Sustainable Green Infrastructure

Ekarizan Shaffie · Rohana Hassan ·
Ahmad Kamil Arshad ·
Norshariza Mohamad Bhkari ·
Muhd Norhasri Muhd Sidek
Editors

Sustainable Green Infrastructure

Materials and Technologies

Springer

Editors
Ekarizan Shaffie
Institute for Infrastructure Engineering
and Sustainable Management (IIESM)
Universiti Teknologi MARA
Shah Alam, Selangor, Malaysia

Rohana Hassan
Institute for Infrastructure Engineering
and Sustainable Management (IIESM)
Universiti Teknologi MARA
Shah Alam, Selangor, Malaysia

Ahmad Kamil Arshad
Institute for Infrastructure Engineering
and Sustainable Management (IIESM)
Universiti Teknologi MARA
Shah Alam, Selangor, Malaysia

Norshariza Mohamad Bhkari
Institute for Infrastructure Engineering
and Sustainable Management (IIESM)
Universiti Teknologi MARA
Shah Alam, Selangor, Malaysia

Muhd Norhasri Muhd Sidek
Institute for Infrastructure Engineering
and Sustainable Management (IIESM)
Universiti Teknologi MARA
Shah Alam, Selangor, Malaysia

ISBN 978-981-96-1485-1 ISBN 978-981-96-1486-8 (eBook)
https://doi.org/10.1007/978-981-96-1486-8

This Springer imprint is published by the registered company Springer Nature Singapore Pte Ltd.
The registered company address is: 152 Beach Road, #21-01/04 Gateway East, Singapore 189721, Singapore

Preface

This book's collection of chapters offers a detailed exploration of innovative solutions and sustainable practices across various industries, emphasising their role in promoting environmental sustainability and enhancing infrastructure development. Each chapter delves into key research areas, presenting findings that align with global sustainability goals and providing valuable insights for researchers, policymakers, and practitioners alike.

The chapter on green energy technology and self-generation assesses the effects of renewable energy investments on electricity savings, offering a strategic framework for optimising energy consumption through self-generation initiatives. Practical guidelines are provided for investors and policymakers to balance business profitability with environmental sustainability.

Another chapter underscores the significance of accurate wood density assessment for sustainable forest management. It compares advanced tools that enhance carbon sequestration estimates, vital for combating climate change and supporting sustainable forestry practices. The role of academic libraries as catalysts for sustainability is also explored, illustrating how they can spearhead environmental initiatives within research institutions and contribute to broader sustainability goals.

Further chapters explore the potential of natural resources, such as *Piper sarmentosum*, in agricultural and medicinal applications and critically evaluate the 2019 European Green Deal, which aims to position Europe as a global sustainability leader. Chapters on sustainable building materials from agricultural waste and innovative construction solutions highlight the sector's transformation by integrating recycled and eco-friendly materials, promoting circular economies, and reducing carbon footprints.

The diversity of research in this book reflects an urgent need for sustainable practices across all sectors—from energy and construction to transportation and waste management. As global challenges like climate change, resource scarcity, and population growth escalate, the solutions and insights presented here will be crucial in guiding future policies and fostering global sustainability efforts.

Subsequent chapters examine strategies to integrate sustainability in construction projects, offering ways to minimise environmental impact, reduce waste, and enhance the lifespan of buildings. Topics include water resource management and incorporating green spaces in airport designs, emphasising using the WEAP model to balance water supply and demand, and enhancing traveller well-being through sustainable design. A chapter on green de-stress relief spaces in Malaysian airports highlights integrating natural elements in high-traffic areas to improve air quality, reduce traveller stress, and contribute to a more sustainable travel experience.

The volume covers innovations in structural design using sustainable materials, Building Information Modelling (BIM), and project management enhancements to avoid construction delays. The insights demonstrate how BIM improves project efficiency, reduces errors, and supports sustainable building practices. Research on value management shows how it can optimise project outcomes. At the same time, additional studies examine the impact of curing compounds on concrete performance and the role of green employee behaviour in advancing organisational sustainability goals.

The book compares global strategies to address transportation-related pollution. It explores recycled materials in road construction, including coal ash in pavement and scrap metal recycling on construction sites, highlighting efficient waste management's role in reducing environmental impact. A study on metal scrap waste management in construction emphasises the importance of efficient recycling processes for minimising industry waste. Another chapter models entry strategies for Malaysian construction firms in global markets, showing how strategic planning can support sustainable growth internationally. The book also presents innovative bamboo-based structural solutions as eco-friendly alternatives to conventional materials, showcasing how bamboo panels reduce the environmental impact of construction. The chapter on cold in-place recycling in Malaysia addresses implementation challenges and benefits, providing practical solutions for sustainable road construction.

Collectively, these chapters illustrate the diverse integration of sustainability across industries. The research in this volume provides critical insights for innovation and adaptation in the face of twenty-first-century sustainability challenges. We hope these contributions inspire further research and encourage stakeholders to embrace sustainable practices that lead to a greener, more resilient future.

Shah Alam, Malaysia — Ekarizan Shaffie
Shah Alam, Malaysia — Rohana Hassan
Shah Alam, Malaysia — Ahmad Kamil Arshad
Shah Alam, Malaysia — Norshariza Mohamad Bhkari
Shah Alam, Malaysia — Muhd Norhasri Muhd Sidek

Contents

The Relationship of Green Energy Technology Investment and Self-generation Production Toward Sustainability in Electricity Saving Consumption 1
Noriza Mohd Saad and Munirah Ramli

Assessing Tree Wood Density: A Comparative Analysis of Resistograph and Increment Borer Measurements 23
J. Joseph

The Green Enclave: Unearthing Strategic Prowess of Academic Libraries as Sustainability Catalysts 37
Haziah Sa'ari, Marlia Idrus, and Mohd Dasuki Sahak

The Chromatographic Profiling of *Piper Sarmentosum* 61
Hannis Fadzillah Mohsin, Kathleen J. Jalani, and Ibtisam Abdul Wahab

2019 European Green Deal: An Ambitious Project with Lights and Shadows 69
Piermauro Catarinella and Suhardi Kram

Sustainable Building Materials from Agricultural Wastes 97
Vicky Kumar, Adamu Sani Abubakar, Tooba Qureshi, Abiola U. Adebanjo, Shamsul Rahman Mohamed Kutty, and Siti N. Abd Razak

Innovative Green Infrastructure Initiatives for Development in the Construction Industry 121
Abiola U. Adebanjo, Nasir Shafiq, Vicky Kumar, Oladele J. Olatoyan, Siti N. Abd Razak, and Nur Fadilah Yusof

Water Availability and Demand Analysis for Kedah River Basin Using Water Evaluation and Planning (WEAP) Model 139
Wei-Koon Lee, Sulaiman Kpakra Faba, and Noor Safwan Muhamad

The Importance of Green De-Stress Relief Spaces in Malaysian Airports: Enhancing Traveller Well-Being and Environmental Sustainability ... 155
Azlan Ariff Ali Ariff, Muhammad Izzat Syafiq Mohd Shahril, Zalina Samadi, and Embong Mohamad

Performance Evaluation of Sustainable Built-Up Composite Cold-Formed Steel Column Filled with Rubberwood Dust Pellet Concrete ... 175
Mohd Syahrul Hisyam Mohd Sani and Fadhluhartini Muftah

Analyzing the Obstacles and Process of Introducing Building Information Modeling (BIM) in the Design Stage of C&S Engineering Consulting Firm Within Malaysia ... 193
Mohd Rashid Bin Ya'acob, Che Maznah Mat Isa, Siti Hamidah Abdull Rahman, and Salmaliza Salleh

Investigation on Factors Contributing to Public Project Delays in Kota Kinabalu Based on Human Errors Using the RII Method: A Pilot Study ... 209
Nazaruddin Abdul Taha, Hazwani Halid, Siti Jahara Matlan, Mohamad Shakri Mohmad Shariff, and Che Maznah Mat Isa

Enhancement in the Application of Value Management Through Project Management in the Malaysian Construction Industry ... 229
Mohd Hilmi Malek, Che Maznah Mat Isa, and Aini Jaapar

The Effect of Paraffin-Based and Acrylic-Based Curing Compounds on Silica Fume Blended Cement Concrete Compressive Strength and Density ... 257
Muhamad Saiful Hafizi Bin Mohd Tauhid, Nur Ilya Farhana Bt Md Noh, JL Ng, Jin Chai Lee, Ruzaimah Razman, Salihah Surol, Wardati Hashim, Mohamad Azim Bin Mohammad Azmi, and Deprizon Syamsunur

Employee Green Behavior at Work: Toward Organizational Environmental Sustainability ... 269
Siti Norashikin Bashirun, Shereen Noranee, and Zuhairah Hasan

Transport and Air Pollution in ASEAN: Lessons Learnt from Countries Practicing a Sustainable Approach ... 285
Roszi Naszariah Nasni Naseri and Nur Asyikeen Kamarudin

Properties of Cement Stabilized Cold In-Place Recycling-Base Course Incorporated with Ground Coal Bottom Ash and Calcium Chloride ... 301
Mohd Izzat Joohari, Ekarizan Shaffie, Nortasha Zulaika Mohd Khalil, and Hanna Natasya Hisham

Management of Metal Scrap from Construction Site for a Sustainable Construction Practice in Malaysia: A Pilot Case Study 315
Umol Syamsyul Rakiman, Raja Zuraidah Raja Mohd Rasi, and Ainol Mardhiyah Rahmat

Strategic Entry Location Decisions by Using of Entry Location Decisions Modeling (ELDeMo) 325
Norizzati Binti Ibrahim and Che Maznah Mat Isa

Experimental and Numerical Investigation of Concrete-Bamboo Sandwich Panel (BSP) as Load-Bearing Wall Under Compression Loading 351
Mohammad Afif Hasnizam, Rohana Hassan, Mohd Hanizan Bahari, Ezahtul Shahreen Ab Wahab, Ali Awaludin, and Ahmad Mazlan Othman

Cold In-place Recycling in Malaysia: Implementation Problems and Solutions 365
Ahmad Kamil Arshad, Ekarizan Shaffie, Mohd Izzat Asyraf Mohamad Kamal, Noor Azreena Kamaluddin, Wardati Hashim, and Mohd Yazip Hj Matori

The Relationship of Green Energy Technology Investment and Self-generation Production Toward Sustainability in Electricity Saving Consumption

Noriza Mohd Saad and Munirah Ramli

Abstract Renewable energy (RE) is a crucial issue for many countries, including Malaysia. People have concerns on climate changes that harm health and social impact in long term. Thus, many developed countries aim to achieve high mix diversification of RE, such as solar PV in green energy technology investment. Focusing on that, this study intends to investigate the relationship of green technology investment and self-generation toward electricity saving consumption. As a consumer, the optimum saving in electricity bill is preferable, because return in Net Energy Metering (NEM) scheme. This study retrieved data from PV Monitoring System (PVMS) from the latest data starting from July 2018 to December 2020 whereby data during Covid-19 pandemic (2021 and 2022) was excluded since biased energy consumption. Thus, a total of 120 observations were considered to run an estimation regression model. The results demonstrate cost, efficiency, and irradiation have strongly significant positive and negative effect to the sustainability of electricity savings consumption. Theoretically, this study enhances literature on social responsible investing and ecological modernization theory in creating awareness of the essential balancing between business profit and environment sustainability. Practically, it provides a general guideline to investor in selecting a suitable install capacity for investment.

Keywords Renewable energy · Green energy technology investment · Electricity saving consumption · Residential · Malaysia

---

N. M. Saad (✉)
Faculty of Business Management, Universiti Teknologi MARA Campus Machang, Bukit Ilmu, 18500 Machang, Kelantan, Malaysia
e-mail: noriza@uitm.edu.my

M. Ramli
College of Business Management & Accounting, Universiti Tenaga Nasional Muadzam Shah Campus, 26700 Muadzam Shah, Pahang, Malaysia
e-mail: munirah12@uniten.edu.my

E. Shaffie et al. (eds.), *Sustainable Green Infrastructure*,
https://doi.org/10.1007/978-981-96-1486-8_1

1 Introduction

Electricity always refers as a technical work, related to engineering, maintenance, and utility providers, for those who supply energy to people for basic essentials. However, consumer is the main player in the energy industry who demand more energy and have created market circulation (Shi et al. 2020; Sulaima et al. 2019). The kilowatt of electricity consumed per capita is part of determinant for the development pace of the country. Consumer urges electricity supply to switch on the electronic, such lighting, refrigerator, and phone charger (Ridzuan et al. 2020). Everything has started from home (Junlakarm et al. 2021). So, considering households among all sectors is the front page who is important actor in electricity industry. It is the time to look up how much household have spent for the utility, and how big the expenses that household potentially saved from solar installation and NEM scheme (Mohd Saad et al. 2023; Jakoplic et al. 2021; Junlakarm et al. 2021; Ur Rehman et al. 2020; Pacudan 2018).

Apart from that, the government put high encouragement to consumer being producer (prosumer) through green technology investment (Mohd Saad et. al. 2022; Taghizadeh and Yoshino 2019). Tax relief is introduced, which are Green Income Tax Allowance (GITA) and Green Income Tax Exempted (GITE) to generate, distribute and consume green energy. The incentives become more attractive investments because the household can reduce income tax at the same time independently generating electricity (Azhar 2019) for the self-consumption and investment as well. The government also predicts that the disposable income for the household will be greater and increase the purchasing power in the future (Xue et al 2021; Yi et al. 2021; Hu et al. 2019). Some of green tax incentives available in world are exemption on the payment of excise for renewable energy electricity, tax incentive in personal income, deduction in taxable profit, value added tax, and effective tax incentives in property tax. Apart from that, the high corporate also reacts positively to the government announcement. Mohd Saad (2021) claims that the stock changes also are quite sensitive for the investor (i.e., stockholder) as the government makes any announcement for the green tax. This implies that all levels of community have reacted positively in pursuing profit with an improvement of environment purpose (Pushpakumara et al. 2018; Wahab et al. 2019).

Next, the crucial in efficiency of solar is concerned to maximize cost saving in electricity bills. The household observed innovation in technology of solar, measuring using capacity factor, as ratio of generation over the fixed capacity. Noted to that, International Renewable Agency reported that the average of capacity factor is approximately 17%. Hence, the coming advanced technology perhaps could boost the performance up (de Santoli et al., 2019; Chatri, 2018). A high-efficiency solar panel is supposed to generate more kWh although the capacity is small, so the solar gives better performance (EIA, 2020). Therefore, people could utilize the gift natural resource for the benefit to the country in the future sustainability (Abdullah et al. 2019; Buur et al. 2019).

Next, it is important for investor to find the best location in installing solar regardless for their house or client's house (Pazikadin et al. 2020; Aliyu and Tekbiyik 2019).

Adding to that, the investment needs to be explored accordingly, to maximize the passive income for the household (Oh et al. 2018). Moreover, the technical study to check heating and temperature is important, so that the household could receive high irradiation (Hu et al. 2019). The solar irradiation is linked to the greater electricity production for self-consumption (Jakoplić et al. 2021).

2 An Overview of Solar Pv as a Green Energy Technology Investment in Malaysia

Renewable energy is one of the national ambitions to cut the resources (i.e., gas and coal) that may harm the nature because take long time to depletion and release CO_2 emissions (Salam and Khan, 2018). According to Malaysia Energy Statistics Handbook, Malaysia heavily relies on coal (43.4%) and natural gas (39.1%) in generating electricity. Both sources are non-renewable, which has limited source and the cost is expensive due to scarce. In addition, International Energy Agency (IEA) stressed up that the advantage of non-renewable requires long period to replenish; thus, it is dangerous to human health and environmental impact. In 2018, government has revised the sustainability target to achieve 20% of mix renewable energy (solar, biomass, biogas, and hydro).

Despite the numerous sources of renewable energy, this study focuses on solar sources especially in rooftop solar using NEM scheme. The rooftop solar PV become popular in ASEAN country, such as in Thailand and Vietnam to boost the mix renewable energy generation. The expansion development of solar industry in Malaysia has triggered the need to strengthen the household's trust in managing electricity bills (Mohd Saad et al. 2023; Azhar et al. 2019; Enongene et al. 2019; Fachrizal and Tang, 2019; Al-Mulali et al. 2013; Bertsch et al. 2017; Comello and Reichelstein 2017; Darghouth et al. 2011; Kahane 1991).

The solar industry which benefits many people and contributes to the economy development (i.e., employment) has expanded to become a leading industry in power generation in Malaysia. Solar is the utilization of sun radiation to generate electricity by converting the sunlight into electricity using the technology, called solar PV. In year 2019, it is noteworthy that Malaysia has exponentially increased renewable capacity from 179 MW in 2018 to 725 MW in 2019 (Annual Report Suruhanjaya Tenaga, 2019). This is in line with national target to achieve renewable energy 20% of the capacity mix in year 2025. Meanwhile, solar energy development was boosted by large-scale solar (solar farm) and rooftop solar. For the household side, the pricing policy offered by government is attractive based on one-to-one mechanism at Tenaga National Berhad, TNB's retail electricity tariff rates, rather than a displaced cost mechanism with lower tariff rate. Then, Suruhanjaya Tenaga recorder 125 licenses for NEM were issued as compared to 37 licenses in 2018. Therefore, the promotion of mix renewable energy has attracted public's attention to invest in green technology, solar PV. Here, investment made by the household refers to the initial cost spend and at the same time takes the risk in installing solar, while aiming for financial and non-financial profit (Kozlova and Collan 2020). The important of effective rate offers by government must compensate to the cost of initial investment (Tran et al. 2020; Taghizadeh-Hesary and Yoshino 2019). On the other hand, the financial profit refers to the cost saving for electricity bill, income tax exemption, and asset value, whereas the non-financial profit is the benefit in long term, where users become part of contributing reducing pollution (Buur et al. 2019).

In order to attract interest from the public investment, it begins with the most technically feasible, making cost-effective and socially acceptable (EIA, 2021). From Fig. 1, International Renewable Energy Agency (IRENA) reported the rapid declining of cost of renewable energy generation via solar PV. Thus, the developing country has a golden opportunity to serve the immerse electricity demand at the cost-effective and sustainable method (Minh 2020; Chaianong et al. 2019). While looking at the mass production and the local manufacturing able to grow, Malaysia has potentially to offer a competitive price of solar installation in the future (Razali et al. 2020). Adding to that, Malaysia is an example growing producer, as the third's world largest producer photovoltaic cell (ACE, 2020).

Besides, the government, Malaysia Development Authority (MIDA) also introduces tax allowance and tax exempted to those purchase green technology (SEDA, 2019). The purpose both Green Income Tax Allowance (GITA) and Green Income Tax Exempted (GITE) is to reduce the burden of acquisition expensive cost while supporting mix renewable energy generation (Ramli et al. 2020; Abidin 2018; Bussic-Sontic et al. 2018). The Ministry of Energy and Natural Resource implements the Renewable Energy Act 2011 (Act 725), to support the national scheme in Fit in Tariff (FiT) and NEM. The promotion of both schemes is eventually to leasing the negative

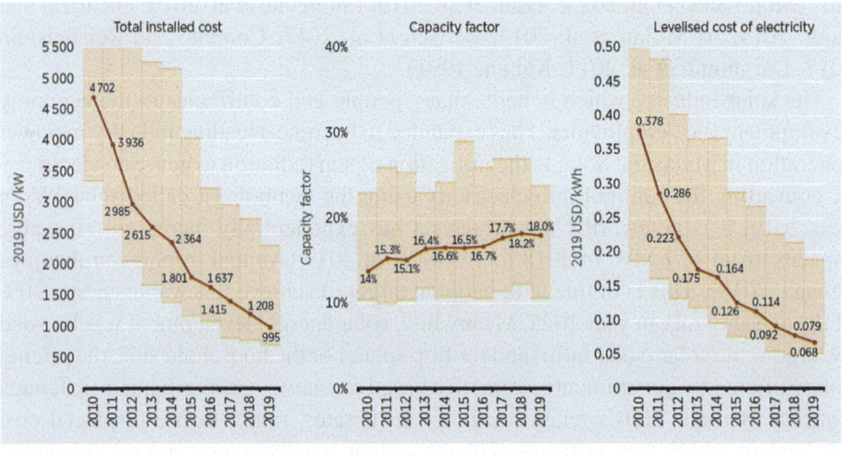

Fig. 1 Global weighted average total installed cost, capacity factors, and levelized cost of electricity (LCOE) for solar PV, 2010 to 2019. *Source* IRENA Renewable Cost Database

impact on the environment using mix renewable energy, like solar source (SEDA, 2020).

However, according to the Institute of Strategic and International Studies (ISIS) Malaysia, the energy policy design is biased toward industry compared to household (ISIS 2014). First, the commercial has enjoyed double tax deduction as they invest in solar panel, while residential is not entitled for tax benefits if they purchased by cash. Second, the quota allocation is bigger to commercial and industry 450 MW and 50 MW, as NEM 1.0 and 2.0 have introduced during year of 2016 to 2020. Thus, the residential as our local user have lesser priority compared to commercial, which will mostly good manufacturing will be exported to international user (Khazanah Research Institute, 2018).

Thus, the new scheme NEM 3.0 in 2021 has been revised to enlarge the opportunity for residential user and increases the quota to 100 mw (residential) and 300 mw (commercial, industry, agriculture, and mining). The scheme of NEM has welcoming more self-generating electricity and built new ventures in electricity market (Xue et al. 2021; Morrisson et al. 2017) whereby government has highlighted the residential as a crucial player in this scheme.

3 Literature Review

3.1 Electricity Consumption by Residential in Malaysia

Malaysia experiences a rapid increasing in electricity consumption, due to macroeconomic development. The economy has been growing and population size expanded ever since 2010 (Khazanah Research Institute 2018). As reported in Malaysia Energy Information Handbook (MEIH, 2019), the final consumption increases gradually, from 5263 ktoe to 13,152 ktoe with the industrial sector exhibits the highest consumption indicating Malaysia is highly energy intensive. Residential electricity consumption represents half of industrial and 20% of total nation's consumption.

Malaysia has designed a few energy policy to address the significant issue in energy market. Apart from the energy policy, Green Technology Master Plan Malaysia (GTPM) was launched by government involving six sectors; (1) energy, (2) manufacturing, (3) transportation, (4) building, (5) waste, and (6) water. The Master Plan intends to plan the green technology development and create low-carbon and resource-efficient economy. Also, as moving toward become a high-income nation, the strategy must be in line with the international trends, especially in terms of climate change. According to Abidin (2018), Malaysia should put priority in addressing issues in population growth, non-renewable resource depletion, and overdependence on conventional or fossil oil. Thus, the green technology strategy intends to support the sustainable economic growth to underline the climate change and the environmental harms. Therefore, one of the schemes to support the agenda is Net Energy Metering (NEM).

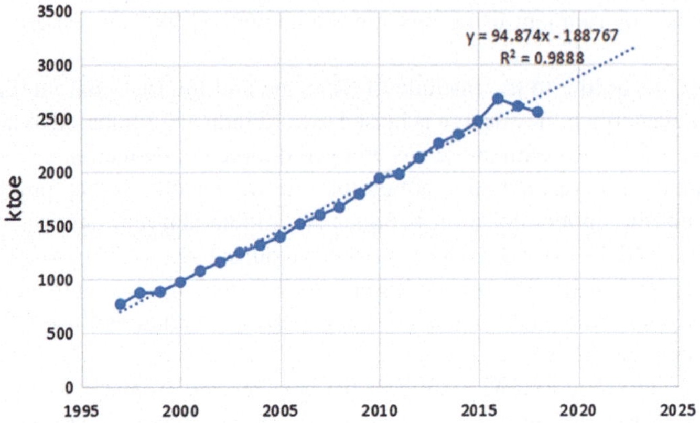

Fig. 2 Annual final consumption for residential sector (ktoe). *Source* Malaysia Energy Information Hub (MEIH), Energy Commissions

Figure 2 shows the annual final consumption for residential sector (in' ktoe) from 1997 to projection year up to 2025 in Malaysia. Throughout of this period, it shows an increasing in pattern as represented by exponential line of final consumption year by year. With respect to the value of r-square, it is reported by estimation model is 98.88 percent explains a strong relationship of the variation in final consumption for residential sector.

3.2 Net Energy Metering (NEM)

Basically, the idea of NEM was introduced in 1983, originally applied in the USA (Razali et al. 2020). In Malaysia, the renewable energy generation began in 2011, as the authority has launched the Renewable Energy Act 2011 (Act 725) to welcome more participation among user to be producers in generating, distributing, and selling to others SEDA (2011). Meanwhile, the NEM scheme was introduced in 2016, replacing Feet-in Tariff (FiT) in 2011, because the government aims to induce the user consume the green energy instead of targeting profitability from solar PV installation (Vaka et al. 2020; Mohd Saad 2022). In 2020, the NEM scheme successfully achieved commercial by 500 MW, including 50 MW for residential and 450 MW for industrial and commercial (SEDA, 2020).

Figure 3 shows the timeline for the development scheme in solar scheme, revised by the government to attract more investors. Basically, NEM scheme has three different tariff rates offered to consumer to offset the surplus of electricity.

By definition, NEM is an electricity bills mechanism that allows user to generate electricity, then energy will be self-consumed inside the building first, and then if there is shortage, the grid will distribute the electricity to user for own consumption. Meanwhile, if there is surplus, or over solar energy generating, the user is eligible

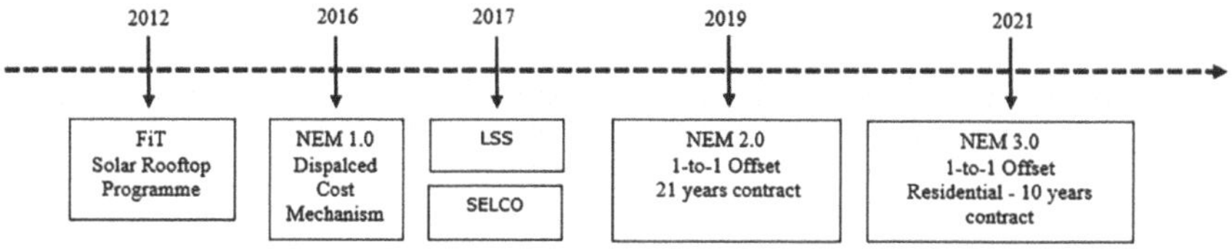

Fig. 3 Timeline for solar scheme. *Source* Author's compilation information for the timeline Illustration

to store the energy at the grid storage (SEDA, 2021). Chua and Oh (2012) point out the NEM allows user to install two-way meters, as deducting the solar electricity generated with the consumption from the building. Hence, the user is entitled a saving in monthly bill after deducting all self-generating.

3.3 RE Investment in Solar Technology

He et al. (2019), point out the green investment as the financial form that properly allocates for natural resource and manages accordingly the capital flow for low pollution, high energy efficiency, and environment friendly. Adding to, the financial resource consists of green financing, green bond, and green insurance to establish win–win situation, resulting the environment protection and business profitability (Table 1).

According to Comello and Reichelstein (2017), the cost of solar PV is affordable as long as the tariff rate is above the threshold of break-even. Adding to that, the government must consider the capacity factor or the ability of solar PV while adjusting the tariff rate. Razali et al. (2020) criticized the small capacity installed as not competitive cost compared to the large size. The study found that the cost of solar PV is expensive; thus, the solar generation is not capable to compensate with the investment. Thus, the investor must know the profile load of consumption before

Table 1 Comparison between residential and the other sectors

Sector	Residential	Commercial, industrial
Voltage	Low voltage (1–12 kWp)	Low voltage Medium voltage High voltage
Equipment component	Solar panel, inverter, meter generation, consumption meter reading	Solar panel, inverter, main switch board (MSB), meter generation, switch boards grid
Net NEMAS	No	Yes—If more than 72kWp

Source Author's Compilations

Table 2 Literature review matrix

Author	Objective of solar investment	Factors contributing to solar investment profitability	Country
Camello (2016)	Cost competitiveness (LCOE)	Household profile, capacity, solar generation, tariff rate	California, Nevada, Hawaii
Xue et al. 2021	Installed capacity (kWp)	Initial cost of PV, information, and awareness	Norway
Razali et al. (2020)	Annualized electricity cost	Profile load, solar irradiation, and capital installment	Malaysia
Ospino-Castro (2020)	Profitability	Irradiance, technical size, and financing assessment	Colombia
Tran et al. (2020)	Investment engagement	Legal framework, financial resource, awareness, and consumer behavior	Vietnam
Aliyu and Tekbiyi-Ersoy (2019)	Optimum capacity	Best technology and location	Nigeria
Taghizadeh-Hesary and Yoshino (2019)	Investment cost	Induce private participation in green financing by green credit guarantee scheme (GCGS)	Japan

deciding to install green technology, like solar rooftop (Ospino-Castro et al., 2020). The summary of the past studies relevant to this study is shown in Table 2.

3.4 Underpinning of Theories; Ecological Modernization, and Social Responsibility Investment

The theory of ecological modernization was pioneered by Joseph Huber in 1982, inspired by the environment policy for greener country (Pushpakumara et al. 2018). The introduction green tools mainly is to cure the issue of climate safety. The innovation policy measures must be explored by combining the environment and economic feasibility. Zahari and Ramayah (2017) study the performance of business based on the green innovation. Without the changes of process in green innovation, the environment would not benefit from it. Green innovation thus involves the consideration of investment to expand the ability and performance. Furthermore, Sidhu and Gibhon (2021) observed that the green innovation is able to reduce the cost of environment, resulting in reducing the cost maintenance, enhancing the revenue, and improving the quality of production by adopting the green technology, such as solar PV (Darghouth et al. 2011). Solar PV is one of the alternatives while using natural sources instead of coal and gas. The incentive has educated people to be creative while managing the source of pollution (Mastrangelo and Aguiar 2019). The environment and development have been linked together, but the regular policy rules the direction toward positive or negative impact.

Social responsible investing (SRI) theory is referred as sustainable investing, responsible investing, and impact investing, describing the behavior and motivation of the investor in making decision. Initially, the investment mode is toward non-profitable and corporate social responsibility (CSR). The development of SRI began in late 1960s and 1970s, as people were concerned to human life. Starting in 1980, sustainability is popular among the nation, and the need to mitigate climate change become a priority. The issue of environment protection has concerned, integrating the ethical to under behavior of environment, social, and governance (ESG) as a strategic planning in balancing the return. Malaysia has announced the official SRI establishment in National Budget's 2014 (25th October, 2013). As a result, the budget has been allocated for those who show high performance in the SRI index (Ministry of Finance Malaysia, 2013). Wahab et al. (2019) state that sending money for solar installation investment is part of example of SRI, while people desire to balance between profit return and environment protection. Tran et al. (2020) believed that the green investment made to improve the environment while reducing air pollution and greenhouse emission without affecting the consumption and production.

Thus, this study was expected to contribute the significant theoretical and practical contributions to the existing knowledge especially in solar industry and household expenditure in Malaysia. Despite its significance between in household consumption, literature has revealed limited studies on the relationship between solar investment and the electricity consumption (Takeda et al. 2019; Zahari and Ramayah 2017). Also, the social responsibility theory enhances the existing literature on the consideration of household looking the making investment decision (Camilleri 2020). The theory is used in this study to focus the household that involves an interaction with the environment to minimize the cost in electricity bill. The successful of the solar investment leads to better environment protection and enhances the financial expenditure, resulting in better self-consumption. Additionally, the theory explains the relationship between green energy and the potential saving for household expenditure. This study successfully proved that the household decision contributes to sustainability performance. Adding to that, the best performance of solar panel is estimated to give the greater impact to the monthly electricity bills. As the electricity export high could save more (Takeda et al. 2019; Wahab et al. 2019).

4 Data and Methodology

Secondary data was utilized for the study. The data was purchased and retrieved from SEDA portal for PV Monitoring System (PVMS) and NEM Calculator since this portal monitors all renewable energy, such as solar quota approval and production in Malaysia. The PVMS begins with 150 grid-connected solar PV participating on voluntary basis. Here, there are two limitations that was identified. First, data for large population could not be obtained to represent Malaysia Solar industry, because out of 150 users registered, there are only 40 users from residential. Besides, the cost of purchasing data is an expensive; therefore, four plants of solar PV were chosen as a sample study that has completed data required. With respect to the sample period, the timeline of solar generation available has also limited since the PVMS was established, which is started on July 2018. Thus, 30 months of solar generation have been observed from July 2018 to December 2020. Data in years of Covid-19 pandemic, 2021 and 2022, was excluded due to mass changes in electricity consumption as well as production and does not represent the normal situation to influence the independent variables.

Particularly, this study refers to two types of report in PVMS database that are solar energy generated (SEG) and irradiation. Data SEG is utilized to calculate the electricity export (EX) and capacity factor (CF) based on formula explained in Eq. (1). Next, data profile was collected, compiled, and then matched for electricity consumption and initial investment cost of solar installation using NEM.

The study also identified there are limitation to have fixed cost of installation, since the price of solar is uncontrollable by authority and it depends on PV Service Provider solely. Firstly, let say, that if the selecting sample has been used four years before, the panel solar price is no more relevant to the current study. Second, the installation cost indicates a decreasing in trends and fluctuates depending on the efficiency of plant, technology changes, size of production (KWh), and research development. Besides, the applicants for solar PV, which includes four main sectors; residential; commercial; industrial; and commercial with a total of 3072 applicants for 500 MW capacity. Out of four sectors, residential sector is the priority given by government in NEM 3.0 scheme, where quota capacity has increased from 50 to 100 MW in 2021. The chance given is to increase more participation in investing in renewable energy, as residential only installed 4–12 kWp compared to others (SEDA, 2020). Besides, the study considers all solar PV plants from poly-crystalline type only. Thus, the specific user profile used in this study can be referring in Table 3.

Based on the previous study, the electricity bills were measured in each country's currency (Razali et al. 2019; Pacudan 2018; Comello and Reichelstein 2017) whereby Ringgit Malaysia (RM) used by government to charge the utility from Malaysian's household. With respect to the dependent variable, net electricity bills (RM) represent an electricity saving consumption for the monthly expenses after investing in green energy technology for solar panel (Canelas et al. 2020). Small amount imported from grid indicates that more saving user are made. The equation for net electricity bills can be derived as follows (Table 4):

Table 3 User profile

Types of solar plant	Capacity size (kWp)	Average electricity monthly consumption (kWh)	Total investment cost* in Malaysian Ringgit (RM)
A	4	750	18,275.00
B	8	1100	33,669.00
C	10	1300	44,978.00
D	12	1500	43,308.00

Note: * Estimate Minimum Upfront Cost
Source: Author's Calculation using NEM Calculator

$$\text{Net Electricity Bill} = \text{Total Electricity Imported} - \text{Total Electricity Exported} \quad (1)$$

Here, the total electricity consumption is described as electricity imported. Electricity is the basic energy for the household to use the electronic tools (Canelas et al. 2020; Ur Rahmen et al. 2020). The usage of electricity starts from the small voltage, such as remote control television up to the big voltage, such as air conditional. Shi et al. (2020) point out that the needs of electricity depend on how many users in house, type of applicants, and the lifestyle. However, this analysis assumed that the other factors that influence the electricity consumption are constant, ceteris paribus, because the study focuses on solar impact only.

Then, the total solar electricity generation (self-consumption) was represented by electricity exported (Darghouth et al. 2011). The concept of NEM promotes user to consume the self-generating electricity first (SEDA, 2016). The regulator puts a limitation for maximum solar generating due to prevention of power overload and high penetration in the grid. In addition, the household does not have the storage for over electricity generated and needs to transfer in the grid. The regulators are

Table 4 Variables for the study

No	Notation	Description of variable	Measurement proxy	Source
Dependent variable (green energy technology investment)				
1	EC	Net electricity bills for the NEM solar user	Electricity imported deducted by electricity exported	PVMS database and NEM calculator
Independent variables (sustainability in electricity savings consumption)				
1	LCOE	Initial investment	Levelized cost of electricity	NEM calculator
2	EFF	Efficiency of solar panel	The ratio of generating electricity divided by installed capacity	PVMS database
3	IR	Solar irradiation	Sun irradiation efficiency	PVMS database

concern if all solar energy generated is transferred to storage while overlapping to the other coal and fuel generation (Ur Rahmen et al. 2020). Noted to NEM scheme, there are assumptions listed by the authority: (1) The electricity consumption exceeds self-generating than household imported from grid. (2) The electricity consumption is less self-generating, then household export to grid for storage. The electricity will be used for the next month. (3) The electricity consumption is equal self-generating.

Since, this study finds out that the quota given to residential user is below than monthly usage (NEM Guideline, 2020). Any applicants can apply for quota of solar capacity based on 75% maximum demand (MD) of electricity averaging in three months consumption. Thus, the power distributor only allows user to import electricity from grid to meet the shortage demand. The measurement for electricity import is measured in RM. Unlike the electricity import, the rate for the user is at regressive rate. The beginning stage rate is at the fourth tier (RM0.5460 per kilowatt). The household will calculate at higher rate. Thus, it could benefit more. Besides, the assumptions of electricity for numbers 2 and 3 for less and equal of self-generating are not applicable in this study because we would like to control the electricity consumption, regardless the household does not stay inside the house or the house is rented to single person. The choosing of first assumption is to generalize the electricity consumption and to reflect the encouragement of authority to self-consumption green energy for the long sustainability (Ridzuan et al. 2020) in green energy technology investment.

As regards to independent variables, the cost-competitiveness of asset evaluation can be measured based on the levelized cost of electricity (LCOE) (Pacudan 2018). The user is able to evaluate the ratio of cost investment in solar installation to actual return electricity generated from the particular asset. Azhar et al. (2019) discussed the LCOE is to compare the operational cost and to increase the penetration of renewable energy for the long-term sustainability. Therefore, LCOE can be expressed as Eq. (2).

$$LCOE = \left(\sum_{t=1}^{n} \frac{I_t+M_t}{(1+r)^t}\right) \Big/ \left(\sum_{t=1}^{n} \frac{E_t}{(1+r)^t}\right) \tag{2}$$

where

I_t = investment expenditure in the year (including financing cost),
M_t = the cost of operation and maintenance,
E_t = the electricity generation,
r = the weighted average cost of capital (WACC),
n = the lifetime of the system.

Based on IRENA Renewable Cost Data, generally LCOE is calculated based on weighted average cost of capital (WACC) at 5%.

Next, the efficiency of solar panel appears to be major factor in determining cost saving in electricity bills in most of study (Pacudan 2018). National Renewable Energy Laboratory (NREL) refers to solar capacity as the ability of solar to meet demand. Progress. Progressive capacity factor ratio reflects lower net electricity bills that allow household to consume electricity from self-generation (Kozlova and Collan

2020; Malala and Tsuyoshi 2020; Fachrizal and Tang 2019; Celik 2002; Kahane 1991).

Based on the availability data in PVMS database, this study used solar energy generation in kWh, which indicates the output of solar panel for the specific plant and then divided with installed capacity. However, due to solar capacity is in terms of kWp refers to kilowatt Peak, it indicates the maximum electricity generated can reach one hour a day. Thus, the solar capacity needs to multiply with 24 h and 30 days to ratio with monthly solar energy generated.

The Eq. (3) is derived for capacity factor (CF):

$$\mathrm{CF} = \frac{\sum \mathrm{SEG}}{\sum \mathrm{SC}} \tag{3}$$

where: CF = Capacity factor,

SEG = solar energy generated,

SC = solar capacity (KWp*24 h*30 days).

Then, with respect to the high irradiation of solar, it represents the higher potential of household to generate a greater electricity production in kWh through the solar panel. Solar irradiation is the amount of the panel received the radiation from sun per unit area of the size. Energy Information Administration (EIA) is defined as the rate at which specific area of surface received the radiant energy during a specific time interval. Aliyu and Tekbiyik (2019) found a strong relationship of solar irradiation as the best location to generate electricity, besides of having best technology. The previous study also supported that the worth investing in high solar irradiation will impact to less electricity imported from the grid. This raw input is averaged in daily basis for the specific month. In this study, the irradiation is measured by (w/m^2) log of the total watt divided by space of solar panel (lnIR) on average of the month (Jakoplic et al. 2021; Pazikadin et al. 2020; Zheng et al. 2011).

4.1 Method Analysis—Panel Data Regression

In this study, panel data regression is utilized to conduct the analysis. According to Baltagi (2003), panel data is described as a longitudinal data that covers multi-dimensional data and also measurement over time. The advantage of using panel data is it is capable of controlling heterogeneity, more information, and reduces the risk of multicollinearity. The data structure for this analysis is macropanel, defined as time series panel (T > N). T refer to the time of study, which is month basis covering a period of July 2018 to December 2020 consisting of 30 times observations. While, N represent the cross-section for four types of solar plants, A, B, C, and D based on different capacity size (i.e., 4 kWp, 8 kWp, 10 kWp, and 12 kWp) respectively. Therefore, this panel data covers medium length of time series ($T = 30$) for relatively a few individuals ($N = 4$) for balanced panel data.

Next, there are three regression estimation models were developed with added additional model for verification, robustness model. Here, model 1 is for pooled ordinary least square (OLS), model 2 for random effect (RE), model 3 for fixed effect-within estimation (FE-WE), and model 4 is for robustness model by including individual of solar plants (types of solar PV as a dummies).

The model regression Eq. (4) for OLS, model 1 is expressed as follows:

$$\text{In EC}_{\text{it}} = \beta_{\text{OLS}} + \beta_{\text{OLS}_1}(\text{In LCOE}_{\text{it}}) + \beta_{\text{OLS}_2}(\text{In EFF}_{\text{it}}) + \beta_{\text{OLS}_3}(\text{In IR}_{\text{it}}) + \varepsilon_{\text{it}} \quad (4)$$

where:

In EC_{it} = the household monthly electricity bills in RM for the month, *i* in the year, *t*,

β_{OLS} = the coefficient estimates in OLS of the explanatory variables,

$\text{In LCOE}_{\text{it}}$ = the ratio of levelized cost of electricity for the month,

$\text{In EFF}_{\text{it}}$ = the ratio of efficiency of the solar panel for the month,

In IR_{it} = the average of irradiation for the month,

ε = the standard error of estimation.

Then, FE happens when the specific effect λ_i is correlated with explanatory variables. According to Yao et al. (2019), FE for model 5 has the ability to remove the effects of time-invariant characteristics. So, by applying RE, we are able to evaluate the net predictors on the outcome variable. Furthermore, the eerror term for fixed effecstimator has not correlated between time invariant with the other individual characteristic. Thus, the model 6 is unique as it can be used within variation in data only and allows for endogeneity of regression. In contrast, if there is correlation in error terms, then RE is appropriate.

$$\text{In EC}_{\text{it}} = (\beta_{\text{fe}} + \lambda_i) + \beta_{\text{fe}_1}(\text{In LCOE}_{\text{it}}) + \beta_{\text{fe}_2}(\text{In EFF}_{\text{it}}) + \beta_{\text{fe}_4}(\text{In IR}_{\text{it}}) + u_{\text{it}} \quad (5)$$

where:

β_{fe} = the coefficient estimates in fixed effect of the explanatory variables,

$(\beta_{\text{fe}} + \lambda_i)$ = the intercept for fixed effect,

u_{it} = the error term for fixed effect.Structure of the Malaysian

$$\text{In EC}_{\text{it}} = (\beta_{\text{re}} + \lambda_i) + \beta_{\text{re}_1}(\text{In LCOE}_{\text{it}}) + \beta_{\text{re}_2}(\text{In EFF}_{\text{it}}) + \beta_{\text{re}_4}(\text{In IR}_{\text{it}}) + u_{\text{it}} \quad (6)$$

where:

β_{re} = the coefficient estimates in random effect of explanatory variables,

$\lambda_i + u_{\text{it}}$ = the error term decomposed two composite error terms for unobserved heterogeneity and it is identically and independently distributed.

4.2 *Model Selection and Diagnostic Check Test*

Since the study developed three main model equations in order to run panel data regression, thus model selection needs to be run to choose the appropriate model among OLS, RE, and FE. Here, Breusch-Pagan Lagragian Multiplier (BP-LM) and Hausman test applied in choosing the most appropriate model for forecasting. Originally, the method was developed by Breusch and Pagan (1980) to detect the heteroscadicity and distinguish the random of pooled OLS effect. In case if it has heteroscadicity, it turns out toward the standard areas which statistically software programs reports are actually wrong. There are biased we actually need to correct for so that is why it is important to test for heteroscadicity. The decision rules imply that if estimation coefficients differ significantly, the study needs to use Hausman test (to distinguish between RE or OLS).

5 Results and Discussion

5.1 *Multicollinearity Results*

Multicollinearity occurs when two variables are not independently and related each other. There are two ways to detect the multicollinearity problem in this study that are pairwise coefficients of correlation matrix test and variance inflation factor (VIF). As reported in Table 5, the results of p-value in the correlation matrix for all variables below than 0.8 indicate a risk-free from multicollinearity problem among all variables. Reconfirm test was also shown by the results of VIF which is reported less than 5 with mean value for 1.18 implying that less collinearity exists among variables as well. Thus, the tests confirm that the variables used in this study are not multicollinearity and are appropriate to run regressions in order to achieve the objectives of the study.

Table 5 Pairwise correlation matrix and variance inflation factor (VIF) results

	EC	LCOE	EFF	IR	VIF	1/VIF
EC	1					
LCOE	0.2624	1			1.17	0.8538
EFF	−0.7739	−0.3299	1		1.28	0.7822
IR	−0.2687	−0.1694	0.5265	1	1.10	0.9082
Mean VIF					1.18	

Table 6 Descriptive statistics results

Variable	Mean	Std.Dev.	Min	Max	Skewness
EC	186.15	107.83	42.57	522.09	−0.9186
LCOE	0.79	0.05	0.74	0.91	1.4955
EFF	0.12	0.03	0.04	0.17	−0.9186
IR	3.27	0.83	0.17	4.99	−0.5420

5.2 *Descriptive Statistic Results*

The result from Table 6 for the high standard deviation (RM107.83) for EC shows the value spread out over a wider range from mean (RM186.15) with minimum of RM42.57 to maximum of RM522.09 per month. The size of installed capacity has influenced electricity bills to spread from small to big size. The pattern has created heterogeneity reflecting the individual characteristic which is consistent to panel data strength. The wide variance also allows this study to explore the boundaries of having different sizes installed capacities. Next, the results for LCOE, EFF, and IR show a smaller standard deviation compared to EC. For instance, LCOE spreads from ratio of 0.74 to 0.91 RM/Kwh indicate smaller LCOE ratio would carry less cost of return to user. Capacity factor for solar shows the maximum ratio is at 17% with average of 12%. The finding is consistent to the reference in IRENA Renewable Cost Database in 2019, only 18% recorded for the average of global capacity factor. The solar generation also has low efficiency compared to conventional electricity generation (i.e., coal and fuel), because the effective rate for solar generating electricity is approximately 4.3 h compared to 24 h for coal. The means of EFF imply that the solar panel in Malaysia has a medium performance, associated with high irradiation solar. The irradiation is measured based on daily averaged solar panel. The average of solar panels receiving the radiation is 3.27 kWh/m^2, indicating that most of plant gets the peak irradiation ($0.17 < IR < 4.99$). The higher irradiation will lead a greater solar electricity generated.

5.3 *Model Selection (BP-LM and Hausman) Results*

As reported in Table 7, Hausman test concluded that model 2 for FE is the most appropriate estimator based on model selection test tested compared to OLS and RE. Specifically, Hausman test shows fixed effect outcome is more appropriate in explaining the relationship between green energy technology investment with energy consumption since the statistical test is significant at 95% confident level for $p < 0.05$. It is noted that, although we run pooled OLS and random effect, we found that the similar results have been published.

Table 7 BP-LM and Hausman test results

Panel	Model	Test				Model selection
		BP-LM		Hausman		
		Chibar-squared	Prob.	Chibar-squared	Prob.	
All ($N = 120$)	OLS	0.000	1.000	–	–	FE
	RE					
	FE	–	–	1184.92	0.000	

5.4 *Multivariate Panel Regression Result (MPRR)*

Table 8 shows that all four models are valid model in explaining the relationship between green energy technology investments by solar PV with electricity consumption proxies. This can be seen by looking at their significant value of F-test and Wald chi2 with 99 percent confident level, respectively. However, the best-fit model belongs to FE, therefore, the relationship between green energy technology investments with electricity saving consumption relies on this model, model 2.

With respect to the R-square and adjusted R-square, the results explain that the selected model (model 2 for FE) has not so strong relation between dependent and independent variables whereby the R-square and adjusted R-squared are at 47.1% and 44.3%, respectively, justifying that only 47.1% of the changes in dependent (green energy technology investment) by solar PV was explained by independent variables, energy consumption proxies. However, once dummy was considered in the model equation; the R-squared and adjusted R-squared were increased to 72.7% and 71.3%, respectively. Means that, Solar PV plant as a dummy is considered another change factor in explaining the variations in dependent by independent variables.

Table 8 Regression results

Variable	Model 1	Model 2	Model 3	Model 4
	POLS	FE	RE	FE-WE
Intercept	1.141	1.517	1.141	1.219
ln LCOE	−0.456	−1.409***	−0.456	−1.409***
ln EFF	−1.718***	−1.394***	−1.718***	−1.394***
ln IR	0.162	2.348***	0.162	2.348***
Plant B				0.007
Plant C				4.634***
Plant D				7.259***
R-squared	0.491	0.471		0.727
Adj. R-squared	0.478	0.443		0.713
F-Test	37.25***	33.6***		50.25***
Wald chi2			111.8***	

There are two out of three variables indicate a negative relationship between green energy technology investment and electricity consumption, LCOE and EFF, as shown by p-value at − 1.409 and − 1.394, respectively. The negative impact of electricity import is consistent with Comello and Reichelstein (2017). Since the LCOE shows a significant relationship, the technology assessment of electricity production is always preferred in computing the financial cost, especially in initial investment (Ur Rahmen et al. 2020; Dale 2013). As reported, the increase of 1% cost RM/Kwh leads to decrease in 1.40% of electricity bills. Consistent with the theory of ecological modernization, the urban people tend to be aware of the preservation of nature; thus, green technology will become new alternative as shifting toward renewable energy as they have ability to self-generating electricity using solar. However, the irradiation variable demonstrates a positive significant relationship at p-value of 2.347 with 99 percent confident level. The positive relationship in cost and efficiency has supported the finding of Pacudan (2018), claiming that the cost will reduce the net electricity bills significantly.

After including individual dummies, the model becomes more appropriate and improvises the relation between dependent and independent variables as represented by R-squared and adjusted R-square where it provides a better results compared to normal MPPR (Pooled OLS, RE, and FE). This shows that 72% of the independent variables influence the monthly net electricity bill. The results show that the plant-specific effect, two solar PV plant, C and D have high significant results at 99% confident level. However, plant B is insignificant. This result explains that the high level income of household in residential customer with high install capacity for solar tends to consume more electricity.

6 Conclusion and Policy Recommendation

The study provides key outcomes on specific focus area to identify self-generation electricity for self-consumption and investment in Malaysia. This subsequently implications to the sustainability of household expenditure in terms of economic and social responsibility toward supporting government national agendas in achieving mix renewable energy by 25% in year 2025. Besides, the findings revealed a significant effect of RE investment by solar PV toward energy consumption for cost saving.

Herewith, there are several recommendations proposed. First, extending NEM rate for long term in encouraging household purchasing green technology, solar panel. In particular, the government can analyze the effective rate offered while household export and import electricity while reflecting to the cost of initial investment. Previously, the high cost bear by the household has been doubting as government has changed contract FiT to NEM 1.0, NEM 2.0, and NEM 3.0.

Second, the outcome of the analysis brings that the innovation and research in upgrading technology of solar should be optimized (Peng and Liu 2018). As can be seen, the average of solar efficiency measured by capacity factor is around 17%. Thus, the coming advanced technology perhaps could boost the performance up to

25% at least. A good solar panel should generate more kWh although the capacity is small, so the solar is more efficient (EIA, 2020). Adding to that, prioritizing research in innovation will potentially lead to the rapid growth in modern economic lifestyle. With regard to technology and innovation investment, the household is suggested to consider investing in battery storage. The additional investment aims to improve the technical and economic benefit in the future. For example, NEM 2.0 offers a long contract to pay high cost for 21 years, but for the new investor in NEM 3.0, the government has shorten the contract become ten years begin in 2021. Thus, the exceed electricity generation could be transferred into batteries storage, and thus, it could cut the monthly electricity bills as well in the future (Deotti et al. 2020).

Third, it is important for investor to find the best location in installing solar regardless for their house or client's house. For instance, the Supply Agreement of Renewable Energy (SARE) Malaysia program must be allocated at high rate of solar irradiation, high temperature, and heating point area (Aliyu and Tekbiyik-Ersoy 2019). Besides, in the future, techno-economic analysis could be included in the analysis that involves simple payback period or internal rate of return. The techno-economic analysis will improve more analysis to capture real cost and financing of the solar panel.

Last but not least, this study has a short timeline from middle year 2018 to the end of year 2020. To expand the investigation, the researcher has divided the timeline into monthly basis to have 30 months period observations. It should be noted that the efficiency and depreciation of solar panel may influence the performance of solar electricity generating along the lifetime. Therefore, there may be unexplored elements influencing the enhancement of the relationship between self-generating of green energy technology investment toward electricity saving consumption. Going forward, it is suggested to establish a support in financial banking, such as insurance and bank loan to encourage the investor participating in generating green energy.

References

Abdullah WSW, Osman M, Ab Kadir MZA, Verayiah R (2019) The potential and status of renewable energy development in Malaysia. Energies 12(12):24–37

Abidin MZ (2018) Driving green economy for Malaysia through green technology and green culture. J Soc Automot Eng Malays 2(1)

Aliyu A, Tekbiyik-Ersoy N (2019) A novel framework for cost optimization of renewable energy installations: a case study of Nigeria. Energies 12(22):4311–4321

Al-Mulali U, Fereidouni HG, Lee JY, Sab CNBC (2013) Exploring the relationship between urbanization, energy consumption, and CO_2 emission in MENA countries. Renew Sustain Energy Rev 23:107–112

Azhar NAHM, Hock GC, Shaari SA, Yunus B, Kiong TS (2019) Feasibility study of renewable energy using levelized cost energy. In: Proceedings of IEEE 7th conference on systems, process and control (ICSPC) Dec 2019, p 43–47

Baltagi BH, Bresson G, Pirotte A (2003) Fixed effects, random effects or Hausman–Taylor?: A pretest estimator. Econ Lett 79(3):361–369

Bertsch V, Geldermann J, Lühn T (2017) What drives the profitability of household PV investments, self-consumption and self-sufficiency? Appl Energy 204:1–15
Breusch TS, Pagan AR (1980) The Lagrange multiplier test and its applications to model specification in econometrics. Rev Econ Stud 47(1):239–253
Buur L, Pedersen RH, Nystrand MJ, Macuane JJ (2019) Understanding the three key relationships in natural resource investments in Africa: an analytical framework. Extr Ind Soc 6(4):1195–1204
Camilleri MA (2020) The market for socially responsible investing: a review of the developments. Soc Responsib J 17(3):412–28
Canelas E, Pinto-Varela T, Sawik B (2020) Electricity portfolio optimization for large consumers: Iberian electricity market case study. Energies 13(9):22–49
Chaianong A, Bangviwat A, Menke C, Darghouth NR (2019) Cost–benefit analysis of rooftop PV systems on utilities and ratepayers in Thailand. Energies 12(12):22–65
Chatri F, Yahoo M, Othman J (2018) The economic effects of renewable energy expansion in the electricity sector: A CGE analysis for Malaysia. Renew Sustain Energy Rev 95:203–216
Celik AN (2002) Optimisation and techno-economic analysis of autonomous photovoltaic–wind hybrid energy systems in comparison to single photovoltaic and wind systems. Energy Convers Manage 43(18):2453–2468
Chua SC, Oh TH (2012) Solar energy outlook in Malaysia. Renew Sustain Energy Rev 16(1):564–574
Comello S, Reichelstein S (2017) Cost competitiveness of residential solar PV: the impact of net metering restrictions. Renew Sustain Energy Rev 75:46–57
Dale M (2013) A comparative analysis of energy costs of photovoltaic, solar thermal & wind electricity generation technologies. Appl Sci 3(2):325–337
Darghouth NR, Barbose G, Wiser R (2011) The impact of rate design and net metering on the bill savings from distributed PV for residential customers in California. Energy Policy 39(9):5243–5253
Deotti L, Guedes W, Dias B, Soares T (2020) Technical and economic analysis of battery storage for residential solar photovoltaic systems in the Brazilian regulatory context. Energies 13(24):6517–6528
Enongene KE, Abanda FH, Otene IJJ, Obi SI, Okafor C (2019) The potential of solar photovoltaic systems for residential homes in Lagos city of Nigeria. J Environ Manage 244:247–256
Fachrizal M, Tang J (2019) Forecasting annual solar PV capacity installation in Thailand residential sector: a user segmentation approach. Eng J 23(6):99–115
He L, Liu R, Zhong Z, Wang D, Xia Y (2019) Can green financial development promote renewable energy investment efficiency? A consideration of bank credit. Renew Energy 143:974–984
Hu J, Harmsen R, Crijns-Graus W, Worrell E (2019) Geographical optimization of variable renewable energy capacity in China using modern portfolio theory. Appl Energy 253:11–36
Jakoplić A, Franković D, Kirinčić V, Plavšić T (2021) Benefits of short-term photovoltaic power production forecasting to the power system. Optim Eng 22(1):9–27
Junlakarn S, Kittner N, Tongsopit S, Saelim S (2021) A cross-country comparison of compensation mechanisms for distributed photovoltaic in the Philippines, Thailand, and Vietnam. Renew Sustain Energy Rev 11:8–20
Kahane A (1991) New perspectives for energy efficiency and system optimization. Energy Policy 19(3):199–201
Khazanah Research Institute (2018) Structure of the Malaysian economy: an input-output analysis.
Kozlova M, Collan M (2020) Renewable energy investment attractiveness: enabling multi-criteria cross-regional analysis from the investors' perspective. Renew Energy 150:382–400
Malala ON, Tsuyoshi A (2020) Portfolio optimization of electricity generating resources in Kenya. Electr J 33(4):67–77
Mastrangelo ME, Aguiar S (2019) Are ecological modernization narratives useful for understanding and steering social-ecological change in the Argentine Chaco? Sustainability 11(13):3593–3599
Minh KP, McKenna R, Fichtner W (2020) A cost-effective and transferable methodology for rooftop PV potential assessment in developing countries. Energies 13(10):2501–2511

Morrison C, Ramsey E, Bond D (2017) The role of social entrepreneurs in developing community resilience in remote areas. J Enterprising Communities: People Places Glob Econ 11(1):95–112
Oh TH, Hasanuzzaman M, Selvaraj J, Teo SC, Chua SC (2018) Energy policy and alternative energy in Malaysia. Renew Sustain Energy Rev 81:3021–3031
Pacudan R (2018) The economics of net metering policy in the Philippines. Int Energy J 18(3)
Pazikadin AR, Rifai D, Ali K, Malik MZ, Abdalla AN, Faraj MA (2020) Solar irradiance measurement instrumentation and power solar generation forecasting based on artificial neural networks. Sci Total Environ 715:36–48
Peng H, Liu Y (2018) How government subsidies promote the growth of entrepreneurial companies in clean energy industry: an empirical study in China. J Clean Prod 188:508–520
Pushpakumara H, Atan H, Khatibi A, Azam SF, Tham (2018) Contribution of green orientation for the organizational performance: a review of stakeholder relationships & ecological modernization perspectives on sustainability. Int J Bus Manag Rev 6(9):56–72
Ramli M, Saad NM, Abdullah Z (2020) Encouraging solar PV investment as one of RE in Malaysia. Glob Bus Manag Res 12(4)
Razali AH, Abdullah MP, Hassan MY, Hussin F (2019) Comparison of new and previous net energy metering (NEM) scheme in Malaysia. ELEKTRIKA-J Electr Eng 18(1):36–42
Razali AH, Abdullah MP, Said DM, Hassan MY (2020) Annualized electricity cost of residential solar PV system under Malaysia's NEM scheme. ELEKTRIKA J Electr Eng 19(1):50–54
Ridzuan AR, Kamaluddin M, Ismail NA, Razak MIM, Haron NF (2020) Macroeconomic indicators for electrical consumption demand model in Malaysia. Int J Energy Econ Policy 10(1):16–24
Mohd Saad N, Mohamed EF, Mohd Arshad MT, Mohayiddin AL (2023) Electricity tariff changes and consumer sentiment on household consumption expenditure in Malaysia. J Quant Econ 20(1):175–191
Mohd Saad N, Ishak I, Jaafar AH, Laton Z (2021) Relationship between installed capacity with total installation cost on solar PV among prosumer NEM 2.0 in Malaysia. Linguist Cult Rev 5(S1):1467–1479
Mohd Saad N, Ibrahim J, Ramli M, Arshad MTM, Mohayiddin AL (2022) The impact of optimized electricity tariff changes towards renewable energy and non-renewable energy in Malaysia. OPEC Energy Rev 46(2):127–138
Salam MA, Khan SA (2018) Transition towards sustainable energy production–a review of the progress for solar energy in Saudi Arabia. Energy Explor Exploit 36(1):3–27
Shi Y, Ren X, Guo K, Zhou Y, Wang J (2020) Research on the economic development pattern of Chinese counties based on electricity consumption. Energy Policy 147:11–18
Sidhu AM, Gibbon J (2021) Institutionalisation of weak conceptions of sustainability in the United Nations clean development mechanism: empirical evidence from Malaysian organisations. Account Audit Account J 34(5):1220–1245
Sulaima MF, Dahlan NY, Yasin ZM, Rosli MM, Omar Z, Hassan MY (2019) A review of electricity pricing in peninsular Malaysia. Renew Sustain Energy Rev 110:348–367
SEDA (2011) Sustainable energy development authority (SEDA) of Malaysia. Guideline of solar photovoltaic installation of NEM scheme. Available: http://seda.gov.my
Taghizadeh HF, Yoshino N (2019) The way to induce private participation in green finance and investment. Financ Res Lett 31:98–103
Takeda S, Keeley AR, Sakurai S, Managi S, Norris CB (2019) Are renewable as friendly to humans as to the environment?: A social life cycle assessment of renewable electricity. Sustainability 11(5):53–70
Tran T, Do H, Vu T, Do N (2020) The factors affecting green investment for sustainable development. Decis Sci Lett 9(3):365–386
Ur Rehman W, Bhatti AR, Awan AB, Sajjad IA, Khan AA, Bo R, Oboreh SO (2020) The penetration of renewable and sustainable energy in Asia: a state-of-the-art review on net-metering. IEEE Access 8:170364–88
Vaka M, Walvekar R, Rasheed AK, Khalid M (2020) A review on Malaysia's solar energy pathway towards carbon-neutral Malaysia beyond Covid'19 pandemic. J Clean Prod 12:28–34

Wahab MZ, Asmadi MN (2019) Malaysian initiatives to support sustainable and responsible investment (SRI) especially through sukuk approach. J Emerg Econ & Islam Res 7(3):1–11
Xue Y, Lindkvist CM, Temeljotov SA (2021) Barriers and potential solutions to the diffusion of solar photovoltaic from the public-private-people partnership perspective of Norway. Renew Sustain Energy Rev 137:11–36
Yao X, Yasmeen R, Li Y, Hafeez M, Padda IUH (2019) Free trade agreements and environment for sustainable development: a gravity model analysis. Sustainability 11(3):597–611
Yi L, Li T, Zhang T (2021) Optimal investment selection of regional integrated energy system under multiple strategic objectives portfolio. Energy 218:94–109
Zahari FM, Ramayah T (2017) Green innovation and firm performance: the ecological modernization perspective. J Technol Oper Manag 12(1):21–31

Assessing Tree Wood Density: A Comparative Analysis of Resistograph and Increment Borer Measurements

J. Joseph

Abstract Estimating wood density is essential for attaining SDG 13's goal of combating climate change, especially in reducing greenhouse gas emissions and promoting sustainable forest management (SFM). Since biomass is directly related to the amount of carbon stored in trees, accurate estimates of wood density (WD) led to more precise calculations of carbon storage. By understanding the density of different tree species, forest managers can better estimate the total carbon sequestration potential of a forest. This information is vital for forest carbon accounting efforts, which are used to track progress towards climate targets. A Resistograph is a non-destructive device for determining the density of wood, and it provides a reasonably accurate estimate of wood density. This study examined the Resistograph technology by estimating the density of Eucalyptus tree wood using amplitude as the primary variable resistance. Therefore, the predicted WD value from the Resistograph-generated variables was obtained and compared to the actual WD value from tree wood cores. The positive, strong correlations ($r = 0.791$) between the predicted and core WD with significant differences ($p = < 0.05$) indicate that the Resistograph is reliable for estimating WD without significant tree destruction.

Keywords Resistograph · Increment borers · Climate change · Wood density · Sustainable forest management

1 Introduction

Wood is a renewable resource with the potential to substantially contribute to achieving the sustainable development goals (SDGs), including the goal of combating climate change through the promotion of sustainable forest management and the use of wood as a building material. Resistograph-based wood density (WD) evaluation is a crucial tool for attaining these objectives. The objective of sustainable forest

J. Joseph (✉)
Faculty of Plantation and Agrotechnology, Universiti Teknologi MARA UiTM Sabah Branch, Kota Kinabalu Campus, Locked Bag 71, 88997 Kota Kinabalu, Sabah, Malaysia
e-mail: jacqueline@uitm.edu.my

E. Shaffie et al. (eds.), *Sustainable Green Infrastructure*,
https://doi.org/10.1007/978-981-96-1486-8_2

management is to ensure that forests continue to provide ecological, economic, and social benefits over the long term. The Resistograph can play a vital role in sustainable forest management by providing information on the wood's quality and ensuring that only high-quality wood is harvested. In turn, this can help promote the growth and health of the forest, preserve biodiversity, and lessen the environmental impact of logging activities. The SDGs aim to establish a sustainable future for everyone by addressing the world's social, economic, and environmental challenges. One of the primary objectives is to combat climate change through the reduction of greenhouse gas emissions and the promotion of sustainable practices. Precise WD value is essential for converting tree volume into forest aboveground biomass when measuring forest carbon stocks. An accurate estimate of forest carbon storage is required for climate change mitigation strategies. Forests are essential for carbon sequestration and climate change mitigation, according to Mohd Zulkifli et al. (2021).

In this context, determining the WD of tropical tree species using a Resistograph is crucial for several reasons. First, tropical forests are among the most biodiverse and valuable ecosystems on Earth, providing essential ecosystem services, including carbon sequestration, water regulation, and biodiversity conservation. Understanding the timber density of the tree species in these forests is crucial for sustainable forest management practices, such as selective logging and forest restoration initiatives. Secondly, using the Resistograph to investigate WD in tropical tree species can aid in the conservation of endangered species. By understanding the wood properties of endangered species, we can gain a greater understanding of how they function in their environment and aid in preventing overexploitation of their populations.

The investigation of timber density using the Resistograph can ultimately contribute to climate change mitigation. Wood products have a smaller carbon imprint than other building materials and are, therefore, an integral part of sustainable building practices. By comprehending the timber density of tropical tree species, we can promote the use of sustainable wood products in construction, thereby reducing greenhouse gas emissions and contributing to efforts to mitigate climate change. A Resistograph is a non-destructive diagnostic instrument that can be used to evaluate the density and quality of wood products used in building construction. The Resistograph can predict the WD value by producing variables such as drilling resistance, drilling speed, and drilling time, allowing for the evaluation of the mechanical properties and quality of wood products used in construction.

Moreover, as demonstrated by carbon footprint assessments at UiTM Perak and Seri Iskandar, the economic valuation of forest carbon contributes to the reduction of GHG emissions and the mitigation of climate change (Mohd. Shobri et al., 2017). Accurate WD estimates determine the dependability of carbon stock estimates by selecting biomass estimation techniques from the highest tier based on industry best practices. Tier 3 is a method for estimating biomass and carbon stocks that permits multiple methods, including process and based models. According to Egglesto (2006), the proxy issue at Tier 3 is transparent documentation of the validity and completeness of the used data, assumptions, equations, and models.

Because WD measurements are typically only available for a subset of species, scientists must rely on the WD repositories in the ICRAF and Dryad databases.

According to Machado et al (2014), only a few WD-based allometric models can estimate the aboveground biomass of tropical forests. It is crucial to develop general WD prediction models of tropical WD to refine estimates of unknown WD (Flores and Comes, 2011). A fixed WD value in protected forest areas necessitates detrimental, time-consuming, and impractical methods. Micro-drilling devices provide a quicker and more accessible alternative for assessing WD variability in living trees (Silva et al., 2020; Tomczak et al., 2022). Extensive research has been conducted on WD assessment using resistance micro-drilling for a variety of applications, including timber structural analysis, tree health assessment, tree reproduction, and ageing investigation (Icel and Guller, 2016; Isik et al., 2019; Isik and Li, 2003; Kubus, 2009; Rinn et al., 1996). Several investigations have been conducted on the relationship between Resistograph readings (amplitude) and WD (Bouffier et al., 2008; Fundova et al., 2018; Joseph et al., 2022; Karlinasari et al., 2017; Oliviera et al., 2017; Da Silva et al., 2020; Todoroki et al., 2021; Tomczak et al., 2022).

In this study, we observed the relationship between the resistance data (amplitude) and the Resistograph drilling resistance and compared it to the actual WD determined from Eucalyptus wood cores. Eucalyptus has been chosen in this study because Eucalyptus trees hold significant importance in WD assessment due to the following reasons:

- Commercial value: Eucalyptus is widely recognized for its commercial value in the timber industry. It is one of the most widely planted hardwood tree genera globally, known for its fast growth and versatile timber properties. Assessing the WD of *Eucalyptus* sp. helps determine their suitability for various applications, such as construction, furniture production, flooring, and pulp and paper manufacturing.
- Resource management: Eucalyptus plantations are established for economic purposes, including timber production. Accurate wood density assessments of Eucalyptus trees support efficient resource management and plantation planning. Understanding the wood density of Eucalyptus trees helps determine the ideal rotation cycles, pruning schedules, and harvesting techniques for maximum productivity.
- Carbon sequestration and climate change mitigation: Eucalyptus trees play a crucial role in carbon sequestration, helping mitigate climate change. WD is directly related to carbon storage capacity, as denser wood generally contains a higher concentration of carbon. By assessing the WD of Eucalyptus trees, carbon sequestration potential can be estimated, contributing to carbon accounting efforts and strategies for mitigating climate change.

Hence, it is highly envisaged that the findings of this study could justify the use of a Resistograph, which could provide an accurate estimate of tree density without necessitating tree felling.

Fig. 1 **a** Tree diameter measurement, **b** Hollow tree. 2022 *Source* Jacqueline Joseph,

2 Methods and Materials

2.1 Study Area

This study was conducted in Long Mio, Sipitang, Sabah, Malaysia. It is in transition zone of lowland and montane tropical forests and ranges in elevation from approximately 1000 to 1600 m above sea level. It consists of state land and forest reserves, which was previously under the jurisdiction of Sabah Forest Industries Sdn. Bhd. and recently relinquished to Sabah Forestry Department.

2.2 Trees Selection

Seventeen individuals of Eucalyptus sp. were sampled and measured using the Rinntech Resistograph® R650-ED (Heidelburg, Germany) from each tree at breast height about 1.30 m above ground level (Fig. 1a).

2.3 Drilling Resistance Measurement and Processing

Prior to drilling, tree diameter and height measurements were recorded. A measuring tape was used to measure the circumference of the tree trunk at breast height (DBH), about 1.3 m above the ground for standardization. Meanwhile, the tree height was measured by using the Laser Range Finder. The Resistograph uses a micro-drilling device with a needle of 3 mm diameter, 50 cm length, and has drill torque mechanism, corresponding to a magnitude of drill difficulty in penetrating the wood. It only creates a very minimal drilling hole about 3 mm after tree drilling operation. The overall health and stability of the tree were assessed before taking any measurements. Trees that are in immediate danger of failure or pose risks (truncated, dead, and hollow) during the measurement process were avoided (Fig. 1b). The profiles were processed using DECOM Scientific 2.38m1 version. Next, the profiles were imported to the Microsoft Excel program and statistically analysed by using the IBM SPSS Statistic Version 26.

2.4 Wood Cores Extraction

A known value of 5.15 mm bark-to-pith increment cores was extracted from each tree in the same manner as Resistograph using power drill-borer, DeWalt DCF899HP2 20 V impact wrench 950 Nm and Haglöf increment borer (Fig. 2a–c). An increment Borer is a specialized tool consisting of a handle, a borer bit, and an extractor. Typically, the core is taken at breast height and approximately 10 cm apart from the Resistograph drilling hole. The tip of the borer bit is placed against the tree, and the handle begins to turn clockwise, applying steady pressure. The bit will cut into the tree, creating a narrow, cylindrical hole. The handle is continued turning until the borer bit reaches the desired depth. In this study, the desired depth or the length of the fresh cores is pre-determined based on the partial diameter of the tree. The extractor is attached to the borer bit and removed carefully from the tree, bringing out the wood core. This procedure was described by Joseph et al., 2023.

(a) (b) (C)

Fig. 2 **a** Tree drilling using the Rinntech Resistograph, **b** Tree coring using Haglof increment borer, **c** Tree coring using power drill-borer. 2022 *Source* Jacqueline Joseph,

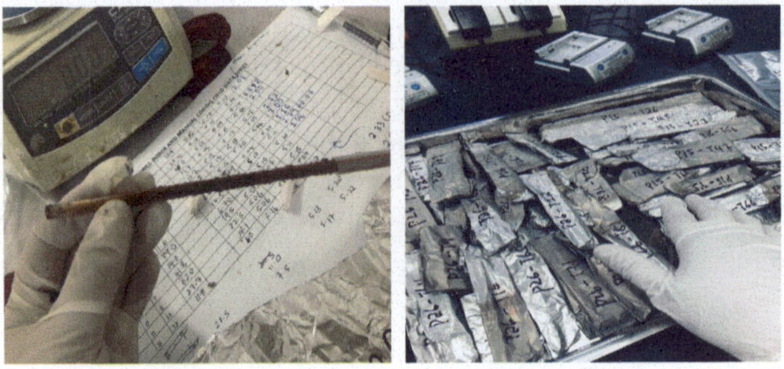

(a) (b)

Fig. 3 **a** Measuring the fresh wood core, **b** wood core ready for air-drying process. 2022 *Source* Jacqueline Joseph,

2.5 WD Determination

Core fresh weight was weighted using digital micro weighing scale after returning from the inventory plots (Fig. 3a) Then, cores were wrapped in aluminium foil, placed into plastic zip-locks, kept in a cooler and transported to the laboratory (Fig. 3b). In accordance with the dimensional methodology described by Chave et al. (2006), the wood density of each sample was calculated.

2.6 Estimation of Tree WD Using Resistograph Generated Variables

Correlations between core WD and variable from resistograph, the amplitude is inspected using Pearson's correlation. An unpair sample t-test is run to observe the mean differences between the predicted WD from the resistograph generated variables and true wood density value from the wood cores (Fig. 4).

3 Results and Discussion

The density of a tree's timber affects its capacity to sequester carbon, store nutrients, resist pests and diseases, and provide wood products with desirable qualities such as strength and durability. In addition, WD can be used as an indicator of forest health, as variations in WD can indicate alterations in forest composition, structure, and function. It is a crucial factor in determining the quality of wood and the mechanical properties of wood products. The Resistograph is a non-destructive diagnostic

Fig. 4 **a** 3 mm Resistograph drilling hole, **b** 5.5 mm Haglof increment borer coring hole, **c** coring hole is sealed with beewax. 2022 *Source* Jacqueline Joseph,

instrument used to determine the WD and quality. The Resistograph produces variables such as drilling resistance, drilling speed, and drilling time that are utilized to predict the WD value. Nevertheless, the accuracy of the predicted WD value from Resistograph-generated variables can vary based on the wood species, age, and growth conditions. In this context, the following discussion sheds light on the results of the predicted WD value derived from Resistograph-generated variables and the true WD value derived from the wood core.

The primary variable, amplitude generated by the Resistograph, is utilized to predict the sample's WD. It is positively correlated with the true wood core density, as shown in Fig. 5. It also indicates that a strong correlation exists between the two variables. In other words, the true wood core density increases as the Resistograph variable increases, and vice versa. In Resistograph measurements, amplitudes refer to the magnitude of the resistance encountered during drilling at different depths within the tree. These amplitudes are directly related to the wood density of the tree being assessed. A positive correlation indicates that the variable generated by the Resistograph accurately predicts the actual wood core WD due to the following reasons:

- Physical properties of wood: WD is primarily influenced by the physical properties of the wood, including the arrangement of its cells and the amount of cell wall material present. Denser wood has a higher concentration of cell wall material and a more compact cell structure. This increase in density leads to greater resistance encountered by the Resistograph during drilling, resulting in higher amplitudes.
- Uniformity of WD: Core WD is typically assessed through a physical core sample extracted from the tree. This method provides a direct measurement of the WD at a specific location. The Resistograph, on the other hand, provides a continuous profile of drilling resistance as it moves through the tree. The amplitudes recorded by the Resistograph reflect the overall WD characteristics along the drilling path, integrating the density variations across different wood zones and regions within the tree.

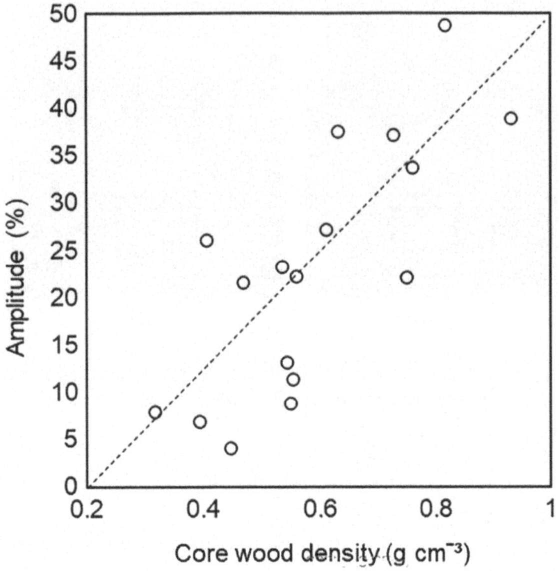

Fig. 5 Correlation between the core WD with the variable generated from Resistograph, the amplitude (%)

- Sampling depth: The drilling depth during Resistograph measurements can be adjusted to penetrate different wood zones within the tree. By drilling more deeply, the Resistograph can reach the core wood, where the highest WD is typically found. Consequently, the amplitudes recorded by the Resistograph tend to correlate strongly with core WD, as they capture the resistance encountered in the denser portions of the tree.
- Calibration and validation: To establish a strong positive correlation between amplitudes and core WD, calibration and validation studies are often conducted. These studies involve comparing Resistograph measurements with corresponding core samples from the same tree or similar trees. By validating the amplitudes against the core WD, a reliable relationship can be established, and the accuracy of Resistograph-based WD estimations can be ensured.

These results also indicate that the Resistograph is an efficient, non-destructive instrument for testing WD. Without the necessity of the destructive testing, the Resistograph-generated variables can provide a fast and efficient method for determining the density and quality of wood. This is especially beneficial in situations where the wood must be evaluated on-site, such as in historical structures or cultural heritage sites. The predicted WD value derived from Resistograph-generated variables may differ based on the Resistograph model employed and the model's calibration. In our research, the true value of WD is determined by destructive testing techniques such as the wood core test. The wood core test entails removing a fresh sample of wood from a standing, living tree and determining its density in a laboratory (Fig. 4b). Figure 4c illustrates tree damage caused by the Haglof increment borer. However, beeswax is used to seal the coring opening during the tree evaluation.

In this study, the true density of wood is used to validate the Resistograph drilling resistance independently. It is considered the gold standard for assessing the quality and mechanical properties of wood products because it involves removing a core of wood from the tree and analysing it in a laboratory.

Table 1 demonstrates that there is a strong correlation ($r = 0.791$) between the predicted WD from the Resistograph variables and the WD from the tree's wood core that was extracted using increment borer. In addition, the correlation shows no statistical difference ($p > 0.05$) between the predicted WD based on Resistograph variables and the WD of cores. These findings provide evidence to support the existence of a meaningful relationship between the two WD values derived from different methods. It also indicates that the two measurements are closely aligned or have minimal variation. This implies that the predicted WD is a reliable estimate or an accurate representation of the core WD. The similarity between the predicted WD value and the core WD suggests that the predicted WD value can be used as a substitute or proxy for the actual measurement of core WD. This can be particularly useful when obtaining core WD measurements, which is challenging, time-consuming, or costly. The close agreement between the two values provides confidence in the accuracy and validity of the prediction. Several studies have investigated the relationship between the predicted value of WD from Resistograph-generated variables and the actual value of WD from the wood core. The results are consistent with these findings. In a study conducted by Bouffier et al. (2008), for instance, the Resistograph-generated variables were used to predict the WD of samples of maritime pine. The predicted values of WD were compared to the true values of WD determined by the wood core test. The correlation between the predicted and true WD values was substantial, indicating that the resistograph accurately predicted the WD of the samples.

Table 2 allows for a straightforward comparison between the WD of different samples and their corresponding resistance measurements, which can be useful in identifying patterns or correlations in wood properties. Generally, higher wood densities tend to correspond with higher amplitude values, which indicates greater resistance. For example, the highest density of 0.93 g/cm^3 has an amplitude of 38.90%, and the second-highest density of 0.82 g/cm^3 has an amplitude of 48.67%. The WD ranges from 0.32 to 0.93 g/cm^3, indicating a diverse set of samples in terms of density. Lower WD tends to have lower amplitude values. For instance, a density of 0.32 g/cm^3 corresponds to an amplitude of 7.80%, and a density of 0.40 g/cm^3 corresponds to an amplitude of 6.80%. The general trend of higher density corresponding to higher amplitude suggests that denser wood typically offers greater resistance when

Table 1 Unpaired sample correlations

	N	Correlation	Sig
Cores WD (from the increment borer)—Predicted WD (from the Resistograph)	17	0.791	0.055

drilled. This is expected as denser wood usually has more compact cellular structures, making it harder for the drill to penetrate.

Table 2 Data of true WD from the fresh wood core (extracted using an increment borer) and amplitude (resistance data from a Resistograph)

Wood Density (g cm^{-3})	Amplitude (%)
0.75	22.00
0.73	37.07
0.45	4.00
0.82	48.67
0.40	6.80
0.32	7.80
0.54	23.20
0.76	33.67
0.61	27.10
0.56	22.15
0.55	7.00
0.47	21.53
0.63	37.45
0.41	26.00
0.55	13.10
0.93	38.90
0.56	11.28
0.75	22.00
0.73	37.07
0.45	4.00
0.82	48.67
0.40	6.80
0.32	7.80
0.54	23.20
0.76	33.67
0.61	27.10
0.56	22.15
0.55	7.00
0.47	21.53
0.63	37.45
0.41	26.00
0.55	13.10
0.93	38.90
0.56	11.28

Figure 6 depicts a linear regression model of core WD and predicted WD, which explains 76.8% of the variation in the predicted WD. This indicates a strong relationship between the two variables. Additionally, the statistical significance (p = < 0.5) suggests that the observed relationship is unlikely to have occurred by chance and reinforces the reliability of the model's prediction. These findings provide evidence to support the predictive capability and reliability of the linear regression model. The linear regression model demonstrates a strong relationship between core WD and predicted WD, explaining a substantial proportion of the variation in the predicted WD.

Similarly, the Resistograph-generated variables were used to predict the WD of Norway spruce samples in a study conducted by Içel & Güler (2019). The predicted values of WD were compared to the true values of WD determined by the wood core test. The results revealed a moderate correlation between the predicted and true values of WD, indicating that the variables derived by the Resistograph were less effective at predicting the WD of the samples. The accuracy of the predicted WD value based on Resistograph-generated variables is contingent on several variables, including the timber species, age, and growth conditions. In some instances, the variables generated by a resistograph may not accurately predict the true value of WD, resulting in inaccurate evaluations of the mechanical properties and quality of wood products.

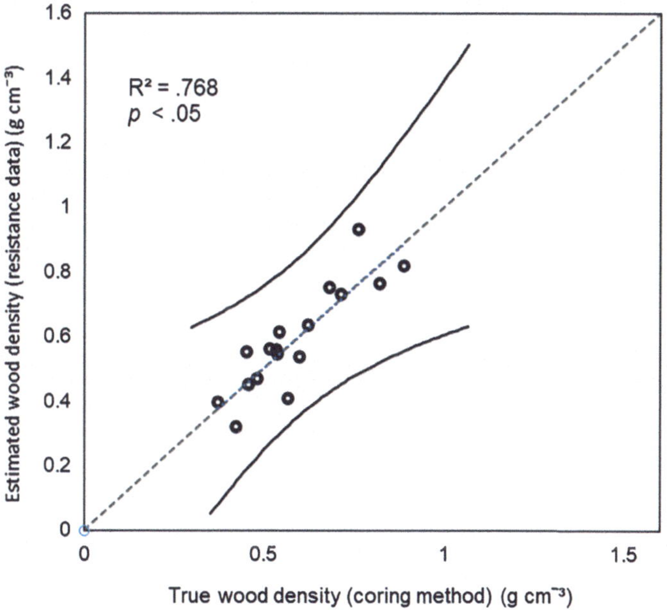

Fig. 6 Correlations between the predicted WD from resistograph-generated variables and true WD from wood cores

Although the Resistograph-generated variables are beneficial for non-destructive testing of wood, the accuracy of the predicted WD value can vary depending on various factors, including the wood species, age, and growth conditions. Therefore, it is necessary to calibrate the Resistograph model for each specific wood species and growth condition to predict the WD value accurately.

Understanding the significance of WD for SFM is crucial for preserving the integrity and productivity of forest ecosystems while meeting societal requirements. Based on this study's findings, the Resistograph has been shown to facilitate the WD assessment of sampled trees without significant defects.

4 Conclusion

In conclusion, using Resistograph technology to estimate WD in Eucalyptus trees offers several advantages over the traditional coring method. Resistograph technology allows for a non-destructive assessment of WD. Unlike coring, which involves extracting a core sample from the tree, the Resistograph measures the drilling resistance as it penetrates the wood. This non-destructive approach eliminates the need for invasive procedures that can potentially harm the tree. The Resistograph provides real-time measurements of WD along the entire drilling path. It generates a continuous profile of drilling resistance, allowing for a more detailed and comprehensive assessment of WD distribution within the tree.

In contrast, coring provides a single sample, which may only partially capture the variability in WD throughout the tree. It is important to note that while the Resistograph technology offers several advantages, it is not without limitations. The accuracy of WD estimation using Resistograph may vary depending on factors such as tree species, moisture content, and other tree characteristics. Therefore, it is crucial to validate the results obtained from Resistograph measurements by comparing them with coring or other established methods. By considering its strengths and limitations, the Resistograph can be a valuable tool for WD assessments and related research on economically significant trees.

Acknowledgements This project was funded by Ministry of Education Malaysia (Project: FRGS/1/2019/WAB07/UMS/01/1). I would like to thank Sabah Biodiversity Centre for providing the Access Licence (JKM/MBS.1000-2/2 JLD.12 (61) and permission to sample trees was provided by Sabah Forestry Department.

References

Bouffier L, Charlot C, Raffin A, Rozenberg P, Kremer A (2008) Can wood density be efficiently selected at early stage in maritime pine (*Pinus pinaster* Ait.)? Ann for Sci 65:106–113. https://doi.org/10.1051/forest:2007078

Chave J (2006) Measuring wood density for tropical forest trees: a field manual sixth framework programme (2002–2006). PAN-AMAZONIA Project, pp 1–6

Da Silva CES, Pace JHC, Gomes FJB, de Carvalho PCL, de Azevedo Reis et al (2020) Comparison between resistograph analysis with physical properties of the wood of Brazilian native tree species. Floresta e Ambient 27:1–7. https://doi.org/10.1590/2179-8087.005219

Egglesto HS (2006) 6 IPCC guideline for national greenhouse gas inventories. Institute for Global Environmental Strategies, Hayama, Japan

Flores O, Comes AD (2011) Estimating the wood density of species for carbon stock assessments. Methods Ecol Evol Br Ecol Soc. https://doi.org/10.1111/j.2041-210X.2010.00068.x

Fundova I, Funda T, Wu HX (2018) Non-destructive wood density assessment of Scots Pine (*Pinus sylvestris* L.) using Resistograph and Pilodyn. PLoS ONE 13:1–16. https://doi.org/10.1371/journal.pone.0204518

Içel B, Güler G (2016) Nondestructive determination of Spruce lumber wood density using drilling resistance (Resistograph) method. Turk J Agri For 40:900–907. https://doi.org/10.3906/tar-1606-76

Isik F, Li B (2003) Rapid assessment of wood density of live trees using the Resistograph for selection in tree improvement programs. Can J for Res 33:2426–2435. https://doi.org/10.1139/x03-176

Genetic variation in acoustic time of flight and drill resistance of juvenile wood in a large loblolly pine breeding population. For Sci 65:469–482. https://doi.org/10.1093/forsci/fxz002

Joseph J, Mui-How P, Debully-Maripa R, Wong Vun Chiong W, Korom A, James, D, Chai L (2022) Assessing intraspecific wood density variations of *Syzgium* sp. in tropical forest of Southwest Sabah. IOP Conf Ser: Earth Environ Sci. https://doi.org/10.1088/1755-1315/1053/1/012014

Joseph J, Maripa RD, Phua MH (2023) Envisaging the potential use of resistance micro drilling on wood density assessment: a review. In: Hashim UR et al (Eds) Green infrastructure. Springer, Singapore. https://doi.org/10.1007/978-981-99-7003-2_20

Karlinasari L, Danu MI, Nandika D, Turjaman M (2017) Drilling resistance method to evaluate density and hardness properties of resinous wood of agarwood (*Aquilaria malaccensis*). Wood Res 62(5):683–690

Kubus M (2009) The evaluation of using resistograph when specifying the health condition of a monumental tree Notulae Botanicae Horti Agrobotanici Cluj-Napoca 37:157–164. https://doi.org/10.15835/nbha3713112

Machado JS, Louzada JL, Santos AJA., Nunes L, Anjos O, Rodrigues J, Simões, RMS, Pereira H (2014) Variation of wood density and mechanical properties of Blackwood (*Acacia melanoxylon* R. Br.). Mater Des 56:975–980. https://doi.org/10.1016/j.matdes.2013.12.016

Mohd Shobri N, Hj Wan Ali W, Mt Akhir N, Md. Sakip S (2017) carbon footprint assessment at Universiti Teknologi MARA Seri Iskandar Campus, Malaysia. Malays J Sustain Environ (MySE) 2(1):59–72. https://doi.org/10.24191/myse.v2i1.5576

Mohd Zulkifli S, Noorazlan Ong N, Sheikh Zain R (2021). The impacts of Malaysian economic development on CO_2 emission. Malays J Sustain Environ (MySE) 8(1):105–118. https://doi.org/10.24191/myse.v8i2.13241

Oliveira JT, Wang X, Vidaurre GB (2017) Assessing specific gravity of young Eucalyptus plantation trees using a resistance drilling technique. Holzforschung 71:137–145. https://doi.org/10.1515/hf-2016-0058

Rinn F, Schweingruber FH, Schar E (1996) Resistograph and X-Ray density charts of wood comparative evaluation of drill resistance profiles and x-ray density charts of different wood species. Holzforschung 50:303–311. https://doi.org/10.1515/hfsg.1996.50.4.303

Todoroki CL, Lowell EC, Filipescu CN (2021) Wood density estimates of standing trees by microdrilling and other non-destructive measures. N Z J For Sci 51:1–12. https://doi.org/10.33494/nzjfs512021x74x

Tomczak K, Tomczak A, Jelonek T (2022) Measuring radial variation in basic density of Pendulate Oak: comparing increment core samples with the IML Power Drill. Forests 13(4):589. https://doi.org/10.3390/f13040589

The Green Enclave: Unearthing Strategic Prowess of Academic Libraries as Sustainability Catalysts

Haziah Sa'ari, Marlia Idrus, and Mohd Dasuki Sahak

Abstract This chapter delves into the intricate strategies employed by academic libraries to drive sustainability within esteemed Malaysian research universities. By drawing on an extensive literature review and conducting in-depth interviews with top management and academic librarians, the study unveils their pivotal roles in campus sustainability. The focus lies on knowledge and awareness, advocacy and leadership, and engagement and participation domains. Empirical evidence from the UI Green University Ranking situates the universities within the global sustainability context, affirming their commitment. The findings underscore academic librarians as powerful change agents in sustainability, offering valuable insights for researchers and practitioners in academic library sustainability. This chapter highlights academic libraries' contributions to sustainability, impacting higher education decision-making and the global discourse on nurturing sustainability in academic settings.

Keywords Academic libraries · Sustainability initiatives · Green library movement · Campus sustainability · Catalysts

H. Sa'ari (✉)
School of Information Science, College of Computing, Informatics and Mathematics, Universiti Teknologi MARA, Kampus Rembau, Rembau, Negeri Sembilan, Malaysia
e-mail: haziah095@uitm.edu.my; azie.saari@vuw.ac.nz

School of Information Management, Victoria University of Wellington, Wellington, New Zealand

M. Idrus
School of Information Science, College of Computing, Informatics and Mathematics, Universiti Teknologi MARA, Kampus Puncak Perdana, Shah Alam, Selangor, Malaysia
e-mail: marlia945@uitm.edu.my

M. D. Sahak
Sultan Abdul Samad Library, Universiti Putra Malaysia, Serdang, Selangor, Malaysia
e-mail: dasuki@upm.edu.my

E. Shaffie et al. (eds.), *Sustainable Green Infrastructure*,
https://doi.org/10.1007/978-981-96-1486-8_3

1 Introduction

1.1 Background

The growing environmental crises emphasize the crucial role of societal institutions in promoting sustainability (Cobbinah et al., 2019; Tong and Ebi, 2019). Academic libraries, as centers of knowledge, have emerged as important players in fostering ecological awareness, leading to the development of "green libraries" (Aulisio, 2013; Kulkarni, 2021). However, integrating sustainability into these institutions faces various challenges, including resource management and economic and ecological considerations (Jankowska & Marcum, 2010). To advance green libraries, an integrative framework that includes sustainable practices and addresses challenges identified by Fedorowicz-Kruszewska (2023) is necessary. The upcoming collection in Library Hi Tech, edited by Chiu and Ho (2022), highlights the role of green technology and librarians' attitudes toward eco-friendly practices (Soroya et al., 2022). By incorporating these insights, academic libraries can strengthen their sustainability leadership.

In Malaysia, academic libraries are not just repositories but key players in the knowledge economy, driving significant societal changes. These institutions, situated at the nexus of knowledge preservation and community engagement, bear the responsibility of spearheading sustainability initiatives. As Malaysia faces environmental challenges like deforestation and increasing carbon footprints, its commitment to sustainability, aligned with the 11th Malaysia Plan and Sustainable Development Goals (SDGs) (Department of Environment, Ministry of Natural Resources, Environment, and Climate Change, 2023), creates a unique opportunity for academic libraries to integrate sustainable practices. However, various barriers, including resource limitations and the lack of tailored frameworks, pose challenges to their role in the green library movement. To overcome these challenges, focused research on enhancing sustainability in Malaysian academic libraries is essential. This study aims to provide context-specific insights and strategies that resonate with the unique socio-academic landscape of these institutions. With this knowledge, librarians, policymakers, and the academic community can work together to advance the green library movement in Malaysia.

This chapter aims to highlight the significant efforts made by academic libraries to promote sustainability. This research provides valuable insights by focusing on operational practices and architectural considerations, supported by a thorough literature review and interviews with leading librarians in Malaysia's top research institutions. These insights will contribute to the broader discussions among librarians, policymakers, and scholars, enhancing the role of libraries in environmental stewardship and informing policy development in this area.

2 Literature Review

2.1 Pioneering Sustainability: Strategic Approaches and the Green Library Movement in Academic Libraries

The emergence of the green library paradigm marks a significant shift in academic libraries, highlighting the importance of sustainability. This evolution, rooted in an awareness of libraries' environmental impact, demonstrates a strong commitment to sustainable practices. Green libraries serve as leaders in promoting eco-friendly practices across their operations, services, and infrastructure, including energy efficiency, waste reduction, and the adoption of innovative, eco-friendly technologies (Aldrich et al., 2013). Beyond physical changes, green libraries focus on comprehensive environmental stewardship. These institutions extend their influence beyond their walls, engaging in discussions about climate challenges and conservation efforts. They collaborate with community stakeholders to strengthen their commitment to sustainability (Antonelli, 2016). The movement encompasses site selection, resource conservation, eco-friendly construction, indoor air quality, and the use of renewable energy (Shahet et al., 2015). Green libraries not only embrace sustainable practices but they also guide broader strategic goals by adopting a visionary approach.

As leaders in sustainable practices, green libraries refine and direct strategic initiatives. Academic discourse highlights libraries' efforts to reduce energy use and promote recycling. This focus on sustainability is evident in their services, teaching, and community engagement, embedding an eco-friendly ethos in all aspects of their operations. By integrating sustainability into their strategies, academic libraries address complex ecological challenges. Their commitment aligns their operations with broader environmental goals (Bell, 2018). The following sections will explore the specific strategies these institutions use to advance sustainability.

2.2 Quantification of the Green Movement and Sustainability

Assessing sustainable architectural design in libraries is essential to understanding their role in supporting sustainability. This goes beyond basic functionality, aiming to create spaces that minimize environmental impact. Key elements include maximizing natural light, implementing energy-efficient systems, and selecting eco-friendly materials (Edem and Atu, 2016). Modern architectural designs also emphasize flexibility to meet changing user needs, preventing the construction of unnecessarily large spaces. It is also important to enhance biodiversity through green landscaping and strategic site choices that reduce carbon emissions from transportation (Rasooli and Weingaertner, 2020). Libraries should aim to design spaces that promote user well-being and intellectual growth, becoming models of sustainable design and institutional commitment.

Brown's (2003) principles of sustainable design, which include energy efficiency, green materials, and improving indoor environmental quality, align with the metrics used to assess progress in the green movement. These principles are consistent with the Greenmetric UI evaluation categories, developed in 2015 to assess sustainability trends in academia, such as energy conservation, waste reduction, water conservation, and educational initiatives (Khairul Anuar, 2017). Many libraries have embraced green practices not just as a responsibility but as a transformative shift to enhance indoor environments (Reitz, 2004; Pan, 2012; KU Eco Library, 2012). Academic librarians play a key role in this transformation, integrating elements like optimal lighting, effective ventilation, and outdoor reading areas. These initiatives are consistent with green libraries' role as leaders in sustainable change. We evaluate library sustainability using key criteria such as budget allocation, energy efficiency, water conservation, paper reduction, and institutional eco-initiatives. The UI-Greenmetric World University Ranking's comprehensive approach to measuring environmental stewardship (UI Greenmetric, 2023) serves as the model for these criteria.

2.3 *Academic Librarians: Catalysts for Sustainable Change*

In this study, we emphasize the critical role of academic librarians in advancing sustainability within educational institutions. Drawing on the conceptual frameworks of scholars like Buchanan and Boddy (1992), Kanter (2000) and Rogers (2003), librarians emerge not just as participants but as leaders in driving sustainability forward. Within academia, librarians play multiple roles in promoting sustainability. They not only share their expertise, but they also serve as catalysts, advocates, and leaders of sustainable reforms. Their adaptability and broad skill set allow them to integrate sustainable practices into library operations, services, and curricula. They work collaboratively with faculty, scholars, and administrators, ensuring that their contributions extend beyond knowledge dissemination to embedding sustainability across the institution. Visser and Crane (2010) describe the varied roles of change agents, a concept perfectly embodied by the efforts of academic librarians. Their roles, from sustainability experts to advocates, are impacting not only within the institution but also across the academic community.

2.4 *Justification and Rationale for Selected Elements in Assessing Academic Librarians as Change Agents for Sustainability*

Based on key arguments, specific evaluation criteria are selected for assessing academic librarians as change agents for sustainability. First, it is essential to explore librarians' awareness of sustainability issues, especially in light of the pressing need to enhance their environmental knowledge (Tribelhorn, 2023; Kousar et al., 2022; Calculli et al., 2021). Second, assessing librarians' involvement and dynamism in sustainability efforts is crucial, given their unique potential to drive significant changes, as noted by Appleton and Woolley (2023). Finally, it is crucial to examine the advocacy and leadership demonstrated by academic librarians in the sustainability field, given its close connection to the implementation of transformational strategies (Hamad & Al-Fadel, 2022).

This three-part approach—focusing on knowledge and awareness, engagement and participation, and advocacy and leadership—creates a comprehensive framework that highlights the importance of academic librarians in promoting sustainability. With this approach, institutions can align their environmental efforts with foundational goals and empower librarians to lead transformative initiatives toward a more sustainable future. As illustrated in Fig. 1, this conceptual framework captures the diverse contributions of librarians, showcasing their pivotal role in advancing environmental stewardship and promoting sustainable practices.

Fig. 1 The nexus of academic librarians as sustainability change agents (developed by authors)

3 Method of the Study

This study embarks on an exploration of the pivotal roles undertaken by academic librarians in catalyzing sustainability within two prominent Malaysian public universities, both of which have garnered recognition as Research Universities (RU) by the Malaysian Ministry of Higher Education. These universities have demonstrated a fervent commitment to sustainable practices, as evidenced by their participation in various sustainability initiatives and rankings (Fadzil et al., 2012; Rahman, 2014; University of Malaya Eco-Campus Blueprint (UMECB), 2016; Khairul Anuar, 2017). The research employs a nuanced case study approach, qualitative data analysis, and a deliberate inquiry into quantitative sustainability initiatives, providing a comprehensive view of academic librarians' perspectives and their significant contributions to sustainability.

The selection of these two distinguished public universities is rooted in their RU status, which underscores their leadership in research and innovation, making them exemplary cases for studying the role of academic librarians in advancing sustainability. Their high-standing in sustainability rankings (UIGM 2016, 2017, 2018) further justify a closer examination of how academic librarians contribute to the broader sustainability goals within these academic institutions. The study engages four academic librarians, carefully selected based on their roles as key drivers of sustainability initiatives within their respective universities. The selection process involved purposive sampling, focusing on chief librarians, deputy chief librarians, and Green Library Managers, ensuring that the insights gathered are both rich and representative of the broader sustainability efforts.

This selection is also informed by the unique challenges these librarians face, particularly regarding the handling and sharing of data. Some data essential to sustainability efforts is restricted for internal use only due to institutional policies and confidentiality agreements, presenting a significant challenge that this study acknowledges and seeks to explore. The insights gathered from interviews are analyzed through content-based analysis for quantitative data and thematic analysis for qualitative data. These methods are aligned with established content analysis techniques (Cavanagh, 1997; Berg and Latin, 2008; Zhang and Wildemuth, 2009), providing a rigorous framework for interpreting the data. Ethical considerations guide this research, with protocols in place to ensure informed consent, anonymization of informants, and the confidentiality of sensitive data, particularly in light of restrictions that limit the sharing of certain information beyond internal stakeholders. The study acknowledges its limitations, particularly the focus on only two universities and the selective nature of the informants. These factors, while enhancing the depth of the study, also limit the generalizability of the findings, a trade-off that is recognized and addressed within the scope of scholarly rigor.

4 Findings and Discussions

This section presents the findings and discussions derived from our investigation, structured into two main parts. Part I focuses on the insights obtained from interviews with top management and academic librarians, particularly regarding their knowledge, advocacy, and engagement in sustainability. Part II highlights the latest green initiatives implemented by Case A and Case B, showcasing their commitment to environmental conservation and the integration of sustainable practices within their institutions (Table 1).

4.1 Knowledge and Awareness

We carefully designed the interview questions to explore how academic librarians promote green sustainability awareness among staff and the broader campus community. Through their strategic efforts, academic librarians play a crucial role in fostering a culture of ecological awareness within the university setting.

4.1.1 Proactively Facilitating Sustainability Engagement Through Knowledge Transfer

The responses from our informants, representing their respective universities, consistently emphasized their endorsement of their institutions' green sustainability agendas. The active involvement of academic librarians in sustainability programs,

Table 1 Demographic profile

Elements of sampling	Case A	Case B
Status of institution	Research University	Research University
Geographical location	Centre of Peninsular	Southern of Peninsular
Participants	Chief Librarian/Deputy/Green Library Manager Jusa C/S48/S44	Chief Librarian/Deputy/Green Library Manager Jusa C/S48/S44
Experience	Over 15 years	Over 15 years
Available sustainability achievements of libraries and institutions	√	√
Number of participant	2	2
Years of experience in green library	More than 5 years	More than 5 years
Specialized department responsible for green library	Library quality department	Green library department

managed by the Campus Sustainability Office, illustrates the library's significant role in promoting awareness and knowledge about green sustainability. This proactive involvement highlights the transformative influence academic librarians have as they guide their institutions toward greater environmental consciousness.

The creation of dedicated Green Managers, supported by the Library's Green Office Steering Committee, further reinforces the library's proactive role in sustainability efforts. As one informant, Academic Librarian 1, noted, "The systematic integration of sustainability updates during biannual library committee meetings has greatly enhanced knowledge and awareness. We consistently inform our staff and stakeholders about ongoing sustainability initiatives through this approach (CSB, Q1, Academic Librarian 1). This transparent reporting mechanism not only fosters accountability but also encourages active participation, allowing stakeholders to clearly see the impact of their contributions.

Additionally, the library's annual report submission to the Campus Sustainability Officer underscores its ongoing commitment to advancing sustainability within the university. This report serves as a platform to highlight the library's achievements and contributions to sustainability within the broader institutional context. This alignment with the university's comprehensive sustainability goals reflects a shared commitment to ecological responsibility. Through their efforts to engage both staff and stakeholders in sustainability initiatives, academic librarians extend their influence beyond mere awareness, nurturing a culture of environmental mindfulness. The knowledge and insights gained from their involvement in sustainability programs drive behavioral changes and promote sustainable practices that reach beyond the library's physical space. The library, as a leader in promoting green sustainability awareness, plays a pivotal role in galvanizing collective efforts toward ecological responsibility and sustainable practices.

4.1.2 Proactive Initiatives Through Nurturing Sustainability Awareness

To understand the proactive measures undertaken by academic librarians to foster sustainability awareness within their academic communities, we conducted in-depth, semi-structured interviews that provided invaluable insights. These insights vividly highlight the dynamic role of academic librarians as catalysts for sustainable change and the diverse strategies they use to embed a culture of environmental stewardship within their institutions.

Academic librarians, operating under the Campus Sustainability Office, actively participate in a dedicated sustainability program. Their enthusiastic involvement reflects the library's deep commitment to sustainability and its critical role within the broader institutional sustainability framework. As one librarian articulated, "The active participation in the sustainability program underscores the library's commitment to fostering awareness and propelling sustainable actions" (CSB, Q1, Academic Librarian 2).

A key element of the library's success in promoting sustainability awareness is its practice of transparent reporting. The library shares biannual updates and progress reports during Parent Committee Meetings to keep stakeholders, including staff, informed about ongoing sustainability initiatives. This culture of openness fosters accountability and encourages active engagement, as noted by one informant: "The regular reporting fosters a sense of accountability among stakeholders, motivating their active engagement in positively contributing to sustainability endeavors" (CSB, Q1, Academic Librarian 2).

The library's approach to sustainability not only impacts its internal operations but also influences the broader campus community. By participating in the sustainability program, academic librarians gain insights into innovative practices and pioneering approaches to sustainability. This knowledge extends across the campus, creating a ripple effect that surpasses the library's immediate reach. As one librarian observed, "Our proactive engagement leads to the dissemination of knowledge that surpasses our immediate community, influencing behavioral norms and propagating sustainable practices at a campus-wide scale" (CSA, Q1, Academic Librarian 1).

Academic librarians, in their role as sustainability leaders, inspire collective commitment to environmental responsibility. Their proactive initiatives not only emphasize the importance of sustainability but also foster a culture of environmental awareness. This proactive stance has transformative impacts, as another informant noted: "The library's initiatives not only inspire emulation but also initiate a broader institutional shift towards sustainability, nurturing a culture that embraces ecological stewardship" (CSB, Q1, Academic Librarian 2). The strategic efforts of the library extend beyond its walls, influencing and motivating action across various segments of the institution. As one librarian aptly stated, "Our actions reverberate throughout the campus, creating a ripple effect that magnifies the scope and impact of sustainability consciousness" (CSB, Q1, Academic Librarian 2).

4.2 Advocacy and Leadership: Pioneering Transformative Change

This section examines the active and visionary roles of academic librarians in both Case A and Case B, highlighting their crucial contributions to advancing sustainability within their institutions. These librarians, as key advocates and leaders, play a significant role in fostering environmental responsibility and implementing sustainable practices, guided by the Campus Sustainability Officer, who spearheads the institution's sustainability efforts.

4.2.1 Dynamic Advocacy for Sustainable Practices

The introduction of green managers in various departments, including academic libraries, has proven to be a powerful force in enhancing the effectiveness of sustainability initiatives. An academic librarian observed that the grassroots level effectively realizes sustainability initiatives, fostering a bottom-up approach to sustainable practices (CSB, Q2, Academic Librarian 1). The combined efforts of the Green Managers and the Campus Sustainability Officer significantly strengthen advocacy and leadership in promoting sustainability. Within the library, the Green Manager provides guidance and mentorship, empowering staff to adopt sustainable practices and drive transformative initiatives. Collaboration with the Campus Sustainability Officer fosters coordinated decision-making and facilitates the exchange of expertise on sustainability strategies. This collaborative approach cultivates a shared sense of responsibility among staff and enhances their commitment to achieving sustainability goals. Another academic librarian emphasized that engaging both responsible entities aligns them with the institution's thought process and systemic outlook (CSB, Q1, Academic Librarian 1).

In Case A, one academic librarian elaborated on their comprehensive approach: "Our strategy is grounded in the Public Sector Conducive Ecosystems (EKSA) Implementation Guidebook, which guides our strategic appointments and the implementation of programs that promote green sustainability practices" (CSA, Q2, Academic Librarian 1). These efforts include eco-friendly visits to external organizations, establishing a strong recycling program, cultivating indoor greenery, and offering presentations on "Kitar Semula" (recycling) and "Bengkel Penjagaan Pokok Dalaman" (indoor plant care workshop). Additionally, the continuous reinforcement of the EKSA framework is central to their strategy. The library plays a pivotal role in raising awareness about green sustainability practices, aligning with the broader institutional goals of sustainability.

4.2.2 Pioneering Transformative Change Through Proactive Leadership

Academic librarians, as champions of sustainability, lead through specialized programs and transformative initiatives designed for both staff and the broader campus community. A key platform for these efforts is the annual gatherings organized by the Campus Sustainability Officer, where green managers from various departments, including the library, convene to foster effective communication and strategic coordination of sustainability initiatives. These collaborative meetings serve as a vital space for exchanging innovative ideas and best practices, cultivating a collective commitment to shared sustainability goals, and thereby enhancing the overall impact of these initiatives.

Case B's library exemplifies leadership through its unwavering commitment to continuous improvement and advancement in sustainability. The library's participation in the prestigious "Anugerah Kelestarian Sempena Majlis Citra Karisma" in 2020, which includes a comprehensive sustainability assessment, underscores its dedication to rigorous evaluation and the consolidation of sustainability efforts. This assessment forms the foundation for tracking progress and identifying areas for further improvement, providing a robust framework for data-driven decision-making and the development of a comprehensive sustainability management system. Additionally, the active involvement of academic librarians in the Academic Library SDG 2022 Match, particularly their participation in the Sustainability Award, highlights their proactive role in sustainability leadership. This engagement fosters a strong sense of ownership and responsibility among staff, aligning them with sustainability practices and outcomes.

Moreover, academic librarians have taken on leadership roles tailored to address the specific needs of special groups. A notable success is the Light Up My Heart Project: Special Learning Tool Kit (SLTK) Braille Materials, which won first place in the Academic Library SDG 2022 Match. This project, aimed at supporting visually impaired students at the Princess Elizabeth Johor Bahru PRS, aligns seamlessly with key Sustainable Development Goals (SDGs). It showcases the library's innovative approach, which includes the creation of 15 titles of Special Learning Toolkits (SLTK) in Braille format, complemented by audio formats accessible via QR codes. The recognition Case B's Library received for its outstanding sustainability efforts in the Academic Library SDG 2022 Match serves as an inspiration, motivating other academic institutions to intensify their sustainability efforts. This dedication to initiatives that resonate with the SDGs, coupled with a spirit of innovation, reflects the library's enduring commitment to leading and advocating for sustainability within the campus community.

Furthermore, the library's focus on inclusivity, particularly in meeting the needs of students with special requirements, exemplifies its commitment to fostering a sustainable learning environment that accommodates diverse user needs. Through these initiatives and achievements, academic librarians have set a high standard for sustainable leadership in higher education. Staff and the broader campus community reflect their legacy through elevated sustainability knowledge and awareness.

4.3 Engagement and Active Involvement as Sustainability Advocates

This section highlights the robust engagement and proactive participation of academic librarians in advancing sustainability initiatives within their institutions. Through their unwavering dedication and leadership, academic librarians play a pivotal role in driving sustainability across the campus. As champions, mentors, and change agents, they embody the university's commitment to addressing environmental challenges and promoting sustainable practices. Their dynamic involvement not only ensures successful implementation of initiatives but also fosters a collective responsibility for sustainable outcomes.

4.3.1 Proactive Involvement as Sustainability Advocates

Insights gathered from informants reveal that academic librarians view sustainability not merely as an obligation but as an integral aspect of their organizational identity. Their engagement goes beyond compliance; they see themselves as advocates for sustainability, driven by a genuine responsibility to cultivate an eco-conscious campus. This proactive stance positions them as influential proponents, inspiring others to join the journey toward sustainability. Through frequent interactions with students, faculty, and staff, academic librarians naturally integrate sustainability principles into their daily activities.

As exemplars of sustainable behavior, they serve as powerful role models, motivating the broader campus community to adopt environmentally conscious practices. Moreover, their active engagement seamlessly integrates sustainability initiatives into the core of library operations. They utilize their expertise and insights to contribute to sustainability planning and decision-making, which results in initiatives tailored to the specific needs of stakeholders. This heightened level of involvement enhances the relevance and effectiveness of sustainability efforts, promoting their broader acceptance within the campus environment. One academic librarian observed, "Our engagement in sustainability stems from a genuine sense of duty as academic librarians. Beyond institutional mandates, we are deeply committed to fostering a greener campus. Our proactive participation not only aligns with our institution's sustainability vision but also empowers us to drive positive change among library users and the broader campus community" (CSB, Q3, Academic Librarian 2).

Furthermore, librarians' active participation in sustainability initiatives cultivates a shared sense of ownership and commitment to sustainable practices within the library. Their dedicated efforts contribute to a collective commitment to sustainability, fostering a culture of continuous improvement and innovation that guides the library toward a more environmentally conscious future.

Despite the challenges of balancing core responsibilities with sustainability commitments, academic librarians recognize the importance of meeting top management's expectations. "By actively engaging in sustainability efforts, we ensure that we remain integral to our institution's dedicated path toward sustainability. We, as academic librarians, understand the critical importance of maintaining our role as a key knowledge hub within the university. Embracing sustainability is not just a choice; it is a commitment to maintaining our relevance and indispensability within the evolving academic landscape" (CSB, Q3, Academic Librarian 2). "Our university ranks 25th in the UI-GreenMetric World Ranking, establishing it as Malaysia's leading institution in this domain," stressed another librarian. As committed university members, supporting and actively participating in initiatives that enhance our green campus distinction is a shared responsibility. Libraries play a central role by aligning with university agendas, contributing to the achievement of SDGs, and leading green campus initiatives" (CSA, Q3, Academic Librarian 1).

4.3.2 Proactive Participation as Sustainability Advocates

The dedication of academic librarians to sustainability extends beyond merely meeting institutional mandates or fulfilling Key Performance Indicators (KPIs). It embodies a profound ethical commitment, underscoring their determination to foster a more sustainable campus environment. These librarians are not just fulfilling duties; they are fervent advocates for sustainability, integrating these values into their professional identities and finding satisfaction in driving meaningful, positive change for the broader community. One informant expressed that their engagement in sustainability initiatives stems from a profound sense of duty to environmental stewardship. It transcends the boundaries of institutional obligations, embodying a loftier aspiration to enact palpable change within our campus community and beyond" (CSB, Q3, Academic Librarian 1). For instance, in 2019, their proactive commitment led to the presentation of a seminar titled "The Green Library Initiative at Our University" during the ICoLIS conference, effectively spreading awareness of their efforts (CSA, Q3, Academic Librarian 1).

The findings highlight the significance academic librarians place on proactive participation, as they actively encourage and inspire one another to engage in sustainability efforts. The Green Office Committee, comprising both librarians and support staff, fosters this collective motivation. This committee plays a crucial role in collaborative decision-making, enabling librarians to share ideas, voice concerns, and propose innovative solutions, thereby directly influencing the course of sustainable practices within the institution.

Additionally, the library's strategic investment in enhancing librarians' expertise in sustainability reflects its commitment to sustainability. By prioritizing professional development and training focused on sustainability, the library not only advances its sustainability initiatives but also equips librarians with the necessary knowledge, skills, and capacity to address complex sustainability challenges. "The library's commitment to enhancing our understanding and proficiencies in sustainability acts as a catalyst for adopting leadership roles in its promotion," another informant noted. It marks a transformative journey, transitioning us from passive participants to proactive advocates and catalysts for transformative change" (CSB, Q3, Academic Librarian 1).

Furthermore, academic librarians extend their efforts beyond the institutional boundaries by collaborating with external stakeholders, including non-governmental organizations (NGOs) focused on sustainability. One informant emphasized, "Our roles as academic librarians entail an ethical obligation to transcend the confines of our institution. Engagement with external NGOs in the realm of sustainability broadens our horizons, granting access to diverse expertise, knowledge, and resources. These collaborations resonate as harmonious notes within a larger symphony of interconnected sustainability endeavors" (CSB, Q3, Academic Librarian 1).

4.4 Challenges in the Implementation of Sustainability

The initiatives toward sustainability within academic libraries is defined by both steadfast commitment and inherent challenges. A significant obstacle lies in the complex interweaving of activities that often span various campus departments (PTJs), necessitating coordination that goes beyond surface-level efforts. Enhancing communication channels between PTJs and the Green Office is crucial to preventing redundancy, harnessing synergies, and fostering a unified sustainability strategy.

Despite the establishment of a dedicated task force committee, achieving seamless alignment and coordination remains an ongoing challenge. This highlights the need for a cultural shift that instills a shared sense of ownership and collective purpose among stakeholders. Strategic communication campaigns can initiate such a transformation by promoting active involvement and emphasizing the role of individual contributions in driving sustainability initiatives. Furthermore, academic libraries are well-positioned to lead interdisciplinary collaboration, serving as hubs of knowledge exchange and supporting a comprehensive approach to sustainability across the academic institution.

Resource allocation emerges as a complex issue requiring careful consideration. The sustained success of sustainability initiatives depends on adequate financial support and institutional backing. Substantial funding empowers academic librarians to conceive and implement impactful projects that drive transformative change. Therefore, institutions must recognize sustainability as an investment that yields long-term benefits for both environmental conservation and the institution's reputation. In this effort, regular evaluations and impact assessments are critical. They provide academic libraries with the tools to measure the effectiveness of their initiatives, identify areas for improvement, and capitalize on successful strategies. This data-driven approach aligns sustainability efforts with measurable outcomes, enhancing transparency and accountability.

4.5 Evaluation of Recent Sustainability Initiatives (Up to 2023)

In Part II, we conduct a thorough assessment of the accomplishments and influence of the sustainability initiatives undertaken by Case A and Case B Libraries up to the current year. This evaluation addresses the five key elements previously outlined, providing a comprehensive view of the libraries' strategic approaches. Table 3 systematically categorizes the initiatives, showcasing the libraries' cohesive efforts toward sustainability. These examples highlight the libraries' commitment to energy-efficient practices and other sustainable initiatives, offering clear benchmarks for continuous improvement and progress toward greater environmental sustainability (Table 2).

Table 2 Assessment of green sustainability for Case A and Case B

Case A	Case B
Budget efficiency	
√	√
Budget allocation for conservation programs is determined by the Green Office Committee. If a program requires financial resources, budget allocation requests are presented during the Library's Executive Meeting as needed	
Descriptions	
1. Appliance Efficiency: These are only a few examples that illustrate the commitment of both libraries to energy-efficient practices. For instance, they have invested in appliances such as refrigerators, air conditioning, and fans that adhere to environmentally-friendly standards 2. Optimized Cooling: Similarly, both Case A and Case B libraries have efficiently managed cooling systems through centralized and split-type air conditioning methods, demonstrating a shared commitment to energy efficiency 3. Awareness Campaigns: These examples shed light on the awareness campaigns initiated by both libraries to promote energy-saving practices and the adoption of LED lighting technology, further highlighting their dedication to enhancing energy efficiency	
Water conservation	
√	√
Descriptions	
1. Water-Saving Notifications: In both Case A and Case B libraries, these examples underscore the implementation of notifications in user stalls, surau (prayer rooms), and staff pantries, indicating their commitment to responsible water usage 2. Collaborative Pipe Replacement: Through collaborative efforts between the libraries and the Project and Maintenance Section, these instances highlight the replacement of conventional water pipes across the campus, demonstrating their joint commitment to reducing water wastage	
Paper reduction/paperless	
√	√

(continued)

Table 2 (continued)

Case A	Case B
Descriptions	
1. Digital Collections: These examples serve as evidence of the expansion of digital collections in both libraries, providing electronic alternatives to traditional print materials and actively contributing to reducing paper usage 2. Innovative Digital Services: The introduction of innovative digital services such as online research assistance, virtual reference desks, and digital interlibrary loan systems are further instances that showcase both libraries' dedication to minimizing the reliance on paper documents 3. Technology Infrastructure: The investments in advanced technology infrastructure, including high-speed internet access, modern computing systems, and secure cloud-based storage, serve as further examples of both Case A and Case B libraries' commitment to supporting digitization efforts and reducing paper consumption	

4.6 Proposed Framework: CASAL—Catalyzing Academic Librarians as Sustainability Leaders for a Greener Future

Building upon these insights, we introduce the "CASAL Framework: Catalyzing Academic Librarians as Sustainability Leaders for a Greener Future." This framework empowers librarians to proactively promote environmental stewardship and sustainability. It encompasses eight key components: Visionary Leadership, Strategic Engagement, Influential Advocacy, Continuous Professional Development, a culture of Environmental Consciousness, Transparency and Accountability, Resource Management and Partnerships, and Evaluation and Innovation (Fig. 2).

5 Conclusions

This study has revealed the significant and transformative role those academic librarians play in driving sustainability within Malaysian higher education institutions. Through visionary leadership, strategic engagement, and unwavering advocacy, these librarians have emerged as key agents in embedding sustainable practices into the fabric of their institutions. Their actions extend beyond merely implementing policies; they actively shape and guide the sustainability agenda, ensuring alignment with both institutional aspirations and national sustainability goals.

The analysis of sustainability initiatives across two case studies, Case A and Case B, underscores the effectiveness of these libraries in fostering environmental stewardship. Both libraries exhibit a strong commitment to sustainability, demonstrated by efficient budget management and strategic planning. The systematic allocation

Table 3 A comparison of recent initiatives to enhance sustainability and user services in Case A and Case B

Cluster	Case A	Case B
Accessibility enhancement	1. Three handicapped or special needs group parking spots	1. Braille Format Transliteration: Fostering reading interest particularly those with less visual effort 2. Special Learning Tool Kit (SLTK) Project: Creating a productive reading ecosystem and nurturing reading interest among visually impaired students 3. Library's Terangi Hatiku Project: Library's Terangi Hatiku Project: Braille Materials won first place in the 2022 Academic Library SDG Competition
User services enhancement	1. Implementation of Green Office Committee budget allocation requests for conservation programs 2. Provision of safer and more comfortable bicycle parking to promote health and wellness	1. Librarian Online Chat: Providing comprehensive referral services to meet user needs through face-to-face or online methods based on user preference 2. Virtual Gallery Development: Allowing visitors to virtually explore and browse the rich heritage of the gallery's collections and artifacts
Sustainability awareness	1. Providing water-saving labels to create awareness about water conservation 2. Go Green Screen Saver is an initiative to raise environmental awareness 3. Label QR Code Tanaman di Taman Herba PSAS to enhance user experience and knowledge sharing 4. Program Go Green Kitar Semula—for waste management education 5. Tips from the Kelestarian Hijau campaign for energy conservation 6. Anugerah Go Green Terbaik encourages sustainable ideas and efforts	1. Knowledge Sharing Programs on Sustainability: Raising awareness about sustainability practices among staff and the campus community 2. Monthly Mutual Help: Raising awareness about sustainability practices among staff and the campus community

(continued)

Table 3 (continued)

Cluster	Case A	Case B
Renewable energy provision and establishment	1. Installation of Solar PV System at Engineering and Architecture Library 2. Water Recycle Initiative for sustainable irrigation in the library Herb Garden	1. Online Fee Payment Module through UTMSmart: Offering a more effective and convenient payment method
Innovation and efficiency	1. Establishment of the Green Sustainability Committee 2. QR Code Plant Labels in the Library Herb Garden for disseminating information 3. Creativity through upcycled furniture production 4. Google Drive Log for upcycled goods sales records 5. Plant the date tree to enhance the esthetic value and green areas 6. Creativity through sustainable furniture production 7. Mutual help program for collaborative involvement	1. KIK (Kolaborasi, Inovasi, dan Kreativiti) under Library Smart Pro Group/ Lejar Management 2. Project: Promoting collaboration, innovation, and creativity and improving work productivity by addressing the issue of time
Health and wellness	1. Water recycling initiative: Reusing excess air conditioning water for Taman Herba irrigation	1. Sustainable Kayaking Program: Encouraging the adoption of a healthy lifestyle

of funds through a shared governance model ensures that resources are directed toward impactful conservation programs, reflecting a well-integrated approach to achieving environmental objectives, in line with SDG 12: Responsible Consumption and Production.

In terms of energy efficiency, both libraries have made significant strides by investing in energy-efficient appliances and optimizing cooling systems. Their proactive promotion of LED lighting and energy conservation awareness campaigns further illustrates their dedication to reducing energy consumption and fostering a culture of sustainability on campus, contributing to SDG 7: Affordable and Clean Energy. Water conservation efforts are another area where both libraries have excelled, implementing strategies such as installing water-saving notifications and collaborating on infrastructure improvements to reduce water wastage. These coordinated efforts reinforce the libraries' role as leaders in sustainable water management, aligning with SDG 6: Clean Water and Sanitation.

The move toward digitization highlights the libraries' commitment to reducing paper consumption and enhancing accessibility, supporting SDG 9: Industry, Innovation, and Infrastructure. By expanding digital collections and introducing innovative

Fig. 2 CASAL framework—catalyzing academic librarians as sustainability leaders for a greener future (developed by authors)

digital services, the libraries not only contribute to sustainability but also improve user convenience and accessibility. Investments in technology infrastructure, including high-speed internet and cloud-based storage, further support these paperless initiatives, showcasing a forward-thinking approach to library services. Both libraries have also prioritized sustainability awareness and education, engaging in campaigns and programs designed to educate the campus community on environmental stewardship. These initiatives, such as the Go Green Screen Saver in Case A and the Knowledge Sharing Programs in Case B, foster a culture of sustainability that extends beyond the library, influencing the broader academic environment, in support of SDG 4: Quality Education. Moreover, the integration of health and wellness into sustainability strategies reflects a holistic approach adopted by both libraries. Initiatives like Case A's water recycling for herb garden irrigation and Case B's Sustainable Kayaking Program highlight the interconnectedness of physical well-being and environmental health, reinforcing the libraries' comprehensive approach to sustainability, supporting SDG 3: Good Health and Well-being.

The comparison of recent initiatives between Case A and Case B reveals a clear pattern of proactive and innovative approaches to sustainability. While both libraries share common goals, such as enhancing accessibility and user services, they have also developed unique initiatives tailored to their specific contexts. This diversity of approaches not only fortifies their individual sustainability efforts, but also enhances our understanding of how to effectively integrate sustainability into academic library services. The CASAL framework, as discussed in this study, serves as a robust tool guiding academic libraries in their pursuit of sustainable excellence. This comprehensive framework encapsulates the dynamic roles academic librarians assume—ranging from acting as catalysts for change, advocating for sustainable paradigms, devising strategic pathways, implementing impactful actions, and exercising transformative leadership. Academic libraries in both public and private institutions can effectively use the CASAL framework as an indicator or parameter to measure the effectiveness of sustainability initiatives. This adaptability underscores its potential to foster congruence between institutional aspirations and sustainable practices, resonating on both national and global scales, particularly in relation to SDG 17: Partnerships for the Goals.

6 Recommendations

6.1 Enhanced Coordination and Communication

The study underscores the need for improved coordination across campus departments (PTJs). Given the complexity of sustainability initiatives that span multiple units, establishing more effective communication channels is essential. A central digital platform for sustainability projects and regular meetings between departments could cut down on duplication, make the most of synergistic opportunities, and help create a more cohesive sustainability strategy, all of which would help reach SDG 17: Partnerships for the Goals.

6.2 Increased Transparency in Resource Allocation

Governmental policies and confidentiality requirements constrain the transparency of resource allocation, despite the libraries' demonstrated budget efficiency. Libraries should provide general summaries of budget allocations in public reports. These summaries should outline areas of expenditure while respecting confidentiality, enhancing transparency without breaching policy constraints. This approach aligns with SDG 16: Peace, Justice, and Strong Institutions.

6.3 Strategic Investment in Sustainability Training

Continuous professional development is crucial for sustaining the momentum of sustainability initiatives. Institutions should invest in interdisciplinary training that equips academic librarians with the advanced knowledge and skills required to navigate the complexities of sustainability challenges and lead initiatives that are both innovative and impactful, supporting SDG 4: Quality Education.

6.4 Strengthened Institutional Support and Recognition

Institutional support is vital for the sustained success of sustainability initiatives. Universities should formally recognize the contributions of academic librarians in this domain, integrating sustainability goals into their performance evaluations and offering incentives for sustainability-related achievements. Such recognition would not only motivate librarians but also amplify their influence within the institution, aligning with SDG 8: Decent Work and Economic Growth.

6.5 Active Student Involvement in Sustainability Initiatives

Engaging students in sustainability projects led by librarians can significantly expand the reach and impact of these initiatives. Integrating sustainability education into the curriculum and offering experiential learning opportunities through library-led projects will foster a new generation of environmentally conscious leaders, in line with SDG 4: Quality Education.

6.6 Regular Evaluations and Impact Assessments

Regular evaluations and impact assessments are necessary to ensure the effectiveness of sustainability initiatives. These assessments will gage the results of sustainability programs, pinpoint areas for enhancement, and guarantee the expansion of successful strategies. An evidence-based approach to sustainability will enhance transparency and accountability, ensuring that these efforts contribute to tangible environmental and institutional benefits, in support of SDG 12: Responsible Consumption and Production.

6.7 Leveraging Sustainability to Enhance Reputation and Trust

Successfully implemented sustainability initiatives can significantly enhance the reputation, integrity, and trust of academic libraries within their institutions and the broader community. By leading in sustainability, academic libraries not only contribute to environmental goals but also strengthen their standing as vital, forward-thinking institutions. This, in turn, can attract students, faculty, and funding, furthering the library's mission and impact, aligning with SDG 11: Sustainable Cities and Communities.

Acknowledgements We extend our gratitude to Universiti Teknologi MARA (UiTM) and the Ministry of Higher Education (MOHE) for funding this research. Special thanks to Puan Haslina Othman and both the library green teams for their valuable cooperation and insights.

References

Aldrich RS, Benton S, Schaper L et al (2013) Sustainable thinking. Passageway to better buildings, budgets and beyond. In: Hauke P, Latimer K, Werner KU (eds) The green library. The challenge of environmental sustainability = Die Grüne Bibliothek: Ökologische Nachhaltigkeit in Der Praxis. De Gruyter Sau, Berlin and Boston, MA, pp 15–37

Antonelli M (2016) The role of academic libraries in promoting sustainability. Libr Manag 37(6/7):343–358. https://doi.org/10.1108/LM-12-2015-0126

Aulisio GJ (2013) Green libraries are more than just buildings. Libr Philos Pract (E-J) 919. Retrieved from https://digitalcommons.unl.edu/libphilprac/919

Bell S (2018) The role of sustainable library architecture in creating healthful environments. Libr Leadersh Manag 32(1):1–8

Berg KL, Latin RW (2008) Research methods in health, physical education, exercise science, and recreation, 3rd edn. Lippincott Williams and Wilkins, Baltimore

Brown J (2003) Sustainable architectural design for libraries. Library Hi Tech 21(3):292–299

Buchanan D, Boddy D (1992) The expertise of the change agent: public performance and backstage activity. Prentice Hall

Calculli C, D'Uggento AM, Labarile A, Ribecco N (2021) Evaluating people's awareness about climate changes and environmental issues: a case study. J Clean Prod 324:129244. https://doi.org/10.1016/j.jclepro.2021.129244

Cavanagh S (1997) Content analysis: concepts, methods and applications. Nurse Res 4(3):5–16

Chiu DKW, Ho KKW (2022) Editorial: 40th anniversary: contemporary library research. Library Hi Tech 40(6):1525–1531. https://doi.org/10.1108/LHT-12-2022-517

Cobbinah PB, Asibey MO, Opoku-Gyamfi M, Peprah C (2019) Urban planning and climate change in Ghana. J Urban Manag. https://doi.org/10.1016/J.JUM.2019.02.002

Department of Environment, Ministry of Natural Resources, Environment and Climate Change (2023) Sustainable development goals. Retrieved from https://www.ekonomi.gov.my/en/sustainable-development-goals.

Edem AN, Atu FE (2016) Sustainable library building: a conceptual approach to integrating energy efficient designs in libraries. Int J Libr Inf Sci 8(2):18–27

Fadzil ZF, Hashim HS, Che-Ani AI, Aziz S (2012) Developing a campus sustainability assessment framework for the national university of Malaysia. Int J Environ Ecol Geol Mar Eng 6:44–48

Fedorowicz-Kruszewska M (2023) Green libraries: barriers to concept development. Libr Manag. https://doi.org/10.1108/LM-04-2022-0041

Hamad F, Al-Fadel M (2022) Advocacy of the sustainable development goals in Jordanian academic libraries. IFLA J 48(4):492–509

Jankowska MA, Marcum JW (2010) Sustainability challenge for academic libraries: planning for the future. Coll Res Libr 71(3):235–254. https://doi.org/10.5860/crl-75r1

Khairul Anuar MN (2017) UPM is ranked world's top 27 and Asia's top 3 according to the UI-Greenmetric world university ranking. Retrieved from https://upm.edu.my/news/upm_is_ranked_worlds_top_27_and_asias_top_3_according_to_the_ui_greenmetric_world_university_ranking-37075

Kousar S, Afzal M, Ahmed F, Bojnec Š (2022) Environmental awareness and air quality: the mediating role of environmental protective behaviors. Sustainability 14(6):3138. https://doi.org/10.3390/su14063138

Kulkarni PP (2021) Green library: concept, features and elements. Int J Recent Technol Eng, 9(2):9349–9351. https://doi.org/10.35940/ijrte.B1508.1182R22021

Pan S (2012) Sustainable architectural design in libraries: the case of national library, Singapore. Libr Manag 33(8/9):601–612

Rahman S (2014) UMCares Annual Report 2013/2014

Reitz JM (2004) Green libraries: Minimizing negative impact and maximizing indoor environment quality. J Environ Libr 30(1):22–34

Rogers EM (2003) Diffusion of innovations. Free Press

Shah L, Kumar S, Shah MK (2015) Green libraries in academic institutions: the need of the hour. Int J Res. https://doi.org/10.29121/GRANTHAALAYAH.V3.I9SE.2015.3129

Soroya SH, Mahmood K, Shahid Soroya M, Hussain S, Ilyas A (2022) Green computing intent and behavior of Pakistani academic librarians: PLS-SEM analysis. Libr Hi Tech 40(6):1996–2012

Tong SY, Ebi KL (2019) Preventing and mitigating health risks of climate change. Environ Res. https://doi.org/10.1016/J.ENVRES.2019.04.012

Tribelhorn SK (2023) Preliminary investigation of sustainability awareness and activities among academic libraries in the United States. J Acad Librariansh 49(3):102661. https://doi.org/10.1016/j.acalib.2022.102661

UI Greenmetric (2023) Retrieved from https://greenmetric.ui.ac.id/about/service

University of Malaya Eco-Campus Blueprint (UMECB) (2016). Sustainability science research cluster. University of Malaya Press, University of Malaya

Universitas Green Metric World University Rankings (UIGM) (2016). Retrieved from http://greenmetric.ui.ac.id/overall-ranking-2016/

Universitas Green Metric World University Rankings (UIGM) (2017). Retrieved from http://greenmetric.ui.ac.id/overall-ranking-2017/

Universitas Green Metric World University Rankings (UIGM) (2018). Retrieved from http://greenmetric.ui.ac.id/overall-ranking-2018-2/

Visser W, Crane A (2010) Corporate sustainability and the individual, understanding what drives sustainability professionals as change agents. Soc Sci Res Netw. Retrieved from http://ssrn.com/abstract141559087

Zhang Y, Wildemuth BM (2009) Qualitative analysis of content. In: Wildemuth B (ed) Applications of social research methods to questions in information and library science. Libraries Unlimited, Westport, CT

The Chromatographic Profiling of *Piper Sarmentosum*

Hannis Fadzillah Mohsin, Kathleen J. Jalani, and Ibtisam Abdul Wahab

Abstract *Piper sarmentosum* Roxb. is well known as kadok. It is a small plant that grows only to a height of half a meter. It is utilized as a medicinal herb. Growing and newly planted samples of *P. sarmentosum* are also observed as landscape plants. The species is reported to contain flavonoids. In this study, an automated instrumentation was selected to observe the chemical profile by using High-Performance Liquid Chromatography (HPLC). The on-site method was differentiated with the published procedure, of the national herbal monograph. The mobile phase was opposed to the gradient elution of water and methanol. Instead, the solvent consisted of 0.3% formic acid and acetonitrile. In the findings, the HPLC chromatograms displayed five acceptable peaks, one of them was considered as the minor component. In comparison with the online data, five biomolecules were possibly detected in the methanolic extract, comprising caffeic acid, amurensin, a flavonoid glucoside, rutin, p-coumaric acid, and quercetin. It is hoped that the research would contribute to the sustainability of this traditional plant.

Keywords Chromatography · Herb · Monograph · *Piper* · Sustainability

1 Introduction

Piper sarmentosum (Fig. 1) is an herbaceous plant from the Piperaceae family (Hussain et al., 2012). It is a small and spreading species, growing in Northeastern India, Malaysia, and China (Waman et al., 2019). Articles were published regarding the *Piper sarmentosum*'s research and its role as an antioxidant (Ismail et al., 2018; Rahman et al., 2016; Sun et al., 2020). Short and comprehensive reviews of *Piper sarmentosum* were available recently (Ibrahim et al., 2022a, b). It is also known as betel vine, the wild leaf wraps being one of the most popular Thai foods (Khanthapok and Sukrong, 2019). A domestic sample of *P. sarmentosum* can be seen in Fig. 1.

H. F. Mohsin · K. J. Jalani · I. Abdul Wahab (✉)
Faculty of Pharmacy, Universiti Teknologi MARA (UiTM), Selangor Branch, Puncak Alam Campus, 42300 Bandar Puncak Alam, Selangor, Malaysia
e-mail: ibtisam@uitm.edu.my

E. Shaffie et al. (eds.), *Sustainable Green Infrastructure*,
https://doi.org/10.1007/978-981-96-1486-8_4

Fig. 1 A domestic sample of *Piper sarmentosum*, showing its cylindrical white inflorescence (Pinya et al., 2022)

This plant flowers around the year and is plentiful in the rainy season (Wiart, 2006). Meanwhile, recent phytochemical analyses of the chloroform extract of *P. betle* L. leaves led to the isolation of a new phenylpropanoid analogue, for the first time (Atiya et al., 2023). Chavibetol or *meta*-eugenol is the major and potent phytotoxin in essential oil from the betel leaves (Kemprai et al., 2023).

Betel vine cultivation may face some problems due to the existence of its synonyms. Similar betel plants may be cultivated with different names, and in some regions, different betel plants are cultivated with the same names (Das et al., 2016). In the meantime, the use of *Piper* roots in obesity management (Gaur et al., 2023) and the leaves for antibacterial studies (Valle et al., 2016) were published. There is a need to compile and disseminate the knowledge to exploit this underutilized resource for agricultural sustainability. On the other hand, some *Piper* species are reported to be overharvested for their medicinal and commercial values. This overharvesting can lead to population declines and even extinction.

The traditional uses of *P. sarmentosum* encompass a wide spectrum of biological activities, including anticancer, anti-osteoporotic, antibacterial, antioxidant, antihypertensive, anti-inflammatory, and anti-diabetic characteristics (Adib et al. 2024; Sun et al. 2020). While the potential application of *P. sarmentosum* in green infrastructure has materialized by the growing *P. sarmentosum* (NParks Flora Fauna Web, 2023). These planted specimens exhibit glossy dark greens, and are also used as landscapers by the Shah Alam City Council (Fig. 2).

The alkamides and alkaloids were the bioactive markers in *Piper* species. These compounds were investigated by using High-aPerformance Liquid Chromatography (HPLC), with ultraviolet detection (Hussain et al., 2012; Khalid et al., 2011). Ethanol was able to extract pellitorine, sarmentine, and sarmentosine from the fruits of *P. sarmentosum*, by reflux. These alkaloids (Tuntiwachwuttikul et al., 2006) were separated from the roots too. Although the compounds were eluted by an isocratic mode, the mobile phase was a tertiary solvent system, comprising of methanol, water, and acetonitrile. Later, the leaves' phytocomponents were investigated by using the mass

Fig. 2 A The planted *Piper sarmentosum* is used as a landscaper by a city council, as part of the greenery initiatives for sustainable urban development

data (Bactiar and Nur Azlina, 2019). In this study, the *P. sarmentosum* leaf extract was subjected to a reverse-phase HPLC, by varying the elution modes.

2 Methodology

First, *P. sarmentosum* leaves were obtained from a local market in July 2020. The dried powdered leaves were soaked in methanol and sonicated, prior to the analysis. The extract was filtered through a 0.45 μm Millipore filter, into the vials. The chromatographic profiling of the extract was performed by using Reverse-Phase High-Performance Liquid Chromatography (Agilent 1200 RP-HPLC) (Purba et al., 2021). About 10 μl of the sample was injected for the HPLC, which was run through a column (Thermo C18; 250 mm × 4.6 mm and 5 μm of pore size) as a stationary phase. Two approaches were employed to compare the extract profiles, between a rapid and a non-isocratic, stepwise gradient (Table 1) elutions in RP-HPLC methods. The mobile phase consisted of various compositions of 0.3% formic acid in water and acetonitrile, with a flow rate of 1 ml/min. The data were recorded via dual wavelengths (210 and 360 nm) through diode array detector (DAD) and the column temperature was maintained at 36 °C throughout the experiment. The running time was set at 30 and 50 min, for the rapid and stepwise gradient modes, respectively. The data acquisition was performed by using ChemStation software (Khalid et al., 2011).

3 Results and Discussion

Figure 3 displays the chromatograms of the *P. sarmentosum*. Four major compounds (**1**–**4**) contribute to four individual, significant peaks at retention time, $t_R = 11.772$, 11.845, 12.167 and 12.314 min, respectively. While a minor compound (**5**) was eluted at $t_R = 17.556$ min. When the stepwise mode was applied, the peaks (compound

Table 1 The solvents for the stepwise gradient elution

Time (minutes)	Acetonitrile (%)	0.3% Formic Acid in Water (%)
0	10	90
5	10	90
20	40	60
22	40	60
30	70	30
32	70	30
40	90	10
45	90	10
47	10	90
50	10	90

1–**4**) were retained and could be eluted later (Fig. 4), separately at t_R = 26.878, 27.460, 29.316, and 30.113 min. Later, the minor compound (**5**) was recorded at t_R = 42.859 min. In the absence of the standard compounds, these retention times could be compared. Following the characteristic of the peaks, e.g., intensity and order of the elutions, the compounds could be suggested as caffeic acid (**1**), amurensin, a flavonoid glucoside (**2**), rutin (**3**), *p*-coumaric acid (**4**) and quercetin (**5**) (Fig. 5). Further efforts would include the scale-up of this fractionation, to isolate the pure entities and to confirm whether amurensin could be simultaneously determined in a distinct peak, together with other aglycone (quercetin **5**) (Purba et al. 2021). Furthermore, an earlier phytochemical screening using HPLC technique also showed the presence of rutin **3** and vitexin **6** (apigenin-8-C-β-D-glucopyranoside) as the main flavonoids in aqueous extract of *P. sarmentosum* (Ugusman et al., 2012). The occurrence of both compounds **2** and **6** in the methanol extract is anticipated. The type of solvent used for the extraction of *P. sarmentosum* could influence the outcome of the investigation of its phytocomponents. In terms of HPLC solvent, methanol was utilized and documented by the Malaysian Herbal Monograph (GlobinMed, 2021), to detect the phytosterol, e.g., b-sitosterol. This approach is also able to sustain the utilization of *Piper* samples and their agro-waste for functional food and therapeutic applications (Gaur et al., 2023). The greenness of the reversed-phase LC technique was suggested and described in detail (Alqarni et al., 2021).

As for the flavonoids that could represent the peaks in this study, they can be identified either by performing isolation work or by comparing the retention time (t_R), with the standards (Ugusman et al., 2012). Hussain et al., (2009) commented that there are some internal and external factors that could contribute to the variations in the type of flavonoids found in *P. sarmentosum*. These elements would include the plant's metabolism that could influence its chemical content. Moreover, the growing and environmental conditions (Waman et al., 2019), the age of this natural resource, collection time, and drying and storage conditions would also determine the chromatographic results.

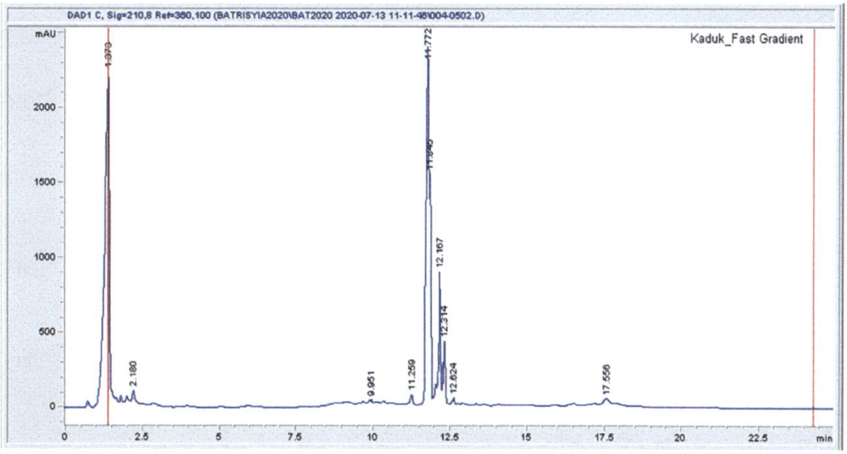

Fig. 3 The chromatogram of methanol extract of *P. sarmentosum* (fast gradient)

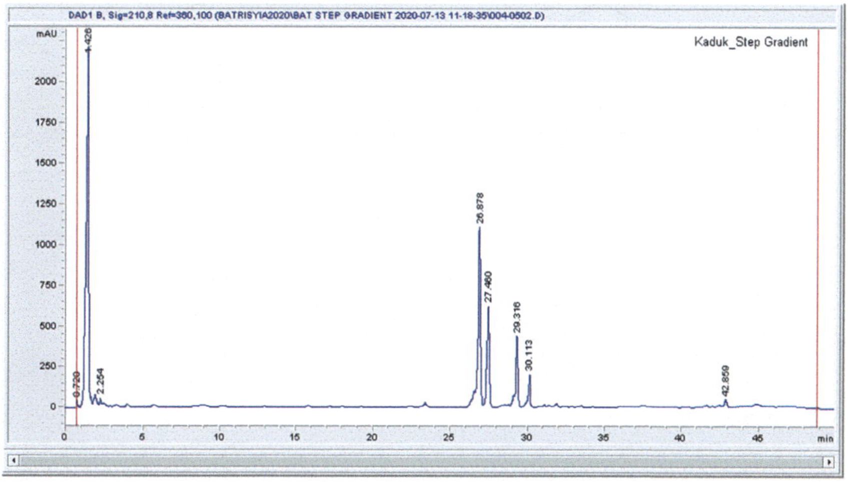

Fig. 4 The chromatogram of methanol extract of *P. sarmentosum* (stepwise gradient)

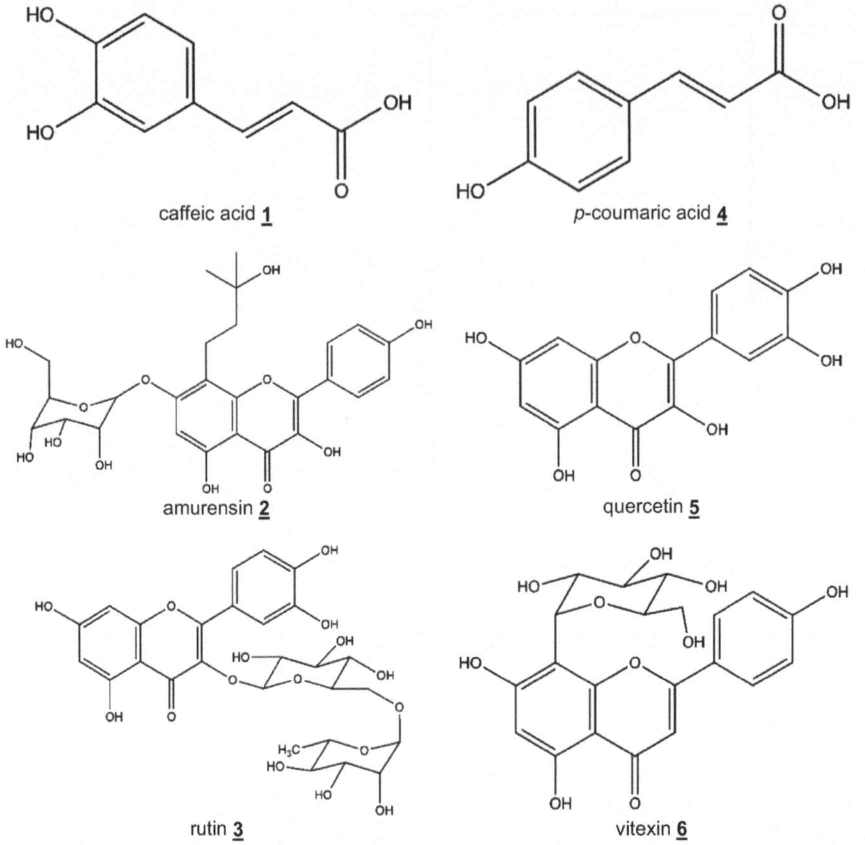

Fig. 5 The chemical structures of compounds (1–6) in *P. sarmentosum*

4 Conclusion

The HPLC chromatograms in this study showed non-identical peaks. The closely eluting peaks in Fig. 3 could be resolved when the stepwise gradient was applied, as seen in Fig. 4. As a result, the retention times between the two chromatographic procedures could be distinguished. Nevertheless, the compounds from both methods would not be similar, since overlapping peaks or shoulders still occur. For example, in compound 1, unsymmetrical signal would suggest other possible entities which could be separated either in aqueous or alcoholic medium, such as rutin 3 and vitexin 6. More research should be encouraged for the optimum utilization of this biomaterial. *P. samentosum* could be a sustainable herbal source of bioactive compounds for development and utilization in food and health products. In summary, the HPLC procedure is effective for a rapid phytochemical assessment and is an analytical tool for quality assurance of *P. samentosum* samples.

Acknowledgements The authors receive partial funding from the Faculty of Pharmacy, UiTM, for this research. The authors would also like to thank Nurmisha Maisarah Yazid and Nur Batrisya Latipulkhobir, for their assistance.

References

Adib AM, Salmin NN, Kasim N, Sui KL, Cordell GA (2024) Ismail NH (2024) The metabolites of *Piper sarmentosum* and their biological properties: a recent update. Phytochem Rev. https://doi.org/10.1007/s11101-024-09930-2

Alqarni MH, Alam P, Foudah AI, Muharram MM, Shakeel F (2021) Combining normal/reversed-phase HPTLC with univariate calibration for the *Piperine* quantification with traditional and ultrasound-assisted extracts of various food spices of *Piper nigrum* L. under green analytical chemistry viewpoint. Molecules 26:732. https://doi.org/10.3390/molecules26030732

Atiya A, Majrashi TA., Yasmin Begum M, Lal UR, Sinha BN, Abdul Qadira SF (2023) A phenyl-propanoid dimer from the leaves of Piper betle. Nat Prod Res 37(9):1550–1556. https://doi.org/10.1080/14786419.2022.2025594

Bactiar CF, Nur Azlina MF (2019) LC-MS Analysis of Phytocomponents in the methanol extract of *Piper Sarmentosum* leaves. Pharmacog J 11(5):1071–1076

Das S, Parida R, Sriram Sandeep I, Nayak S, Mohanty S (2016) Biotechnological intervention in betelvine (*Piper betle* L.): a review on recent advances and future prospects. Asian Pac J Trop Med 9(10):938–946

Gaur P, Gupta P, Tiwari N, Gupta N, Shanker K (2023) Validation and greenness assessment of HPTLC-densitometric method for simultaneous estimation of six bioactive alkamides in *Piper longum* Linn. Root. Microchem J 191:108766

GlobinMed (2021) Malaysian herbal monograph, *Kadok* leaves, *Piper sarmentosum* Roxb. Available via https://www.globinmed.com/medicinal_herbs/piper-sarmentosum-roxb-piperaceae/. Accessed online 11 August 2023

Hussain K, Ismail Z, Sadikun A, Ibrahim P (2009) Evaluation of metabolic changes in fruit of Piper sarmentosum in various seasons by metabolomics using Fourier transform Infrared (FTIR) spectroscopy. Int J Pharm Clin Res 1:68–71

Hussain K, Hashmi FK, Latif A, Ismail Z, Sadikun A (2012) A review of the literature and latest advances in research of *Piper sarmentosum*. Pharm Biol 50(8):1045–1052

Ibrahim MA, Kamarul Arifin NI, Mohd Hasali NH (2022) Total phenolic content in *Piper sarmentosum* using different extraction techniques and solvents: a short review. Matrix Sci Pharma 6:6–10

Ibrahim MA, Zulkiflee NFI, Mohd Hasali NH (2022) A comprehensive review on the phytochemical constituents, antioxidant and anticancer properties of *Piper sarmentosum*. J Pharm 2(2):107–121. https://doi.org/10.31436/jop.v2i2.152

Ismail SM, Kien Hui C, Aminuddin A, Ugusman A (2018) *Piper sarmentosum* as an antioxidant: a systematic review. Sains Malaysiana 47(10):2359–2368

Kemprai P, Bora PK, Saikia SP, Haldar S. (2023) Chavibetol: major and potent Phytotoxin in betel (*Piper betle L.*) leaf essential oil. Pest Manag Sci. https://doi.org/10.1002/ps.7645

Khalid H, Zhari I, Amirin S, Pazilah I (2011) Accelerated stability and chemical kinetics of ethanol extracts of fruit of *Piper sarmentosum* using high performance liquid chromatography. Iran J Pharm Res 10(3):403–413

Khanthapok P, Sukrong S (2019) Anti-aging and health benefits from Thai food: protective effects of bioactive compounds on the free radical theory of aging. J Food Health Bioenvironmental Sci 12(1):54–67

NParks Flora Fauna Web (2023) https://www.nparks.gov.sg/florafaunaweb/flora/2/3/2342 Accessed 9 July 2024

Pinya T, Intharuksa A, Yanaso S, Kamnuan S, Phrutivorapongkul A (2022) Conventional and molecular pharmacognostic characters integrated with chemical profiles of five *Piper* plants in the Thai herbal pharmacopoeia and their admixture/adulteration/substitution situations in Thailand. J Nat Med 76:605–620. https://doi.org/10.1007/s11418-022-01607-4

Purba RAP, Paengkoum S, Paengkoum P (2021) Development of a simple high-performance liquid chromatography-based method to quantify synergistic compounds and their composition in dried leaf extracts of *Piper Sarmentosum* Roxb. Separations 8:152. https://doi.org/10.3390/separations8090152

Rahman SFSA, Sijam K, Omar D (2016) *Piper Sarmentosum* Roxb.: a mini review of ethnobotany, phytochemistry and pharmacology. J Anal Pharm Res 2(5):1–3

Sun X, Chen W, Dai W, Xin H, Rahmand K, Wang Y, Zhang J, Zhang S, Xu L, Han T (2020) *Piper sarmentosum* Roxb.: a review on its botany, traditional uses, phytochemistry, and pharmacological activities. J Ethnopharmacol 263:112897

Tuntiwachwuttikul P, Phansa P, Pootaeng-on Y, Taylor WC (2006) Chemical constituents of the roots of *Piper Sarmentosum*. Chem Pharm Bull 54(2):149–151

Ugusman A, Zakaria Z, Chua KH, Megat Mohd Nordin NA, Abdullah Mahdy Z (2012) Flavonoids Of *Piper Sarmentosum* and its Cytoprotective effects against oxidative stress. EXCLI J 11:705–714

Valle DL, Puzon JJM, Cabrera EC, Rivera WL (2016) Thin layer chromatography-Bioautography and gas chromatography-mass spectrometry of antimicrobial leaf extracts from Philippine *Piper betle L.* against multidrug-resistant bacteria. Evid-Based Complement Altern Med 2016. https://doi.org/10.1155/2016/4976791

Waman AA, Bohra P, Chakraborty G (2019) Vegetative propagation of *Piper Sarmentosum* Roxb.—a medicinally important species. Curr Agri Res 7(1):46–52

Wiart C (2006) Medicinal plants of Asia and the Pacific. In: Culinary and hospitality industry publication services, CRC Press, UK

2019 European Green Deal: An Ambitious Project with Lights and Shadows

Piermauro Catarinella and Suhardi Kram

Abstract In 2019, the European Commission, under the leadership of Ursula von der Leyen, launched the "European Green Deal" (EGD) which was ratified first by the European Council in the same year and subsequently by the European Parliament in 2020. EGD is a comprehensive plan aimed at making Europe the first continent with zero climate impact. This paper explores EGD objectives, which include improving public health, protecting nature and wildlife, and ensuring the well-being of future generations. Specific actions are outlined, such as addressing air and water pollution caused by dangerous chemicals or CO_2 emissions derived from farmers' activity. EGD also focuses on introducing clean technologies, implementing new industrial strategies, and amending waste management legislation to promote a transition towards a modern circular economic model. Various sectors, including automotive, construction (both public and private), food and beverages, energy production, and financial institutions, are involved in this transformative process. Achieving the goals of the European Green Deal requires significant investments, with contributions expected from both European institutions and citizens. While the decision to implement the Green Deal is supported by the coalition behind President von der Leyen, there is not unanimous agreement among member states regarding its validity and effectiveness. This research aims to investigate not only the quality of the Green Deal per se but also if it has solid scientific validity or if, instead, it is the result of an ideological choice, dictated by particular and peculiar partisan interests. How did the European Commission and its politically supporting majority reach this decision? Was it an autonomous deliberation or were there external interferences? Will the EU be able to fully implement EGD? This paper seeks to address these questions by analyzing internal documents from the European Commission and the European Parliament, interviews with prominent politicians and scientists across various media platforms, and recent articles published by Western newspapers.

P. Catarinella (✉)
Academy of Language Studies, Universiti Teknologi MARA, Shah Alam, Selangor, Malaysia
e-mail: piermauro@uitm.edu.my

S. Kram
Faculty of Sports Science and Recreation, Universiti Teknologi MARA Cawangan Sarawak, Samarahan, Sarawak, Malaysia
e-mail: suhardikram@uitm.edu.my

E. Shaffie et al. (eds.), *Sustainable Green Infrastructure*,
https://doi.org/10.1007/978-981-96-1486-8_5

Keywords Climate change · European Commission · European Union · European green deal · National recovery and resilience plan

1 Introduction

In December 2019, the EU Commission presented the European Green Deal, a detailed vision aimed at making Europe the first climate-neutral continent by 2050. This communication outlines the key policies and measures necessary to achieve this ambitious goal. Recognizing climate change and environmental degradation as existential threats, EGD aims to transform the EU into a resource-efficient, competitive, and modern economy. The primary objectives include maintaining Earth's temperature within a 1.5 °C limit, achieving net zero greenhouse gas emissions by 2050, making electricity production cleaner, reducing greenhouse gas emissions by 55% compared to 1990 levels, cutting pesticide use by 50% and fertilizer use by 20% by 2030, planting three billion additional trees in the EU by 2030, investing in environmentally friendly technologies, promoting renewable energy sources, restoring ecosystems, expanding protected areas, supporting sustainable food production, driving innovation in the industry, developing low-impact products, encouraging energy-efficient construction, introducing clean and cost-effective transportation methods, and aiming for 25% of EU agriculture to be organic by 2030 (Misculin 2020).

As part of the European Green Deal, the EU Commission has also introduced a sustainable finance strategy and a Sustainable Europe Investment Plan. These initiatives aim to ensure that the financial system supports businesses' sustainability transition. The Investment Plan allocates at least one thousand million euros of public and private investments over the next ten years, with the goal of mobilizing one trillion euros for sustainable investment through the EU budget and the European Bank for Investments (Neri 2020).

However, in order to achieve these ambitious goals, collaboration among governments, companies, and individuals is necessary. Individuals themselves must contribute through so-called environmental taxes on fossil fuels used for cars and heating homes. The increased expenditure for European taxpayers is estimated at 86 billion euros (Prestigiacomo 2023). Additional resources come from the NRRP—National Recovery and Resilience Plan—originally planned due to the pandemic crisis. There are €723.8 billion (in loans of €385.8 billion and grants of €338 billion) for EU countries to invest, create jobs, and build a green and digital future (Euronews 2022). The EU Recovery and Resilience Plan responds to the urgent need to foster a strong recovery and make European countries' future prosperous and eco-friendly. The reforms and investments in the plan will help all European countries become more sustainable, resilient, and better prepared for future challenges. It will offer opportunities to move towards green and digital transitions. The plan consists of 132 investments and 58 reforms. 37.5% of the whole budget plan will support climate objectives (Brugnara 2022).

The green measures of the plan can be grouped into four categories of interventions: (A) Transport and other green infrastructures (modernization and new construction of public infrastructures with low environmental impact); (B) Energy efficiency of both private and public buildings, through interventions aimed at reducing the use of energy or which encourage the transition from dirty energy to green energy; (C) Investments in renewable energy plants and alternative fuel infrastructure; (D) Environmental prevention works to mitigate the risks of natural catastrophes.

Immediately after the presentation of the EGD, President Ursula von der Leyen, expressed her pride in this project. She stated that it is the most advanced and far-sighted initiative, serving as a virtuous example for all countries worldwide. She is confident that the new EU climate rules will also impact the United States of America and other important nations. This is because EU-based companies and those operating globally and generating significant revenue within the EU will be required to comply with the new norms (Muresu 2023).

Even before 2019, the topic of climate change was globally debated. Following conferences in Berlin, Kyoto, Stockholm, Rio, Bali, Paris, Katowice, and Glasgow, COP27 took place in the Egyptian coastal city of Sharm el-Sheikh in 2022. The Conference of the Parties, also known as "COP", is an annual event that brings together governments that have signed the UNFCCC, the Kyoto Protocol, or the Paris Agreement (Stallard 2022). World leaders, ministers, and negotiators convene to collectively address climate change and its impacts. COP is a highly significant meeting where leaders discuss the planet-warming effects caused by human-generated emissions, primarily from the burning of fossil fuels such as oil, gas, and coal. Unfortunately, although many years have elapsed, no tangible actions have been implemented at a global level, leading some scientists to believe that heads of state from different countries are unable to reach a satisfactory agreement among themselves. Additionally, China and India decided not to participate in COP, which has raised concerns and disappointment, as these two countries are the largest polluters in the world, with China producing approximately ten billion tons of CO_2 annually and India producing over 2.6 billion tons (Caruso 2022).

If the United Nations (UN) has consistently failed to implement tangible measures to address climate change over the years, why should EGD expect a different outcome? In fact, there are numerous concerns regarding the actual feasibility of the EU to successfully execute this project, despite the grandiose statements made by its leaders in Brussels. It is worth noting that the European Parliament, which endorsed the Commission's decision, approved it with 442 votes in favor, 203 against, and 51 abstentions. This demonstrates that there was no unanimous perspective on this matter (Capritti 2021).

There are concerns that this ambitious project may face challenges in its full implementation, considering the EU's history of previous failures in achieving its projects, despite being presented with great enthusiasm and as a problem-solving initiative. Here are a few examples. (1) The issue of migration regulation has consistently resulted in failures. Even today, there is no viable solution to this problem, as the interests of individual member states have always taken precedence (Indelicato 2023). The Malta agreement of September 2019, which aimed to redistribute migrants among all EU countries, has never been effectively enforced. Mediterranean countries continue to bear the burden of the escalating immigration crisis on their own. In 2022, nearly 90,000 migrants arrived on Italian shores, but only 38 were relocated to France, 74 to Germany, and 5 to Luxembourg (Canepa 2022). (2) Following the outbreak of the war between Russia and Ukraine, the EU imposed stringent sanctions on Russia. However, European countries have struggled to reach a consensus on the price of gas imported from Russia (Romano 2022) for an extended period. Each country has primarily prioritized safeguarding its own interests. Even when some form of agreement has been achieved, it has failed to yield positive outcomes for European citizens and has instead highlighted a significant divide among European countries (Alonso 2022). (3) The Global Gateway project, aimed at engaging young people in the realm of politics, proved to be unsuccessful. The project incurred a cost of 387,000 euros and lasted from October 19th to December 15th, 2022 (Mosca 2023). (4) Since 2011, the EU has been endeavoring not only to establish unified regulations for determining the taxable base of companies residing within the EU and branches of companies from non-EU countries operating within the EU but also to achieve legislative harmonization among member states. The goal is to eliminate tax-related distortions that impede fair competition within the EU market (Caggiano 2020). (5) The Schengen Agreement, signed on June 14, 1985, abolished border controls among EU countries and introduced freedom of movement for nationals of the signatory countries. However, this fundamental right has frequently been restricted or suspended on 116 occasions (Bianchini 2020) by individual EU countries for various reasons, including concerns over terrorism, pandemics, uncontrolled immigration, and national security. (6) The President of the European Commission and the President of the European Council often lecture European citizens on how to improve environmental health, yet they themselves set a poor example with their behavior. Even for short distances, they prefer to use private jets (Bianchi 2021). (7) Recently, the reputation of European institutions has been severely damaged by the "Qatargate" scandal. Several politicians have been investigated, and large sums of cash have been discovered in their homes. They are accused of accepting bribes and lavish gifts from wealthy dignitaries from Qatar and other non-European countries. The criminal investigation, initiated by the Belgian judiciary on December 9, 2022, has revealed instances of corruption within the European institutions, where individuals have acted in the interests of non-EU states (Lombardi 2022). (8) An increasing number of European citizens have lost trust in the current EU ruling class to such an extent that the term "euro-bureaucrats" has been coined to refer to EU leaders. Despite holding power for many years, they have failed to establish a European army and develop a cohesive foreign policy. Additionally, they have struggled to foster greater solidarity among member states in economic, fiscal, and labor policies. The EU is perceived as an assembly of states unable to make collective decisions, and its bureaucratic structure is vulnerable to external influences (Minzolini 2023). The European Green Deal is the result of a decision taken by the political majority (a coalition composed of the EPP and the PES, respectively, European People's Party and European Socialists Party) which currently supports President Ursula von der Layen. All European laws have a political connotation, are prone to instability over time, and are subject to continuous changes. The Green Deal, soon after its approval, has undergone multiple reviews due to the occurrence of geopolitical and socio-economic circumstances, which the EU carefully considers (Borrell 2022). Furthermore, between 6 and 7 June 2024, the European Parliament will undergo renewal through elections (Cipolla 2023). The latest polls indicate that political parties opposing the EGD, claiming it undermines the economic foundations of each member state, are projected to emerge as winners (Napolitano 2023).

2 Methodology

This study employs a multi-method approach to investigate the European Green Deal (EGD) from its inception to its current status, with a focus on its scientific validity, political motivations, and implementation feasibility. The methodology consists of three primary components: document analysis, qualitative interview analysis, and media content analysis.

1. **Document Analysis**

A comprehensive analysis of internal documents from the European Commission and the European Parliament was conducted. These documents include official reports, legislative texts, meeting minutes, policy briefs, and correspondence related to the EGD. The aim is to understand the formal decision-making process, identify key stakeholders, and uncover any documented evidence of external influences. The document analysis helps to trace the development of the EGD, assess the rigor of the scientific data used, and evaluate the transparency of the decision-making process.

2. **Qualitative Interview Analysis**

Qualitative interviews with prominent politicians, policymakers, and scientists, sourced from various reputable media platforms, were analyzed to gather diverse perspectives on the EGD. These interviews were selected based on their relevance to the EGD process and their expertise in environmental science, policy, and economics. A thematic analysis was conducted on the interview transcripts to uncover insights into the motivations behind the EGD, the level of scientific consensus, and the political dynamics at play. This analysis helps to identify common themes and divergent viewpoints among the interviewees.

3. **Media Content Analysis**

A media content analysis was performed on recent articles published by leading Western newspapers and media outlets. This analysis focused on the portrayal of the EGD, the framing of its objectives, and the public discourse surrounding its implementation. Articles were selected using a keyword search strategy, focusing on terms such as "European Green Deal," "climate policy," and "sustainable development." The selected articles were then analyzed for their content, tone, and the presence of any bias. This component of the methodology provides insights into the public and media perception of the EGD, as well as the broader societal and political context.

2.1 *Data Synthesis*

The findings from the document analysis, qualitative interview analysis, and media content analysis were synthesized to provide a comprehensive understanding of the EGD. This triangulation approach ensures that the research findings are robust and reflective of multiple sources of evidence. By integrating these different data sources, the study aims to answer the key research questions: the scientific validity of the EGD, the political motivations behind its adoption, and the feasibility of its full implementation.

2.2 *Limitations*

While this multi-method approach provides a thorough examination of the EGD, there are inherent limitations. The document analysis is limited by the availability and accessibility of internal documents. The qualitative interview analysis is constrained by the selection of publicly available interviews, which may not cover all relevant perspectives. The media content analysis might reflect the bias of selected media outlets. These limitations are acknowledged and addressed through careful data triangulation and critical analysis.

3 Problem Statement

Two issues remain insufficiently investigated: firstly, whether the decisions made by the European Commission to address climate change are based on unequivocal scientific foundations, or if they are merely driven by purely political considerations that prioritize the enrichment of influential entities in the green finance sector, to the detriment of citizens. Secondly, it is crucial to assess the feasibility of implementing the plans and projects outlined in the Green Deal within the given timeframe, taking into account the current financial circumstances, future developments, and geopolitical dynamics both within and outside the European Union.

4 Literature Review

The vast scientific literature on climate change and global warming, encompassing their causes and potential solutions, can be broadly classified into three main groups. The first group of scientists asserts that human activities are solely responsible for the environmental issues on Earth, often referred to as "global anthropic warming". The second group argues that climate change is a natural occurrence inherent to the planet's evolutionary processes and cannot be attributed solely to human influence. Meanwhile, a third group remains uncertain and unable to provide a definitive answer due to the lack of conclusive scientific evidence demonstrating the anthropic nature of climate change and global warming. In addition to scientific perspectives, climate change is also examined by non-scientists who approach the subject from political and geopolitical standpoints. This section focuses on reviewing literature from the latter group of authors.

In "A Green and Global Europe" (2022), Nathalie Tocci explains how the unprecedented nature of the current energy transition represents both a unique opportunity and a huge challenge to Europe's future prosperity. Tocci argues that the EU must not act in isolation or ignore the adverse effects of the transition on member states and neighbors. It must also address the global cleavages that may arise with China, the transatlantic relationship, and the Global South as a result of the EU's green agenda. By adopting a truly global approach to the energy transition, Europe can deliver on its responsibilities to people and the planet alike and avoid unleashing social, economic, and security problems that could come back to haunt the Union. The European Union has assumed uncontested global leadership in its ambitious goals to green its economy. The author (Tocci 2022) explores the links between the European Green Deal, domestic European dynamics, and the regional and worldwide geopolitical implications of today and tomorrow in a fascinating way. General readers will obtain a great overview of the European energy and climate policy architecture from this book, while energy and climate specialists will gain new insights into the crucial, but still underexplored, social and political dimensions of this historical transformation.

In his book "Fixing the Climate: Strategies for an Uncertain World", Charles Sabel provides an interesting perspective. The author reflects on past international meetings and debates on climate change, ranging from the Kyoto Protocol to the Paris Agreement. Throughout history, world leaders have consistently failed to find concrete solutions to address the problems plaguing our planet. Despite decades of sustained negotiations, the climate crisis continues to worsen. The power to drive meaningful discussions on climate change lies with ordinary citizens, and only through collaborative efforts, we can hope to achieve positive results. Sabel and his co-authors (Sabel et al. 2022) acknowledge the visionary nature of this book, emphasizing that it is up to ordinary people to take matters into their own hands. Without the active involvement of individuals, world leaders alone will be unable to implement the measures necessary to save the planet.

2023 Sanchez-Reaza's "Making the European Green Deal Work for People" offers an analysis of the challenges posed by the green transition for the European Union. It emphasizes the crucial role of human development in making the transition attainable and sustainable, driving societal changes towards greater environmental consciousness. The green transition entails a significant effort to adopt environmentally sustainable practices on a scale that can make a difference for current and future generations. The authors affirm (Sanchez-Reaza et al. 2023) that a human-centered approach is essential for a just transition to the envisioned sustainable environment under the EGD. This report explores the key role of human development policies in achieving this goal. Firstly, human development policies are indispensable for enabling the green transition. Secondly, they are needed to facilitate societal adaptation to a green economy. Lastly, considering the magnitude of the proposed actions in the EGD, there is a potential for unintended consequences and adverse effects, especially for low-skilled workers and lagging regions, which should be mitigated.

Michael Kaeding's 2023 publication, titled "Climate Change and the Future of Europe", offers illustrative instances of national trajectories leading towards climate neutrality, along with an examination of public attitudes towards the climate crisis. Furthermore, the book presents actionable suggestions for policymakers, delineating the effective implementation of climate policies at both the national and European Union (EU) levels. Notably, the EU has forged an exemplary framework of environmental regulations that rank among the most rigorous globally. These regulations extend beyond conventional environmental concerns such as pollution control and natural resource preservation, encompassing emerging issues of significance like genetically modified organisms (GMOs) and climate change, which profoundly impact daily production and consumption patterns. The EGD sets ambitious objectives to establish Europe as the first climate-neutral continent by 2050. However, the implementation of these objectives varies significantly across member states and EU neighbors due to diverse factors such as local geographies, historical backgrounds, cultural influences, economic conditions, and political landscapes. This book (Kaeding et al. 2023) provides an analysis of the national efforts undertaken by member states and EU neighbors to combat climate change. Subsequently, it draws upon these factors to illuminate the local challenges, tensions, and opportunities encountered on the path towards achieving climate neutrality. Given the inter-country dependencies that have emerged following Russia's war against Ukraine, the book also addresses strategic questions pertaining to EU integration, the transformation of economies, the reduction of energy dependencies, and public perception of these issues. Additionally, the book puts forth concrete recommendations across various policy domains, outlining how individual countries and the EU as a whole should approach the climate crisis.

2022 "Greening Europe" is a multi-chapter book written by different authors and edited by Floriana Cerniglia and Francesco Saraceno. It is recommended for economists, environmentalists, policymakers, and anyone else who wishes to understand the financial and political challenges faced by the EU and its member states in achieving the goals of the European Green Deal. The funds initially allocated by EU institutions are insufficient to facilitate the green transition, as all EU countries are currently experiencing rampant inflation, significantly reducing citizens' purchasing power, while gas, electricity, and heating bills are increasing substantially. The editors (Cerniglia and Saraceno 2022) emphasize that the EU is allocating millions of euros monthly to armaments in support of Ukraine, a country involved in a brutal war that appears likely to continue for many years to come. There is a tangible risk that EU citizens will have to bear more than 70% of the costs of the EGD, leading to their frustration and potentially sparking violent street protests.

A 2023 investigation conducted by the Indian Institute of Technology (IIT) Kanpur challenges the common perception of the environmental benefits of EVs compared to conventional internal combustion engine cars. Contrary to popular belief, the findings of the research by Agarwal et al. indicate that electric cars are not as environmentally friendly as often portrayed. EV batteries emit 15–50% more greenhouse gases than other vehicle categories, raising questions about their true environmental impact. Preeminent studies on electric vehicles conducted before the EGD's approval should have cautioned both the Commission and the European Parliament against adopting impractical measures that could prove detrimental to the economies of individual member states and fail to yield meaningful environmental protection benefits. Notable among these studies *are Electric Vehicles: Present Scenario and Future Perspectives* (European Affairs 2019) and *Environmental Impact: A Comparative Analysis between Electric and Internal Combustion Vehicles* (Farneti 2018). The findings of these investigations underscored the significant challenges associated with electric cars, aligning with the conclusions drawn in subsequent analyses conducted in 2021 and 2023. European engineers have conclusively demonstrated that a substantial increase in the number of electric cars on the roads would not significantly reduce CO2 emissions. These studies recommended that European institutions explore the possibility of postponing the discontinuation of internal combustion engine production and sales, as European states were not adequately prepared for such a radical and abrupt transition.

5 Results and Discussion

5.1 Was EGD Approved Based on Robust and Indisputable Scientific Evidence, or Was It a Political Decision?

When the European Commission approved the Green Deal, it emphasized that this decision was motivated by the grave threat our planet is currently facing. Consequently, addressing the root causes contributing to climate change and detrimental global warming became imperative. The European Commission has embraced the standpoint that climate change is universally attributed to human activities, a consensus supported by the scientific community (Musso 2019). The research conducted by James Powell, a geologist from the United States National Science Board, supports the notion that climate change can be attributed to the rising emissions of greenhouse gases and other factors, all of which are directly linked to human activities (Powell 2017). Prior to Powell's findings, other researchers had already reached a similar conclusion (Cook et al. 2016) that global temperatures could surpass the significant threshold of 1.5 °C by 2026. This escalation in temperatures would lead to the melting of glaciers, a significant rise in sea levels, and an increase in the intensity and frequency of extreme weather events (BAMS 2022). In the past century, the causes of climate change were predominantly attributed to natural processes inherent in the Earth's cycles. However, over the past 150 years, human activity has emerged as one of the primary drivers, owing to the increasing reliance on fossil fuels resulting from technological advancements. This significant shift has led to considerable disruptions in microclimates. Extensive research identifies the excessive release of greenhouse gases into the atmosphere by human activities, alongside the Industrial Revolution, transportation systems, deforestation, urban development, chemical usage, and intensive farming practices, as the primary contributors to climate change (Perinelli 2022). Furthermore, the Intergovernmental Panel on Climate Change (IPCC) corroborates the unmistakable evidence of climate change occurring in the past 70 years, attributing it to anthropogenic factors (IPCC 2023).

However, there exist dissenting opinions among scientists who do not align with the aforementioned arguments. Nobel Prize laureate in Physics 2022, John Clauser, asserts that there is no genuine climate crisis and emphasizes that climate change has always been a part of Earth's history, even dating back thousands of years. Clauser contends that the debates surrounding climate change reflect a worrisome corruption of scientific principles, posing a threat to the global economy and the well-being of billions of people. He further criticizes the sensationalized and distorted portrayal of climate science in mass media, referring to it as a form of pseudo-science (Balsamo 2023).

Similarly, Norwegian physicist and Nobel Prize recipient Professor Ivar Giaever argues against the notion of a climate emergency. Giaever suggests that climate science should be depoliticized while emphasizing the need for climate policies to be grounded in rigorous scientific principles. He points out that Earth's climate has naturally experienced periods of both cooling and warming throughout its existence. Giaever contends that the observed warming has been significantly less than what was predicted by the IPCC based on anthropogenic forcings. Moreover, he asserts that enriching the atmosphere with CO_2 is beneficial rather than detrimental, rejecting the notion that it is a pollutant. Giaever also highlights the lack of statistical evidence linking global warming to the intensification or increased frequency of natural disasters such as hurricanes, floods, and droughts. He advocates for climate policy to be aligned with scientific and economic realities, emphasizing that there is no need for panic or alarm (Micallef 2023). It is worth noting that Ivar Giaever was the first signatory of a petition joined by 1200 scientists and professors from various scientific disciplines worldwide, including Franco Battaglia, Mario Giaccio, Enrico Miccadei, Giuliano Panza, Alberto Prestininzi, Franco Prodi, Nicola Scafetta, among others. This collective effort challenges the scientific premises underlying the implementation of the European Green Deal (Battaglia 2022).

Franco Prodi, an Italian Professor of Physics, Thermodynamics, and Geophysics at the University of Modena and a scientist affiliated with the Italian National Research Council (CNR), as well as a Member of the Cloud Physics Commission of the International Association of Meteorology and Atmospheric Physics (IAMAP), raises doubts about the exclusive responsibility of human activities for global warming. Prodi argues that proponents of anthropogenic climate change fail to consider natural factors such as astrophysical, astronomical, and atmospheric influences on climate change; therefore, global warming is not solely dependent on human activity, contrary to popular belief. He asserts that climate change is an inherent characteristic of the planet and has occurred throughout its history, with an average temperature increase of 0.7 °C per century over the past two hundred years. However, Prodi disputes the notion of a climate emergency. He suggests that since 1970, scientific research has been excessively influenced by political considerations. As an example, he points to the Intergovernmental Panel on Climate Change (IPCC), which he claims has transformed into an international organization comprising members appointed by national environment ministries. Prodi argues that these members lack the freedom to investigate natural phenomena solely from a scientific perspective. Instead, they are accountable to their political superiors and, as a result, may not impartially influence pure research but rather public opinion. Prodi highlights the involvement of government representatives in reviewing and revising scientific papers on climate change and global warming, suggesting that they may seek to exert influence over the final reports destined for policymakers. He contends that it is challenging to deny the political dimension of the IPCC (De Francesco 2022). Professor Prodi opposes the implementation of EGD, arguing that even if fully implemented, it would not yield any ecological benefits. He points out that Europe, with a population of less than 500 million people out of a global population of eight billion, accounts for only 8% of the world's emissions. Prodi questions the notion that Europe can single-handedly set a positive example and save the planet. He asserts that the EGD imposes self-destructive policies on European citizens, while other countries continue to pollute without concern or interference (Senaldi 2023).

Christian Schluchter, an esteemed professor emeritus of paleontology and climatology at the University of Bern in Switzerland, challenges the notion that melting glaciers are solely attributable to human activities. He argues that factors such as the Industrial Revolution and overpopulation began influencing the climate long before recent times. Schluchter highlights that throughout history, there have been eight or nine glaciations, during which glaciers have appeared and disappeared cyclically and he emphasizes the existence of a cyclic pattern within the Earth's system. This viewpoint contradicts the consensus among the majority of climatologists who attribute recent environmental disasters, such as the 2022 drought in Italy and other parts of the world, to human-induced factors. Despite its divergence from prevailing beliefs, Schluchter's thesis holds significant weight (Alberti 2022).

Stating that 99% of scientists agree with the assertion that climate change is caused by human activities is an inaccurate claim lacking scientific evidence. The reports published by the Intergovernmental Panel on Climate Change (IPCC), a United Nations body, advocating for the concept of an exponential increase in temperatures and sea levels, can be deemed misleading from a scientific standpoint. These reports tend to evoke alarm due to the influence of large financial and vested interests driving catastrophism. Given the lack of consensus among scientists regarding the causes of climate change and global warming, it raises the question of why the European Commission has chosen to initiate the European Green Deal, which imposes a series of changes on citizens that are likely to have adverse effects on their lives and the economies of individual member states. Some people believe that the European Commission's policy decision was influenced by the pressure exerted by the green finance lobby, which has experienced a significant expansion of its business.

Green Finance has generated a substantial sum of $41 trillion between 2010 and 2022 (Mediaset 2023). This term encompasses a broad spectrum of financial support for technologies, projects, industries, or enterprises that are environmentally focused. Another interpretation of green finance pertains to financial products or services that are aligned with environmental objectives. This includes various offerings such as green loans, green bonds, green investment funds, green mutual funds, solar bonds, green deposit certificates, and green mortgages. The shift towards a low-carbon economy necessitates substantial investments, which can only be funded through a strong commitment from the private sector. Frequently, the terms green finance and sustainable finance are used interchangeably or seen as interconnected. Nevertheless, the expenses that European citizens are required to bear for the implementation of the EGD may not appear sustainable for them (Barolini 2021).

Currently, China holds a dominant position in both the production and sale of lithium batteries and electric vehicles, establishing itself as a leader in the future of mobility. This situation poses risks to the industrial performance of the European automotive sector. Chinese customs data reveals a substantial growth in exports of lithium batteries, which increased by 66.9% year-on-year in the first quarter of 2023. Additionally, exports of electric cars witnessed a remarkable surge of 122.3%. The proportion of electric cars in the overall automotive market has risen from around 4% in 2020 to 14% in 2022, with projections indicating a further increase to 18% in 2023 (Console 2023). The rise of China's electric car industry indeed poses concerns regarding its potential to dominate the European market. In response, a coalition of eight EU countries, namely the Czech Republic, Italy, Germany, Poland, Hungary, Romania, Slovakia, Portugal, and Austria, has formed a united front. This alliance is currently engaged in a conflict with European institutions, as they seek to protect jobs and businesses and maintain independence from China, which holds a clear advantage in terms of electric mobility but also has higher levels of pollution compared to Europe. The situation highlights the complex balance between economic competition, environmental concerns, and the need for job security within the EU. Certain European ministers have openly criticized what they perceive as a paradox within the European Union (EU), where decisions are imposed from above onto individual member countries, potentially jeopardizing their economies. This critique highlights concerns about the centralized decision-making process and its impact on the economic well-being of member states. It underscores the need for a balanced approach that takes into account the diverse economic circumstances and priorities of each country within the EU (Dama 2023). Within the automotive sector, European industries have voiced their protests against EU decisions. They argue that numerous changes are being demanded within a remarkably short timeframe. It is worth noting that no other non-EU country has mandated a complete transition to zero emissions within just 12 years. Achieving a successful ecological transition necessitates a cohesive synergy between major industries and political actors. The automotive sector plays a vital role in the European economy. However, if the goal is to eliminate internal combustion engine vehicles, significant public incentives are required to make electric cars financially accessible for European citizens. This would involve creating an environment where electric vehicles become affordable and attainable options for the general population (Bonora 2023a, b, c). The European car industry has historically been a global leader in technological advancements related to the development of combustion engines and automobiles. Renowned brands from countries like Italy, Germany, and France have established a rich legacy over a span of more than a century, showcasing the wealth of expertise and know-how accumulated over time. However, there are concerns that, due to political motivations and policy directives, this industry's technological dominance could be relinquished within a short period of time. This raises questions about the adoption of non-European technologies that may rely on raw materials which Europe does not possess in abundance. There are criticisms suggesting that the European Union lacks a comprehensive vision, effective strategies, and a realistic approach to protecting the heritage of the European car industry, particularly in its decision to prioritize electric vehicles as the future. Italian Deputy Prime Minister Matteo Salvini has voiced concerns about potential external influences, suggesting that financial incentives from China may have played a role in influencing the European Commission's decision-making process. Salvini has openly referred to corruption and raised comparisons to previous cases such as Qatargate, which was discussed earlier. These remarks indicate a skepticism towards the motivations behind certain policy decisions and a desire for transparency in the decision-making process (Ciriaco 2023). We can assert that the European Green Deal (EGD) is primarily a political decision for two key reasons. Firstly, the European Commission has consistently disregarded the perspectives of scientists who have studied climate change and global warming, often labeling them as "deniers" because they believe climate change is a natural phenomenon not linked to human activities. Secondly, the EGD appears to prioritize "environmental justice" over "social justice". As a result, while the well-being of European citizens and vital economic sectors are likely to suffer, certain other entities stand to gain substantial profits (Ocone 2024).

5.2 *Was the EGD Approval an Independent Decision or Influenced by External Factors?*

There is a common belief that the decision to implement the Electric EGD by EU authorities was not made freely, as it seems to disregard the economic impact on key sectors in individual member states. Germany, France, and Italy, which have long been engaged in the production and assembly of internal combustion engine cars, are now required to transition to full-scale electric production within a short span of time. This significant shift has raised concerns among automotive unions, as they fear that it could potentially jeopardize over 75,000 jobs in each European country. Such a blow to the automotive sector is viewed as both challenging and burdensome. Consequently, the choices made by the EGD are perceived as driven by ideology and lacking practicality (Angì 2022). European car manufacturers would have preferred the zero emissions target to be postponed until 2040, allowing engineers to determine the most suitable alternative technology, such as hydrogen, rather than solely relying on electricity. The question arises as to why the EU made such a drastic decision exclusively for passenger cars while not addressing the harmful emissions from maritime transport (Società 2023). Cargo ships are recognized as significant contributors to CO_2 emissions in the environment. They primarily rely on high-sulfur fossil fuels, which release substantial amounts of sulfur dioxide (SO_x), nitrogen dioxide (NO_x), and CO_2. These emissions have detrimental effects on the greenhouse effect and contribute to climate change. Furthermore, maritime transport generates other forms of pollution that result in severe damage to marine ecosystems and compromise water quality. To provide some perspective, the emissions produced by a mere fifteen cargo ships are enough to pollute as much as the entire global car fleet.

The EU has consistently disregarded alternative green energy sources for cars and solely focused on electricity. This choice raises questions about why the EU has not allowed for other means besides electric vehicles (EVs) to achieve the goal of "net zero" emissions. It is scientifically important to consider not only the emissions from a car's exhaust pipe but also the environmental impact of electricity production (Robustelli 2023). Electric cars are widely acknowledged to have significantly lower emissions during their usage compared to petrol or diesel-powered vehicles. However, when considering the entire life cycle of an electric car, certain environmental implications arise. This includes the requirement of powerful batteries made from rare minerals, which can result in pollution during the extraction process. Additionally, factors such as deforestation, transportation, and waste disposal further contribute to the overall environmental impact of electric car production (Agizza 2023). Furthermore, electric cars raise important concerns that the EGD fails to address. Due to their inherent characteristics, electric motors operate in a linear and quieter manner compared to combustion engines. However, this comes with the disadvantage of lacking the natural vibrations experienced in traditional engines, which help passengers assimilate the vehicle's movement. Electric cars are also equipped with increasingly large and numerous screens, inundating passengers with excessive amounts of information. These factors contribute to an increased incidence of motion sickness, commonly known as car sickness, whose symptoms are nausea, sweating, pallor, hypothermia, headaches, and vomiting (Emond 2023). Electric vehicles are engineered with safety measures to ensure their compatibility with water. The entire high-voltage system, including the electric motor and battery pack, is constructed to maintain electrical insulation from the chassis, thus minimizing the risk of electric discharges. However, in situations where an electric vehicle becomes partially or fully submerged, particularly during severe floods, there is a genuine concern regarding potential battery fluid leakage or damage to the high-voltage system's electrical components. Such circumstances could lead to short circuits and subsequent fires once the vehicle is no longer submerged in water (Baldini 2023).

In the region of Emilia Romagna, Italy, which was recently struck by severe storms, all-electric cars were parked and isolated for a minimum of 15 days in outdoor areas. A minimum distance of five meters was maintained between each vehicle to mitigate the risk of spontaneous combustion, following incidents where some vehicles caught fire at a dealership (Redazione 2023).

The matter of safety is of utmost importance, necessitating further investigation into electric cars. European citizens were shocked by the news of an electric car catching fire during a routine journey in the streets of Naples, Italy. The occupants of the vehicle, who are currently receiving treatment at Caldarelli Hospital's intensive care unit, suffered severe burns and are now battling between life and death. This tragic incident occurred on June 14, 2023, and the causes of the fire are presently under investigation. Possible factors being examined include battery overheating, electrical defects, spontaneous combustion of the vehicle, or other contributing factors that led to this devastating fire (Popoli 2023). The transition to electric cars has been extensively discussed from various perspectives, but there is one aspect that has received limited analysis thus far: the insurance costs associated with electric vehicles. Car manufacturers currently do not grant access to battery diagnostic data to third parties. As a result, repairers, insurers, and experts are unable to accurately assess the extent of damage to the battery or verify if repair operations after potential accidents have effectively restored all damaged cells. This lack of access to comprehensive information poses challenges in accurately quantifying the repair costs and assessing the overall insurance risks associated with electric vehicles. Due to the limited access to battery diagnostic data, insurance companies often face the necessity of replacing the entire battery, even in cases where the damage is minor. Alternatively, they may choose to declare the vehicle a total loss if the cost of replacing the battery, which can account for a significant portion, often up to 50%, of the electric car's total value, is deemed uneconomical. This situation arises due to the lack of precise information on the extent of battery damage and the inability to accurately assess repair options, leading insurers to take cautious measures to mitigate potential risks and ensure proper coverage. In some European countries, a paradoxical situation has arisen where scratched battery packs, potentially containing intact cells, are being accumulated in landfills. Similarly, electric cars with minimal mileage are being prematurely scrapped. This situation contradicts the ambitions of sustainability that have guided the transition to electric vehicles. Such wasteful practices are not in line with the goals of promoting a more environmentally friendly transportation system. It highlights the need for better strategies and policies to ensure the proper handling, repair, and recycling of electric vehicle components, aligning with the principles of sustainability (Marino 2023). A recent study (Burnham et al. 2021) indicates that a global shift to electric vehicles, encompassing cars, buses, trucks, and vans, by 2030 would result in a temperature decrease of just 0.0001 degrees. However, the study emphasizes that the costs borne by ordinary people would be exorbitant.

Many assert that the powerful Bilderberg group has exerted influence over the European Union's policy decisions regarding environmental protection. This group is often described as a private neoliberal think-tank that engages in discussions on global issues, or even as a "shadow government" where the political and financial agenda of the Western world is allegedly shaped. Participants of the Bilderberg group include heads of state, treasury ministers, influential journalists, European Union politicians, and prominent figures from European and Anglo-American high finance

(Luminari 2020). The 69th closed-door meeting of the Bilderberg Club took place in Lisbon, Portugal, on May 21, 2023, where various topics, including the energy transition, were reportedly discussed. It is not uncommon for representatives of the European Union, including the Commission, to participate in such meetings. Their involvement typically entails attending discussions, sharing insights, and engaging in conversations with other participants. However, the specifics of their participation and the extent to which they disclose information about the discussions are not publicly disclosed. The nature of these meetings often involves confidentiality, and participants generally maintain discretion regarding the details and outcomes of the discussions (Vivaldelli 2023). The Bilderberg group, established in 1954 by Henry Kissinger and David Rockefeller, derives its name from the Hotel De Bilderberg in Oosterbeek, the Netherlands, where the inaugural meeting of the "club" took place. Although the meetings themselves do not result in formal statements or decision-making power, it is important to note that many of the participants hold influential institutional positions that can potentially shape their decisions (Audiello 2023).

The presence of influential members within the Bilderberg group is not without significance, as some of them are also known to be funders of the international network known as #A22. This network comprises young Europeans who engage in acts of civil disobedience to advocate for stronger government actions in combating the climate and ecological crisis. The connection between these funders and the #A22 network suggests a possible alignment of interests and support for grass-roots movements pushing for environmental change. This international network includes various organizations such as "Just Stop Oil" in the UK, "*Letzte Generation*" in Germany, "*Dernière Rénovation*" in France, and "*Ultima Generazione*" in Italy. These organizations have been actively escalating their activities in Europe, taking part in numerous acts of vandalism and other illegal actions to protest against the energy and industrial policies of member states and the European Commission. Notably, they have gained significant media visibility, which allows them to exert a considerable influence on public opinion, particularly among younger individuals. These youth organizations receive significant financial support from notable sources. From this perspective, it is claimed that influential figures in the world of green finance strategically utilize these youth organizations to instill fear in the population. The narrative propagated suggests that without radical and expensive changes, the world is believed to be facing an imminent catastrophe. This approach is seen as benefiting high finance and multinational corporations, which can potentially accumulate immense wealth through the green transition. Additionally, resources are used to propagate panic regarding climate change, linking it to human activities through studies funded with specific objectives while discrediting scientists who hold different viewpoints. Climate catastrophism is viewed as a tool for psychological and political pressure, aiming to accelerate the implementation of green initiatives across all aspects of human life (Romiti 2023).

These influential entities not only have the power to sway the decisions of the European Commission but also to enforce their agendas. Observing the multinationals investing millions of dollars in cultured meat and other lab-grown foods, it is clear that their lobbying efforts yield tangible benefits for their own interests. A

2023 study by the Corporate Europe Observatory demonstrated how powerful multinationals across various sectors have the ability to steer the EU when it comes to making significant political decisions (Sesana 2023).

5.3 Is the European Commission Able to Implement Its EGD After the War Eruption Between Russia and Ukraine?

The European Green Deal was introduced by the European Commission on December 11, 2019, and subsequently ratified by the European Parliament on January 15, 2020, with a majority vote. In order to accomplish its goals, a budget of one thousand billion euros has been allocated over a period of ten years, along with additional funding from the Recovery and Resilience Plan. However, a significant and grave event occurred on February 24, 2022: Russia launched an invasion of Ukraine, initiating a brutal and ongoing conflict. This is not the appropriate context to analyze the causes that led to this war or establish potential violations of international law. Since its inception, the EU has made a political decision to support Ukraine, demonstrating this by welcoming millions of refugees and providing military assistance to the invaded country. The EU could have chosen to remain neutral, as many other countries have done, and assume a mediating role between the conflicting parties to seek a peaceful resolution. However, that was not the case. The EU has actively chosen to engage in the war and consistently emphasizes, in various international forums, that the only viable solution to end the conflict lies in the military victory of the Ukrainian armed forces. A potential Russian triumph, the EU warns, would have catastrophic consequences for the Western nations (Zanellato 2023). The EU has chosen to allocate billions of euros towards supplying Ukraine with an increasing number of replacement weapons, resulting in a diversion of funds from the European Green Deal project due to insufficient resources. The initial allocation of one trillion euros is no longer available, as the entirety of the EU's financial resources is now focused on providing weapons of war to Kyiv. A thorough study conducted by the Stockholm International Peace Research Institute (SIPRI) estimated that military aid to Ukraine in 2022 incurred costs of approximately 2240 billion euros, surpassing the allocation for the EGD. Furthermore, the military expenditure of EU countries for 2023 has risen by 127 billion euros compared to the previous year (Canella 2023). As a result of extravagant expenditures on armaments, the EU is currently facing a scarcity of resources for the European Green Deal and its domestic economy. The overall well-being of European citizens has significantly declined, and all member states find themselves operating within a veritable war economy. The prevailing situation has resulted in a significant surge in the prices of raw materials and essential food items, experiencing a staggering 600% increase since February 24, 2022. Furthermore, the economy is grappling with double-digit inflation rates, which have become a prevalent concern (Grazzini 2022). Funds from the Recovery and Resilience Plan have also been redirected towards Ukraine's weaponry needs. The European Parliament has approved the Commission's proposal to enhance armament production for Ukraine. In the span of two years, the EU has been compelled to set aside its EGD project, which was initiated in 2020, and has increasingly shifted its focus towards warfare, irresponsibly disregarding the environmental catastrophes it is engendering. The ecological devastation caused by the recent conflict in Ukraine is unparalleled in history and will have far-reaching consequences for the entire planet. Over the course of one year, the war in Ukraine has resulted in not only hundreds of thousands of deaths but also extensive environmental destruction. Vast expanses of forests have been decimated, lands have become contaminated, aquifers have been polluted, animal species have faced extinction, and greenhouse gas emissions have surged. The estimated cost of these damages currently stands at approximately 46 billion euros and is expected to rise further. Ukrainian agriculture, devastated by the ongoing war, may suffer repercussions for at least a century. The country has already undergone a similar ordeal following the Chornobyl nuclear disaster, and now, once again, the issue has the potential to become a global concern (Mastrodonato 2023). Numerous areas of land have become unusable due to the presence of landmines and unexploded ammunition, as well as the deposition of toxic substances like white phosphorus. Tank movements and battlefield engagements have resulted in the destruction of additional tracts of land.

The extensive utilization of military resources contributes significantly to pollution. Bombing incidents in both rural and urban areas also play a pivotal role in generating peak emissions. Additionally, the environmental impact of post-war reconstruction represents a separate aspect that necessitates careful assessment. It is estimated that the conflict with Russia has resulted in CO_2 emissions of approximately 120 million tons, equivalent to the amount produced by nearly 27 million gas-powered cars on the road for a year. Furthermore, an additional 23 million tons of emissions have resulted from fires (Paddison 2023). Ukraine stands as one of the most highly industrialized nations in Europe, with its coal mines, chemical plants, and other heavy industries believed to generate billions of tons of liquid waste. Since February 2022, these industries have endured relentless bombardments, resulting in widespread spills that, over time, will find their way to the Black Sea and subsequently to the Mediterranean. Furthermore, the bombings in proximity to the Chornobyl and Zaporizhzhia nuclear power plants have not only posed a severe nuclear catastrophe threat but have also dispersed radioactive dust into the atmosphere (Signorile 2023). Ukraine boasts some of the most fertile soils in Europe, and its cereal exports play a crucial role in meeting the needs of many nations. However, these fertile lands have endured constant bombardment, burning, and pollution caused by metals such as white phosphorus, chromium, nickel, molybdenum, vanadium, cobalt, depleted uranium, and manganese present in the rockets and missiles used during the bombings. The impact of these metals, combined with the chemically toxic dust produced upon missile or rocket impact, has significantly altered the hydrogeological structure of cultivated fields, rendering them infertile. Consequently, the agricultural future of the entire country is at stake, along with the supply of wheat and cereals to numerous countries. In 2022, a substantial quantity of wheat harvested in Ukraine, approximately 29 tons, had to be destroyed due to contamination with highly carcinogenic mycotoxins such as ochratoxin A (OTA) and deoxynivalenol (DON), posing a severe food security issue for countries that rely on Ukraine for agricultural self-sufficiency (Salvatore 2023). On June 6, 2023, a major environmental catastrophe unfolded when the Nova Kakhovka Dam exploded, triggering a blame game between Russians and Ukrainians regarding the incident. This disaster surpassed the magnitude of the Chornobyl incident in 1986, making it the worst environmental catastrophe in Europe over the past three decades. The explosion resulted in the release of 600 tons of oil, contaminating a significant body of water and leading to the destruction of vital ecosystems and national parks. Among the affected areas are Nyzhniodniprovs'kyi, Kam'yanska Sich, and Biloberizhzhia Sviatoslava, the latter being a UNESCO heritage site. United Nations Secretary-General António Guterres bluntly described the ongoing situation as an "ecological catastrophe" (Talignani 2023). For a comprehensive and in-depth overview of the environmental devastation caused by the war and its severe impacts on ecosystems, please refer to the report published by Greenpeace and the Ukrainian Ecoaction NGO (Ansa 2023).

The European Union is unlikely to implement the European Green Deal as it was conceived in 2019 for several reasons. Firstly, the funds initially allocated for the EGD have been redirected to military support for Ukraine, with defense expenditures continuing to rise. On March 8, 2024, the Aspen Institute published a paper titled "Election Year: Transatlantic Public Opinion and Ukraine". The authors, drawing on internal European Commission documents and data from independent think tanks, analyze the financial contributions of the European Union (EU) to the ongoing war in Ukraine. Their findings reveal a significant shift in European spending priorities. According to data from the Ukraine Support Tracker and Stockholm International Peace Research Institute, since February 2022, European contributions to support weaponry for Ukraine have surpassed those of the United States, exceeding €2.2 trillion. Additionally, the EU announced a new €50 billion multi-year support package for Ukraine between 2023 and 2027. As a result, the EU's total financial commitments to Ukraine are now nearly double those of the US. When factoring in contributions from non-EU Western European nations (UK, Norway, Switzerland), the disparity widens further (Aspen 2024).

Additionally, many Members of the European Parliament have shown reduced support for the green transition in light of the upcoming June 2024 European political elections and electoral polls suggest that political parties opposing the EGD are poised to win (Capasso 2023). Substantial support for this argument comes from the 2023 European Environment Agency report, which indicates that the primary objectives of the EGD, ranging from greenhouse gas emissions reduction to biodiversity restoration, are "unlikely to be reached" (EEA 2023).

6 Conclusion

The European Green Deal had a short lifespan, despite being approved by the European Commission in December 2019. However, it was effectively abandoned following the outbreak of the war between Russia and Ukraine. In February 2022, the EU, along with NATO and the USA, made a decision to reallocate the originally designated €1 trillion for the European Green Deal towards the production of powerful and expensive weapons to support Ukraine, one of the warring countries. We strongly disagree with this political decision for two fundamental reasons. Firstly, had the EU chosen to remain neutral, it could have served as a mediator between the conflicting countries and convened a peace conference that could have swiftly ended the fighting within a few weeks. Secondly, this decision could have prevented the war from prolonged devastation, including loss of lives, injuries, destruction, food shortages, and economic crises. Instead, there is now a heightened risk of a nuclear Third World War breaking out (Balestrieri 2023; Chiacchio 2023).

It is worth noting that even if there had not been a war on European soil, the EGD would likely not have achieved the desired effects due to methodological and substantive reasons.

(A) Regarding methodology, it is crucial to highlight that the EGD was determined and enforced by the European Commission without prior dialogue with governments, citizens, entrepreneurs, and labor unions of each European country. Major companies in Italy, Poland, France, and Germany have consistently expressed their concerns that the 2035 deadline, which would halt the production of combustion engine cars, was initially seen as a political mandate. However, the concern is ongoing, as they believe it would be unsustainable unless the intention is to dismantle the entire European automotive sector, leaving twelve million employees in this industry jobless. The European Commission approached the green transition through legislation in an ideological and confused manner, disregarding the diverse economic and social realities of individual European countries, and applying the same regulations uniformly to all member states (Bonora 2023a, b, c).

Climate change, pollution, and global warming are issues that impact the entire planet. It is not logical for these problems to be tackled solely by a geographically limited minority such as Europe, which bears minimal responsibility for these problems. The provided graph indicates that the EU is not the largest contributor to global pollution (Fig. 1).

Therefore, even if the EGD were fully implemented, it would not be sufficient to address the world's problems (Cipolla 2022). The EU should strengthen its relationships with the rest of the world to ensure the inclusion of all nations in the green transition. Rather than proceeding independently and presenting impressive yet unattainable projects, it is essential to consider the aforementioned reasons and collaborate globally. Addressing climate change requires a collaborative effort involving all nations, regardless of their individual contributions to the problem.

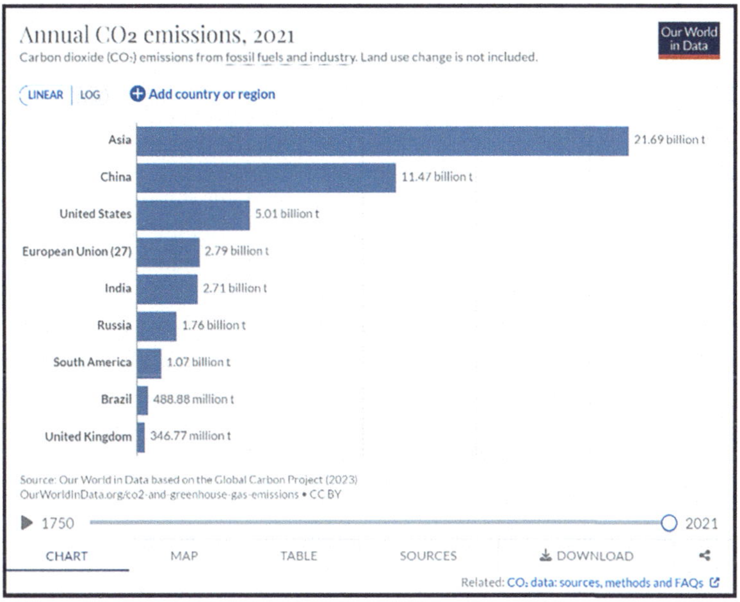

Fig. 1 Illustrates the quantity of CO_2 tons produced by different worldwide regions in 2021 (*source* https://ourworldindata.org/co2-emissions)

(B) Regarding substance, doubts arise due to the fact that sensitive sectors that have a direct impact on millions of European citizens, such as housing and food, were addressed without their direct involvement. In the real estate sector, the European Green Deal has mandated that by 2030, all public and private buildings must meet the energy class E standards. This means that the buildings must have insulated roofs and walls, sun shading devices to limit summer overheating and reduce cooling/ventilation consumption, and energy-efficient windows and doors. To reach this objective, property owners will be required to renovate their homes at their own expense. This politically driven decision is being implemented rapidly, without taking into account the real-life situation, which has been further complicated by the energy crisis and inflation. The potential collateral effects on people's lives and well-being are disregarded. It is crucial to recognize the concerns arising from the EGD's approach, particularly regarding the financial burden placed on property owners and the potential consequences on their quality of life. Balancing the urgent need for environmental sustainability with the practical implications for individuals and society requires careful consideration and inclusive decision-making processes (Ocone 2023). The agricultural sector is also being impacted by the ecological transition. The European Commission has introduced new foods promoted as sustainable and environmentally friendly, but without considering the preservation of culinary traditions in individual countries. As a result, this initiative has sparked strong criticism and reservations from European agricultural producers (Danieli

2023). These two aspects—the real estate and agricultural sectors—pertaining to the green transition will be addressed in an upcoming research study.

It is crucial to recognize that the EGD, as formulated by the political majority supporting President von der Leyen during her tenure from 2019 to 2024, has faced substantial criticism due to the perceived infeasibility of most of its projects. This criticism stems from the EU Commission's steadfast ideological commitment to expediting the transition towards accomplishing the "zero emissions" objective by 2050 and, despite the appeals from both consumers and industrialists to ease the pace of decarbonization, their concerns have been largely disregarded or deemed insignificant (Zaccardi 2023). However, the EU continues to promote the EGD, which raises concerns of hypocrisy, as it neglects to address the fact that public funds have been redirected towards armaments to support Ukraine. The European Green Deal, initially presented as a groundbreaking program with the potential to serve as a global model, represents the fourth debacle over the past five years during Ursula von der Leyen's presidency.

Acknowledgements No financial support was received by the researchers for conducting, writing, and publishing this study.

Conflict of Interest The authors declare that they have no known competing financial interests or personal relationships that could have appeared to influence the work reported in this paper.

References

Agarwal A et al (2023) LCA and TCO analyses of BEVs, HEVs, and ICEVs. https://www.iitk.ac.in/erl/Downloads/LCA%20and%20TCO%20Analyses%20of%20BEVs,%20HEVs,%20and%20ICEVs.pdf

Agizza P (2023) Auto elettriche: la batteria inquina più di un'auto a benzina? Un mito da sfatare. DMOVE. https://www.dmove.it/news/auto-elettriche-la-batteria-inquina-piu-di-una uto-a-benzina-un-mito-da-sfatare

Alberti D (2022) Schluchter (geologo) "Scioglimento ghiacciai non è opera uomo"/ "Prima dell'800...". Il Sussidiario. https://www.ilsussidiario.net/news/schluchter-geologo-scioglime nto-ghiacciai-non-e-opera-uomo-prima-dell800/2398049/

Alonso A (2022) Ok, il tetto è giusto: l'Unione Europea trova l'accordo sul price cap. Euronews. https://it.euronews.com/my-europe/2022/12/19/ok-il-tetto-e-giusto-lunione-eur opea-trova-laccordo-sul-price-cap

Angì N (2022) Anno 2035, fine dei motori a scoppio: cosa accadrà? Al Volante. https://www.alv olante.it/news/anno-2035-fine-dei-motori-scoppio-cosa-accadra-378490

Ansa (2023) Ucraina: Greenpeace, mappa dei danni ambientali della guerra. Danneggiate aree protette e foreste, sparse sostanze tossiche. A&E Inquinamento. https://www.ansa.it/canale_ambiente/notizie/inquinamento/2023/02/21/ucraina-greenpeace-mappa-dei-danni-ambientali-della-guerra_5692edc3-0cdf-4513-8c9b-7977d5b67b7d.html

Aspen Institute (2024) Political and societal analysis. Election year: Transatlantic Public Opinion and Ukraine. Presented on the occasion of the International Dialogue: Assessing Risk, Business in Global Disorder. https://www.aspeninstitute.it/un-anno-elettorale-lopinione-pubblica-transa tlantica-e-lucraina/

Audiello G (2023) Bilderberg: come ogni anno le élite globali si sono incontrate a porte chiuse. L'Indipendente. Esteri e Geopolitica. https://www.lindipendente.online/2023/05/22/bilderberg-come-ogni-anno-le-elite-globali-si-sono-incontrate-a-porte-chiuse/

Baldini E (2023) Auto elettriche e acqua, cosa si rischia? Il Giornale. https://www.ilgiornale.it/news/automotive/auto-elettriche-e-acqua-cosa-si-rischia-2111398.html

Balestrieri B (2023) Putin pronto a dispiegare i missili nucleari. Zelensky: "Non è pronto, ha paura per la sua vita". Huffpost. https://www.huffingtonpost.it/esteri/2023/06/21/news/putin_zelensky-12455041/

Balsamo M (2023) "È pseudoscienza": la lezione del Nobel alle follie degli ambientalisti. Il Giornale. https://www.ilgiornale.it/news/attualit/narrativa-sul-cambiamento-climatico-riflette-corruzione-2149444.html

BAMS (2022) Explaining Extreme Events in 2021 and 2022 from a Climate Perspective. This BAMS special report presents assessments of how human-caused climate change may have affected the strength and likelihood of extreme events. https://www.ametsoc.org/index.cfm/ams/publications/bulletin-of-the-american-meteorological-society-bams/explaining-extreme-events-from-a-climate-perspective/

Barolini A (2021) Che differenza c'è tra finanza etica, sostenibile, verde e solidale. Valori. https://valori.it/differenza-finanza-etica-sostenibile-verde-solidale/

Battaglia F (2022) "L'emergenza climatica non esiste". Basta un Nobel per bocciare Cop27. La Verità. https://www.laverita.info/cop27-petizione-scienziati-2658615260.html

Bianchi A (2021) Un aereo privato per fare solo 50 chilometri, la scelta poco verde della presidente von der Leyen. Europa Today. https://europa.today.it/ambiente/jet-privato-emissioni-von-der-leyen.html

Bianchini N (2020) Cos'è il trattato di Schengen e quante volte è stato sospeso. AGI. Agenzia Italia. https://www.agi.it/estero/news/2020-02-22/coronavirus-trattato-schengen-sospensione-7166660/

Bonora P (2023a) "Auto europea sotto pressione. L'intera industria è nel mirino". Il Giornale. https://www.ilgiornale.it/news/economia/auto-europea-sotto-pressione-lintera-industria-nel-mirino-2170235.html

Bonora P (2023b) Il settore dell'auto alla Commissione Ue: "Parlate con le imprese". Il Giornale. https://www.ilgiornale.it/news/economia/settore-dellauto-commissione-ue-parlate-imprese-2172154.html

Bonora P (2023c) Il mondo dell'auto si inginocchia alla Cina. Il Giornale. https://www.ilgiornale.it/news/economia/mondo-dellauto-si-inginocchia-cina-2173560.html

Borrell J (2022) L'Europa nell'interregno: il nostro risveglio geopolitico dopo l'Ucraina. The Diplomatic Service of the European Union. https://www.eeas.europa.eu/eeas/l%E2%80%99europa-nell%E2%80%99interregno-il-nostro-risveglio-geopolitico-dopo-l%E2%80%99ucraina_it

Brugnara L (2022) 108 misure verdi: cosa fa il PNRR per la transizione ecologica. Osservatorio Conti Pubblici Italiani. https://osservatoriocpi.unicatt.it/ocpi-pubblicazioni-108-misure-verdi-cosa-fa-il-pnrr-per-la-transizione-ecologica

Burnham et al (2021) Regional emissions analysis of light-duty battery electric vehicles. Atmosphere 2021. https://doi.org/10.3390/atmos12111482. https://www.mdpi.com/2073-4433/12/11/1482

Caggiano G (2020) Il cantiere dell'armonizzazione fiscale per il Mercato unico digitale. Annali AISDUE, ISSN: 2723-99. https://www.aisdue.eu/wp-content/uploads/2020/08/Post-Giandonato-Caggiano-Cantiere-Tassazione-digitale-2.pdf

Canella A (2023) Non s'è mai speso tanto per le armi. Comune Info. https://comune-info.net/non-se-mai-speso-tanto-per-le-armi/

Canepa C (2022) Sì, solo 117 migranti sbarcati in Italia nel 2022 sono stati ricollocati in Europa. Pagella Politica. https://pagellapolitica.it/fact-checking/salvini-migranti-ricollocati-francia-europa

Capasso A (2023) Perché il Green Deal va verso il fallimento (anche a causa dell'Italia). Un report prevede che l'Ue non riuscirà a raggiungere gran parte degli obiettivi climatici. Europa Today. https://europa.today.it/ambiente/fallimento-green-deal-contributo-italia.html

Capritti R (2021) Legge UE sul clima: neutralità climatica al 2050 vincolante. InfoBuildenergia. https://www.infobuildenergia.it/commissione-europea-riduzione-emissioni/

Caruso D (2022) COP 27: cos'è e perché è un appuntamento così importante. Leonardo.it. https://leonardo.it/news/cop-27-cos-e/

Cerniglia, Saraceno F (2022) Greening Europe. 2022 European Public Investment Outlook. ISBN: 978-1-80064-905-7. Open Book Publishers. https://www.thirdplacebooks.com/book/9781800649064

Chiacchio G (2023) L'arsenale nucleare russo. Tutte le armi a disposizione di Mosca. Inside Over. https://it.insideover.com/guerra/larsenale-nucleare-russo-tutte-le-armi-a-disposizione-di-mosca.html

Cipolla A (2023) Elezioni europee 2024, quando si vota? Data, legge elettorale e sondaggi. Money.it. https://www.money.it/elezioni-europee-2024-quando-data-legge-elettorale-sondaggi

Cipolla R (2022) COP27: questi sono i 10 Paesi del mondo che inquinano di più (e che potrebbero davvero fare la differenza). Green me. https://www.greenme.it/ambiente/cop27-10-paesi-mondo-inquinano-di-piu/

Ciriaco T (2023) Salvini accusa l'Ue e la Cina: "Vogliono imporci le auto elettriche, forse a Bruxelles troveranno valigette piene di soldi da Pechino". La Repubblica. https://www.repubblica.it/politica/2023/03/31/news/matteo_salvini_cina_auto_corruzione-394439988/

Console S (2023) L'Italia deve avere paura delle auto elettriche cinesi? Today. Attualità. https://www.today.it/attualita/auto-elettriche-cina-italia-salvini.html

Cook J et al (2016) Consensus on consensus: a synthesis of consensus estimates on human-caused global warming. https://doi.org/10.1088/1748-9326/11/4/048002

Dama S (2023) Auto elettrica, la fronda guidata da Salvini si allarga: chi dice "no". Libero Quotidiano. https://www.liberoquotidiano.it/news/politica/35190179/auto-elettica-matteo-salvini-fronda-si-allarga.html

Danieli A (2023) La Ue dice sì ai grilli a tavola. Coldiretti furiosa: "È una follia, ma quale scelta ecologica…". Secolo d'Italia. https://www.secoloditalia.it/2023/01/la-ue-dice-si-ai-grilli-a-tavola-coldiretti-furiosa-e-una-follia-ma-quale-scelta-ecologica/

De Francesco M (2022) Un tema di enorme rilevanza industriale: dubbi scientifici sulla responsabilità umana (CO_2) per il riscaldamento globale. Industria Italiana. https://www.industriaitaliana.it/un-tema-di-enorme-rilevanza-industriale-dubbi-scientifici-sulla-teoria-della-responsabilita-umana-co2-per-il-riscaldamento-globale/

EEA, European Environment Agency (2023) The 8th Environment Action Programme monitoring results 2023. https://www.eea.europa.eu/publications/european-union-8th-environment-action-programme/viewfile

Emond W (2023) Soffri di mal d'auto? Con quelle elettriche e a guida autonoma andrà peggio. In più di un secolo di sviluppo automobilistico il problema del mal d'auto avrebbe dovuto essere risolto. Anzi, alcuni progressi tecnologici, come l'elettrificazione e la guida autonoma, possono aggravare il malessere. La Repubblica. https://www.repubblica.it/tecnologia/2023/05/23/news/mal_dauto_auto_elettriche_guida_autonoma_cinetosi-400823865/?ref=RHRM-BG-P1-S1-T1

Euronews (2022) What is Europe's Recovery Plan and how does it work? Euronews. https://www.euronews.com/next/2022/02/09/what-is-europe-s-recovery-plan-and-how-does-it-work

European Affairs (2019) Electric vehicle: present scenario and future perspectives. Future Electricity. https://www.elettricitafutura.it/public/editor/Press_Room/EF_Studio%20Vettore%20Elettrico_Rev1.pdf

Farneti R (2018) Environmental impact: a comparative analysis between electric and internal combustion vehicles. University of Bologna. Faculty of Mechanical Engineering. https://amslaurea.unibo.it/19088/1/Tesi.pdf

Grazzini E (2022) La guerra in Ucraina mette in crisi l'Euro e l'Unione Europea. MicroMega. https://www.micromega.net/guerra-ucraina-crisi-economica/

Indelicato M (2023) L'UE ribadisce: no alla redistribuzione obbligatoria dei migranti. Inside Over. https://it.insideover.com/migrazioni/l-ue-conferma-la-linea-niente-redistribuzione-dei-migranti.html

IPCC (2023) Reports of the Intergovernmental Panel on Climate Change. https://www.ipcc.ch/

Kaeding M et al (2023) Climate change and the future of Europe. Springer. TEPSA. Trans European Policy Studies Association. ISBN: 978-3-031-23327-2. https://www.tepsa.eu/tepsa-books/

Lombardi A (2022) Qatargate: i personaggi, i soldi, le accuse e le reazioni dello scandalo al Parlamento europeo. La Repubblica. https://www.repubblica.it/esteri/2022/12/14/news/qatargate_cose_e_cosa_e_successo_corruzione_panzeri_eva_kaili-379045314/

Luminari M (2020) Gruppo Bilderberg, la riunione segreta dei potenti del mondo tra cronaca e fake news. Centro Pagina. https://www.centropagina.it/attualita/gruppo-bilderberg-riunione-segreta-dei-potenti-del-mondo-cronaca-fake-news/

Marino G (2023) Auto elettrica e l'allarme batterie: se si danneggiano anche lievemente vanno sostituite. Ecco perché nessuna casa autorizza l'accesso diagnostico e così salgone alle stelle anche i costi assicurativi. La Repubblica. https://www.repubblica.it/motori/sezioni/attualita/2023/06/13/news/auto_elettrica_e_lallarme_batterie_se_si_danneggiano_anche_lievemente_vanno_sostituite_completamente_ecco_perche-404276489/?ref=RHRM-BG-P7-S1-T1

Mastrodonato L (2023) La guerra in Ucraina ha un impatto ambientale devastante. Lifegate Daily. https://www.lifegate.it/guerra-ucraina-impatto-ambientale

Mediaset (2023) Fuori dal coro. Puntata del giorno 30 maggio 2023. https://mediasetinfinity.mediaset.it/video/fuoridalcoro20222023/puntata-del-30-maggio_F312336201002101

Micallef M (2023) "There is no climate emergency"—the Global Climate Intelligence Group led by the Nobel Prize laureate Professor Ivar Giaever. Free Press. https://simonmercieca.com/2023/05/12/there-is-no-climate-emergency-the-global-climate-intelligence-group-led-by-the-nobel-prize-laureate-professor-ivar-giaever/

Minzolini A (2023) Europa senza leader. Il Giornale. https://www.ilgiornale.it/news/politica-estera/europa-senza-leader-2101964.html

Misculin L (2020) Il Green Deal europeo, spiegato bene. Il Post. https://www.ilpost.it/2020/02/02/green-deal-europeo/

Mosca G (2023) Chiude il "metaverso Ue". Il flop della Commissione europea. Il Giornale. https://www.ilgiornale.it/news/nuove-frontiere/commissione-europea-ha-chiuso-proprio-metaverso-2104721.html

Muresu C (2023) Perché gli Stati Uniti dovrebbero seguire l'esempio dell'Ue sulle regole climatiche. Energia Oltre. https://energiaoltre.it/perche-gli-stati-uniti-dovrebbero-seguire-lesempio-dellue-sulle-regole-climatiche/

Musso M (2019) Il 100% degli scienziati crede che il cambiamento climatico sia dovuto agli esseri umani. Wired. https://www.wired.it/attualita/ambiente/2019/11/27/cambiamenti-climatici-consenso/

Napolitano P (2023) "Da Elly il bacio della morte a Sanchez. E i Popolari non governano senza Ecr". Il Giornale. https://www.ilgiornale.it/news/politica/elly-bacio-morte-sanchez-e-i-popolari-non-governano-senza-2159133.html

Neri V (2020) Cosa prevede il piano da mille miliardi di euro per finanziare il green deal europeo. Lifegate/Daily. https://www.lifegate.it/piano-investimenti-green-deal-europeo

Ocone C (2023) L'Ue ci obbliga a ristrutturare? Quando l'ideologia punisce i più deboli. Libero Quotidiano. https://www.liberoquotidiano.it/news/europa/34503432/ue-obbligo-ristrutturare-casa-ideologia-punisce-deboli.html

Ocone C (2024) Trattori in piazza, la sinistra non tutela più i lavoratori e irride i contadini. Fa un certo effetto l'aria di indifferenza, quanto non proprio l'aperta ostilità, con cui a sinistra si guarda alla protesta degli agricoltori. Libero Quotidiano. https://www.liberoquotidiano.it/news/italia/38334619/trattori-piazza-sinistra-non-tutela-lavoratori-irride-contadini.html

Paddison L (2023) Russia's war in Ukraine is undermining global efforts to tackle the climate crisis, new report finds. CNN. https://edition.cnn.com/2023/06/07/europe/ukraine-war-climate-change-impact-intl/index.html

Perinelli C (2022) Cambiamenti Climatici: Cause e Conseguenze in Sintesi. Green Marketing. Riduzione Impatto. https://greenmarketing.agency/cambiamenti-climatici-cause-e-conseguenze/#:~:text=Differenza%20tra%20cambiamenti%20climatici%20e%20riscaldamento%20globale,-Molto%20spesso%20tendiamo&text=I%20cambiamenti%20climatici%20fanno%20riferimento,della%20temperatura%20media%20sulla%20Terra.

Popoli P (2023) Napoli, l'auto esplosa in tangenziale era un prototipo. I familiari della ricercatrice e del tirocinante ustionati: "Vogliamo la verità". La Repubblica. https://napoli.repubblica.it/cronaca/2023/06/25/news/auto_esplosa_tangenziale_cnr_prototipo-405676549/?ref=RHLF-BG-I405681822-P1-S1-T1

Powell J (2017) Scientists reach 100% consensus on anthropogenic global warming. Bull Sci Technol Soc 37(4):183–184. https://doi.org/10.1177/0270467619886266

Prestigiacomo D (2023) La nuova tassa ambientale su auto e case alimentate con i fossili. Europa Today. https://europa.today.it/ambiente/ue-nuova-tassa-case-auto.html

Redazione (2023) Alluvione, la beffa delle auto elettriche: "Mettetele in quarantena". L'onda lunga dell'alluvione assesta un duro colpo al sogno elettrico delle auto. Nicola Porro. https://www.nicolaporro.it/alluvione-la-beffa-delle-auto-elettriche-mettetele-in-quarantena/?utm_source=partner&utm_medium=link&utm_campaign=porro&utm_source=partner&utm_medium=link&utm_campaign=porro

Robustelli L (2023) Auto solo elettriche. Forse si può fare di meglio. Eunews. https://www.eunews.it/2023/03/24/auto-elettriche-italia/

Romano B (2022) Gas, non c'è accordo tra i ministri europei su un tetto al prezzo. Il Sole 24 Ore. https://www.ilsole24ore.com/art/gas-non-c-e-accordo-i-ministri-europei-un-tetto-prezzo-AED5K9yB

Romiti C (2023) La deriva del pensiero unico: il clima sta diventando una religione. Nicola Porro. https://www.nicolaporro.it/la-deriva-del-pensiero-unico-il-clima-sta-diventando-una-religione/?utm_source=partner&utm_medium=link&utm_campaign=porro&utm_source=partner&utm_medium=link&utm_campaign=porro

Sabel C et al (2022) Fixing the climate: strategies for an uncertain world. Princeton University Press. Powell's. ISBN: 0-691-22454-4. https://www.powells.com/book/fixing-the-climate-9780691224558

Salvatore F (2023) Armi ed ecocidio: perché per l'Ucraina può essere la fine del granaio d'Europa. Inside Over. https://it.insideover.com/guerra/armi-ed-ecocidio-perche-per-lucraina-puo-essere-la-fine-del-granaio-deuropa.html

Sanchez-Reaza J et al (2023) Making the European Green Deal Work for People: The Role of Human Development in the Green Transition. World Bank Group. Open Knowledge Repository Beta. https://openknowledge.worldbank.org/entities/publication/7b1fcc3c-f44f-4a11-9828-9ae191a746c6

Senaldi P (2023) Clima, Franco Prodi: "Da cosa dipende la siccità. La Ue si è bevuta il cervello". Libero Quotidiano. https://www.liberoquotidiano.it/news/scienze-tech/35251799/clima-franco-prodi-siccita-da-cosa-dipende-ue-bevuta-cervello.html

Sesana I (2023) Le società di consulenza che influenzano le istituzioni europee per favorire le Big Tech. Altraeconomia. https://altreconomia.it/le-societa-di-consulenza-che-influenzano-le-istituzioni-europee-per-favorire-le-big-tech/

Signorile L (2023) La guerra in Ucraina è anche un disastro ambientale. Per gli ecosistemi, l'agricoltura, le specie a rischio e, sul lungo periodo, per tutto il pianeta. Il Tascabile. https://www.iltascabile.com/scienze/ucraina-guerra-disastro-ambientale/

Società (2023) Azioni dell'UE per ridurre le emissioni di navi e aerei. Attualità. Parlamento Europeo. https://www.europarl.europa.eu/news/it/headlines/society/20220610STO32720/azioni-dell-ue-per-ridurre-le-emissioni-di-navi-e-aerei

Stallard E (2022) What is COP27 and why is it important? BBC Science. https://www.bbc.com/news/science-environment-63316362

Talignani G (2023) Ucraina, il crollo della diga di Kakhovka è una catastrofe ecologica. Contaminazione chimica dei terreni, specie dei parchi nazionali a rischio, mine trascinate dall'acqua e inquinanti che potrebbero arrivare fino al Mar Nero. La Repubblica. https://www.repubblica.it/green-and-blue/2023/06/07/news/ucraina_diga_kakhovka_catastrofe_ecologica_petrolio-403531169/?ref=RHLF-BG-I403548745-P3-S2-T1

Tocci N (2022) A Green and Global Europe. POLITY PRESS. La Feltrinelli. Ibs. ISBN-13: 9781509555178. https://www.ibs.it/green-global-europe-ebook-inglese-nathalie-tocci/e/9781509555185

Vivaldelli R (2023) Dalla Gruber a Gentiloni: ecco chi va al Bilderberg, il meeting dell'élite globale. Il Giornale. https://www.ilgiornale.it/news/politica-estera/gentiloni-gruber-alver-ecco-chi-partecipa-meeting-bilderberg-2153921.html

Zaccardi M (2023) L'Fmi incastra i Verdi: energia alle stelle per i diktat ecologisti. Libero Quotidiano. https://www.liberoquotidiano.it/news/piulibero/36275485/fmi-incastra-verdi-energia-stelle-diktat-ecologisti.html

Zanellato W (2023) "Vittoria russa sarebbe fatale": la profezia di Draghi sull'Ue. Il Giornale. https://www.ilgiornale.it/news/politica-internazionale/kiev-deve-vincere-o-lue-sar-fatale-monito-draghi-sulla-2163538.html

Sustainable Building Materials from Agricultural Wastes

Vicky Kumar, Adamu Sani Abubakar, Tooba Qureshi, Abiola U. Adebanjo, Shamsul Rahman Mohamed Kutty, and Siti N. Abd Razak

Abstract Integrating sustainability and environmental objectives into construction projects during the evaluation stage is crucial. However, this stage is often overlooked. The construction sector faces environmental challenges, such as excessive resource use, necessitating a reassessment of the supply chain of raw materials used in construction activities. This chapter provides an in-depth review of the advantages associated with using sustainable construction materials derived from agricultural by-products. Using agricultural waste in construction can significantly reduce environmental and engineering issues, offering significant benefits over traditional materials. From an environmental standpoint, employing building materials made from agricultural waste offers several advantages. These include minimizing waste, conserving resources, improving energy efficiency, and promoting sustainable land use. From an Engineering perspective, agricultural wastes can improve low density, thermal insulation, fire resistance, and biodegradability. Economic considerations suggest that agricultural wastes' viability depends on their availability and circular economy alignment. Diverse agricultural waste sources, unstandardized manufacturing processes, and lack of quality control procedures pose obstacles despite the benefits. These factors collectively pose challenges for wide-scale adoption in the industry. The integration of agricultural waste-based building materials into the

V. Kumar (✉) · A. S. Abubakar · A. U. Adebanjo · S. R. M. Kutty · S. N. Abd Razak
Department of Civil and Environmental Engineering, Universiti Teknologi PETRONAS, Seri Iskandar, Malaysia
e-mail: vicky_19000193@utp.edu.com

A. S. Abubakar
e-mail: adamu_22000793@utp.edu.my

A. U. Adebanjo
e-mail: abiola_22007022@utp.edu.my

S. R. M. Kutty
e-mail: shamsulrahman@utp.edu.my

S. N. Abd Razak
e-mail: tq1511877@qu.edu.qa; siti_0008995@utp.edu.my

T. Qureshi
Department of Chemical Engineering, College of Engineering, Qatar University, Doha, Qatar

E. Shaffie et al. (eds.), *Sustainable Green Infrastructure*,
https://doi.org/10.1007/978-981-96-1486-8_6

construction sector holds the potential for innovative and sustainable advancements in the future.

Keywords Agricultural wastes · Building and construction · Carbon emissions mitigation · Sustainable materials · Resource conservation

1 Introduction

The construction industry has long been acknowledged as a significant contributor to environmental degradation, accounting for a substantial share of global resource consumption and waste generation (Ding 2004; Olonade and Adebanjo 2021). As the world grapples with the pressing need to combat climate change and reduce the ecological impact of human activities, there is a growing demand for sustainable practices within construction. One of the most promising approaches to achieving eco-friendly building solutions is the utilization of agricultural wastes as raw materials for construction purposes (Sangmesh et al. 2023). The building of structures begins with the selection of suitable materials. An appropriate setup and the synthetic, physical and mechanical properties of the materials determine how strong a structure will perform in service. Accordingly, the foundation of green buildings should begin with selecting and using eco-friendly building materials comparable to or preferable features to conventional building materials.

Building construction and other closely related activities are responsible for many harmful outflows, accounting for 30% of ozone-depleting compounds, with an additional 18% being influenced by material mishandling and carriage. Sixty per cent of the natural stone, rock and sand used annually overall, as well as 30% of the unprocessed timber, are consumed by the building industry. This implies that the structure segment considerably impacts the overall condition from an ecological effect perspective (Kumar Sharma 2020). Cement manufacturing mostly causes carbon dioxide (CO_2) emissions (around 8%). With approximately two-thirds of the world's cement production each, China and India are the two biggest producers. Moreover, approximately 900 kg of CO_2 goes into the atmosphere while manufacturing each ton of cement (Benhelal et al. 2013; Sathiparan 2021). Due to burning fossil fuels, construction has a significant negative influence on both the dwindling of natural resources and the emissions of greenhouse gases (GHG). Regarding its contributions to GHG emissions, the construction industry in the US ranks third among all other industrial sectors. Around the world, it uses up 40% of all primary energy supply, 15% of the world's freshwater assets, produces 25% of waste worldwide, and releases 40–50% of GHG emissions (Ding 2014; Mokhlesian and Holmén 2012).

Despite these challenges, the demand for building materials is continuously increasing and their costs are rising due to a shortage of raw materials (Gejdoš and Potkány 2022). As a result, acquiring natural resources for building materials has become a global issue. Additionally, traditional building materials like cement consume substantial thermal and electrical energy during production, leading to higher construction expenses and the industry's overall lack of sustainability. Therefore, there is a pressing need for more scientific research to develop eco-friendly and sustainable building materials without compromising their quality (Mezher 2011). Some of the feasible initiatives include improving circular economy methods, boosting building energy efficiency and lowering the carbon footprint of construction materials (Sáez-de-Guinoa et al. 2022). To achieve this objective of sustainable construction, advancements in research studies are now primarily focused on the utilization of sustainable building materials in order to reduce the industry's adverse environmental effects. This is one action the construction sector can take to contribute to environmental protection responsibly. The decision concerning the use of building materials is one of the variables that might affect the sustainability of a building project, as indicated by Akadiri in his study (Akadiri 2015).

In developing nations, the proper management of agricultural waste, such as sugarcane bagasse, wheat husks, coconut shells and rice husks, is of utmost importance. For instance, India produces more than 600 million metric tons of agricultural waste annually and with the intensification of farming practices, the amount of waste generated is expected to increase further (Cardoen et al. 2015; Vitali et al. 2013). Conventional waste disposal methods, such as dumping, incinerating and composting, have significant adverse environmental impacts. However, recent studies have demonstrated that agricultural waste and its by-products can solve these challenges (Mortier et al. 2016). By utilizing agricultural waste in the construction industry, the environmental consequences of landfill waste disposal are reduced, thereby mitigating the pollution associated with conventional building materials like cement (Abdel-Shafy and Mansour 2018).

Successful examples include substituting sand with agricultural crop wastes like peanut shells, rice husks, rice straws, and coconut shells in cement block production, meeting the required standards for strength and durability (Prusty et al. 2016). The industry can achieve the revolutionary development it needs to become more sustainable and ecologically benign by employing agricultural waste. Furthermore, using wheat and barley straw fibres to stabilize soil has resulted in bricks with properties conducive to environmentally friendly and safe building materials (Laborel-Préneron et al. 2017). In addition, straw bale construction has been found to create more eco-friendly buildings with improved thermal performance and low embodied energy. This approach aligns with sustainability objectives and helps minimize pollution and other harmful effects (Parisi et al. 2015). The research into using agricultural wastes as building materials began in the early twenty-first century, driven by these waste materials' economic, widespread availability, and cost-efficient nature. They have found applications as supplementary cementitious materials (SCMs), alternative recycled aggregates, and alternative building materials, all contributing to environmental and sustainable development in the construction industry (Adebanjo et al. 2021). Agricultural wastes are viable substitutes in numerous applications because they are plentiful, renewable, and environmentally friendly. By-products like straw, husks, and manure are typically rich in nutrients and organic content, making them ideal for composting and soil enhancement (Koul et al. 2022). They can also be converted into bioenergy, which decreases dependence on fossil fuels and cuts down greenhouse gas emissions. Moreover, using agricultural wastes aids in waste management reduces environmental pollution, and encourages sustainable farming practices. Transforming these wastes into valuable products supports a circular economy and improves resource efficiency.

This chapter seeks to explore the emerging field of sustainable building materials derived from agricultural wastes, providing an overview of the prevailing challenges faced by the construction industry and the urgent need for innovative solutions. Exploring sustainable building materials derived from agricultural wastes represents a transformative step towards a greener, more resilient, and environmentally responsible construction industry (Pansera and Sarkar 2016). Capitalizing on the abundant resources provided by agriculture, can not only mitigate the environmental impacts of construction but also contribute to a sustainable future, balancing human activities with the delicate ecosystems of our planet. Through this work of literature, aim to shed light on the potential of agricultural wastes as invaluable resources for sustainable building solutions and inspire further research and innovation in this promising field. The utilization of agricultural wastes as building materials offers a promising solution to address environmental concerns and promote sustainability in the construction sector. Further research and implementation of these practices can positively impact the environment and contribute to a more sustainable future.

2 Agricultural Wastes as a Source of Sustainable Building Materials

Concrete is currently the most widely applied material for building in the construction sector. In addition to its durability, concrete has other essential qualities such as its adaptability, incombustibility, affordability and ease of acquisition (Hegger et al. 2017). It can also be easily moulded into any shape and is an engineering material that almost always satisfies specified specifications. In today's industrial society, over 10 billion tons of concrete are manufactured annually, highlighting its extensive use. Nonetheless, despite its advantages, the massive production of concrete has considerable adverse environmental effects. In addition, the manufacturing process results in CO_2 emissions due to the substantial number of raw materials needed to make the billions of tons of concrete produced annually worldwide. One ton of CO_2 is discovered to be released into the atmosphere for every ton of Portland cement produced (Aprianti et al. 2015).

Over the past decade, the building material industry has faced significant demand due to the increasing population, leading to a chronic scarcity of construction materials. To keep up with the ever-growing housing needs, there is a tremendous requirement for producing various construction materials like bricks, cement, steel, wood, cladding, and more (Raut et al. 2011). However, conventional production methods for these materials, such as cement, bricks and steel, consume substantial thermal and electrical energy and result in air, water and land pollution. Unfortunately, the importance of using suitable and environmentally friendly building materials has not received enough attention (Raut et al. 2011). Furthermore, according to (Abdel-Shafy and Mansour 2018), the improper disposal of solid waste generated from agricultural and industrial activities poses a severe problem, particularly in many developing countries.

Numerous research studies have looked for substitutes for cement, such as disposable and low-value wastes from industry and agriculture, whose potential advantages can be achieved through reuse, recovery, and rejuvenating programs (Aprianti et al. 2015; Awasthi et al. 2022; Awoyera and Adesina 2020; Kareem et al. 2023; Kolawole et al. 2022; Koul et al. 2022; Narisetty et al. 2022; Pappu et al. 2007; Raut et al. 2011; Sonu et al. 2023; Wang et al. 2016). As a result, scientists have been looking into the usefulness, viability and availability of pozzolanic wastes as alternatives to cement. Agricultural waste, including sugarcane bagasse, rice husk, bamboo, coconut husk and sawdust, is abundant and needs proper management. Reusing such waste materials as sustainable building alternatives presents a promising solution to reduce pollution and address the challenges of landfills and expensive building materials (Pappu et al. 2007). Figure 1 shows the percentage of the production rate of common agricultural wastes, about 998 million tons on a yearly basis. The increasing global population, urbanization, and technological advancements have led to a substantial rise in the quantity and variety of solid wastes generated from various sources, such as industry, mining, households, and agriculture (Pappu et al. 2007). Figure 2 represents the various possible sources of agricultural waste.

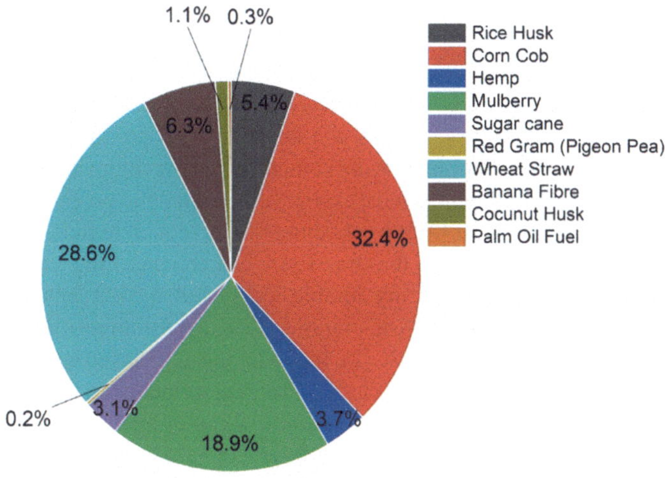

Fig. 1 Annual percentage production per metric tonne of common agricultural wastes. *Sources* Balda et al. (2021), Guna et al. (2021), Jadhav et al. (2020), Korjenic et al. (2011), Lawson et al. (2022), Panyakaew and Fotios (2008), Pinto et al. (2011), Ramezani and Sain (2019), Satpathy and Sahu (2010), and Shyam Kumar et al. (2021)

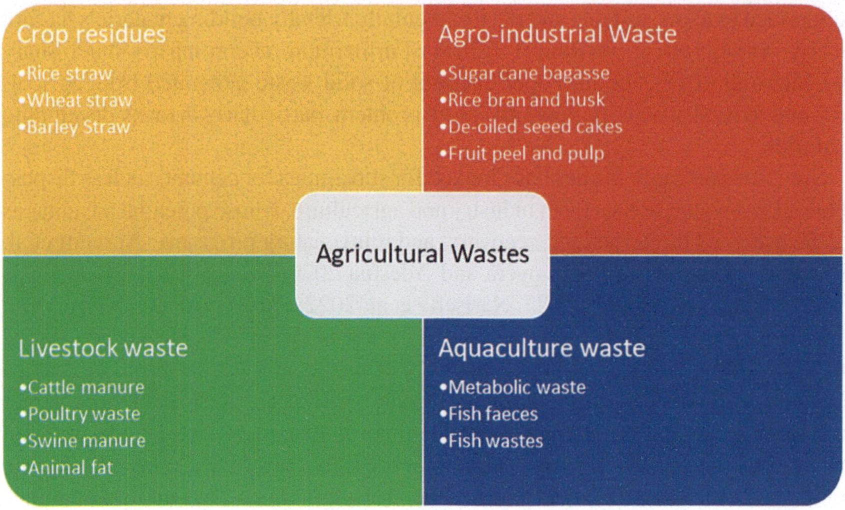

Fig. 2 Categorization of agricultural wastes based on their sources. *Source* Koul et al. (2022)

There as several types of agricultural waste that can be used in construction. Information provided in the succeeding section provides a brief description of some of the agricultural wastes that have been utilized in the construction industry.

2.1 *Rice Husks*

By-products of rice milling, rice husks can be utilized to produce materials like rice husk ash (RHA) or combined with cement to create lightweight and insulating building blocks. Researchers have extensively studied the potential use of rice husk ash (RHA) as a partial substitute for cement in the concrete industry (Nagrale et al. 2012). The findings indicate that fine RHA can enhance concrete's strength, making it a viable option for producing durable concrete alongside cement (Cardoen et al. 2015). However, coarse RHA is employed negatively affects the concrete's strength development (Nagrale et al. 2012). In producing autoclaved aerated concrete (AAC), replacing sand with RHA increases water requirements, reducing compressive strength but yielding lighter AAC (Kunchariyakun et al. 2015). Furthermore, incorporating 5–10% RHA as a cement replacement in self-compacting concrete (SCC) improves its mechanical properties but reduces workability (Adebanjo et al. 2022; Molaei Raisi et al. 2018). These research findings emphasize RHA's potential as an alternative material in concrete production, with effectiveness depending on particle size and specific application.

Fig. 3 Processes to produce sugarcane bagasse ash. *Source* Prusty et al. (2016)

2.2 *Sugarcane Bagasse*

Derived from the remains of crushed sugarcane stalks, bagasse can be processed to produce biodegradable composites and panels for different construction applications. According to Sun et al. when 1 tonne of sugarcane is processed, it generates approximately 280 kg of bagasse during one cycle. Bagasse is the fibrous residue remaining after extracting juice from sugarcane. The worldwide production of bagasse amounts to about 54 million dry tonnes annually (Sun et al. 2004). Utilizing sugarcane bagasse has the potential to replace fossil fuels effectively and can be extensively employed in power generation through boilers, turbines, and furnaces. Figure 3 illustrates the conversion process from raw sugarcane to bagasse and sugarcane bagasse ash.

2.3 *Bamboo*

Bamboo is perennial grass characterized by its large woody stem or culm. Approximately 1662 bamboo species belong to 121 genera (Canavan et al. 2017). However, for construction purposes, only a select few species from the Guadua, Dendrocalamus, Bambusa, and Phyllostachys genera are commonly used (Fahim et al. 2022; Shen et al. 2019; Tahmasebinia et al. 2021; Yor Maikol et al. 2020). The bamboo culm refers to the aboveground part of the plant that contains the majority of its woody material. It is accompanied by a branching system, sheath, foliage leaves, flowers, fruits, and seedlings. The culm grows in a single period, with a well-timed process of cell elongation and some cell division. Unlike trees, bamboo lacks secondary thickening growth, resulting in straight, hollow, and cylindrical culms with node and internode parts. Bamboo culm has gained widespread recognition as an environmentally friendly and sustainable building material, offering an eco-conscious alternative to conventional materials like wood, brick, concrete, and steel (Tahmasebinia et al. 2021). The versatility of bamboo has led to extensive usage in construction across numerous countries, particularly in Asia, Africa, and South America. It is employed for various purposes, including flooring, ceilings, walls, windows, doors, roofs, trusses and rafters in buildings. Additionally, bamboo is a structural material for bridges, water transportation, and skyscraper scaffolding. Its adoption as a building material is considered a step towards sustainable development, given its evident environmental, social, and economic benefits (Shen et al. 2019).

2.4 *Mulberry Stems (MS)*

MS is a sericulture byproduct, that has already been employed as a reinforcing agent in polypropylene (PP) polymers for environmentally friendly building, furniture, and automotive applications. Although lignocellulosic mulberry stems are renewable and sustainable resources, they are rarely used and are often disposed of as garbage by burning or burying. In an effort to replace frequently used non-biodegradable composites, mulberry stems have been tried utilizing a straightforward and hygienic production method. In their study, Mulberry was utilized by Guna et al. (2021) as a reinforcement agent with Polypropylene polymer (PP). According to the results, the best ratio for achieving the highest mechanical strength was 90/10 (MS/PP w/w%). It was also found that density enhancement improved the strength, thermal insulation, sound transmission loss, water stability, and flame-retardant properties (Guna et al. 2021).

2.5 *Sabai Grass*

This has special qualities that make it an excellent material for creating composites, including low density, high lignin concentration, and good moisture absorption. Native to Asia, sabai grass is notably common in the eastern Indian state of Orissa. To investigate the viability of cellulose fiber extraction from sabai grass and ascertain their qualities, a few experiments have been carried out. Sabai grass is also thought to have UV resistance and antibacterial qualities, according to reports (Guna et al. 2019; Satpathy and Sahu 2010).

2.6 *Pigeon Peas*

According to data reported by the Directorate of Pulses Development and the Ministry of Agriculture and Farmers' Welfare, Government of India, pigeon peas (Red Gram) is one of the most ancient, second-most significant crop covering 38.49 lakh hectares and 28.66 lakh tons of manufacturing, extensively farmed in India, which positioned first with a coverage of 79.65% of the land area and a production of 67.28% in the world. The pigeon pea is made up of woody stalks, leaves, and flowers. In the past, the woody stems of the pigeon pea plant were used as temporary structures, fences, and rooftops made of thatch (Shyam Kumar et al. 2021).

2.7 *Wheat Straw*

This is a common agricultural waste that has a lesser commercial value than other lignocellulosic materials. In a study by Rojas, Carlos, et al., a pair of materials for insulation were generated based on natural leftover fibers from the agricultural sector, which happen to be abundant in the south of Chile. Wheat straw (1,510,779 tons of residue on a nationwide scale in 2013–2014) and maize husk (9727 ha of crops in 2014) were the two primarily used in the study. Installing thermal insulation in buildings has been shown to reduce transmission losses, lowering the energy used for heating usage. The remnants from wheat straw and maize husk were chemically analyzed to design prototypes later and subject to physical and mechanical evaluations to demonstrate their effectiveness as insulation materials. The results demonstrated that the thermal conductivity of the prototypes was good enough and that these two materials were suitable for insulating buildings (Rojas et al. 2019).

2.8 *Palm Oil Residues*

The oil palm is a tropical palm tree that is easily cultivated in tropical nations such as Malaysia, Indonesia, Thailand, Africa, and Latin America; three ASIAN countries produce 90% of all palm oil. Due to its high fineness, which is denser and more uniform, POFA (Palm oil fuel ash) can improve the compressive strength of concrete. Furthermore, by having a loss on ignition (LOI) of less than 10%, the usage of POFA as a binder meets the chemical criteria in ASTM C618 as a pozzolanic material. As a result, it may be useful in the production of concrete (Aprianti et al. 2015).

2.9 *Other Agricultural Wastes*

A recent investigation by Weerasinghe, Dakshitha, et al. seeks to fill the gap for utilizing waste by examining the feasibility of employing the naturally degradable bioplastic PLA in conjunction with yarns spun from banana fibres to create a composite panel designed for interim partitioning products utilized by the construction sector. The material for reinforcement was pre-tensioned banana yarns, and the matrix was PolyLactic Acid (PLA). When compared to medium-density fibreboard (MDF), the innovative composite panel generated under ideal circumstances had 52% more impact strength and a whopping 55% greater flexural strength. As a superior-performing material made from renewable resources, the composite panel proposed in this article has the potential to replace MDF (Dassanayaka et al. 2023). In another study, a straightforward and efficient process for producing high strength-high density board materials from entire coconut husks without the use of chemical binders. This material possessed exceptional qualities which were equivalent to, if not superior to, commercial wood-based panels. The compressed husks of coconut boards could be worked on with standard woodworking tools such as drill and chopping, smoothing and finishing. The project's goal was to provide proof of the potential of (fresh) coconut husks as a raw material for the creation of environmentally safe and high-performance building materials. Highly efficient fibre boards, poles, and panels made domestically can be used in low-cost housing and the manufacturing of materials for the packaging industry. The structural and physical characteristics of these boards were assessed in relation to those of conventional MDF and particleboard and found to be good (strength of 50 MPa and stiffness of 5 GPa). This high performance emerges as a commercial opportunity for the development of low-cost building materials (Van Dam et al. 2004) Ilangovan et al. (2022) investigated that; when utilized as reinforcement for polypropylene, sheep wool and chicken feathers have unique qualities that offer great noise, temperature insulation, and flame resistance. Not only are used wool and chicken feathers inexpensively and in great quantities, but they also have unique properties that make them perfect for creating composite materials for a variety of uses. Rajput et al. (2012) recycled paper mill waste and cotton waste to make waste-create bricks which varied the content of cotton waste from 1 to 5% by weight (1–5 wt.%), made and tested according to IS 3495 (Part 1–3): 1992) specifications. The results showed that brickwork was thermally resistant and satisfied the standard (IS 3495 (Part 1–3): 1992). The utilization of these agricultural waste materials for construction offers numerous advantages, including reducing waste, greenhouse gas emissions, and environmental pollution. Additionally, it promotes the use of renewable resources and contributes to more energy-efficient and cost-effective building practices. Ongoing research and development efforts focus on optimizing these materials to meet industry standards and building codes, ensuring their long-term durability and effectiveness in sustainable construction projects. Table 1 demonstrates some of these studies.

3 Properties and Characteristics of Agricultural Waste-Based Building Materials

The use of agricultural waste in construction materials has become a sustainable and environmentally friendly approach to address waste management and resource depletion in the construction industry. Agricultural waste, such as rice husks, straw, bagasse, and coconut shells, possesses unique properties that make them suitable for building materials, offering numerous advantages in terms of performance and environmental impact. One noteworthy property of agricultural waste-based building materials is their lightweight nature. Their low weight makes them easy to handle

Table 1 Other agricultural wastes used for sustainable buildings

Study	Agricultural waste used	Method	Results/ comparison	Benefits
(Balda et al. 2021)	Banana fibres	Composite panel using pre-tensioned banana yarns and polylactic acid (PLA)	• 52% higher impact strength than MDF 55% greater flexural strength than MDF	Superior performance derived from renewable resources potential MDF replacement
(Van Dam et al. 2004)	Coconut husks	High strength-high density board material produced from entire coconut husks without chemical binders	• Qualities equivalent to or superior to commercial wood-based panels strength of 50 MPa stiffness of 5 GPa	Can be worked on with standard woodworking tools. Environmentally safe High-performance Suitable for low-cost housing and packaging industry
(Ilangovan et al. 2022)	Sheep wool and chicken feather	Used as reinforcement for polypropylene	• Great noise, temperature insulation • Flame resistance	Inexpensive and abundant. Unique properties for creating composite materials
(Rajput et al. 2012)	Recycled paper mill waste and cotton waste	Waste-create bricks with varying cotton waste content (1–5 wt.%	• Thermally resistant • Met IS 3495 (Part 1–3): 1992 standards	Reduces waste Decreases greenhouse gas emissions and environmental pollution. Promotes renewable resources More energy-efficient and cost-effective building practices

during construction, reducing the need for heavy machinery and transportation costs (Koul et al. 2022). Additionally, their reduced weight contributes to energy-efficient structures, requiring less energy for heating and cooling, resulting in an eco-friendlier built environment. Another important characteristic of these materials is their exceptional thermal insulation capacity. They possess excellent thermal properties, providing natural insulation to buildings (Abu-Jdayil et al. 2019). This inherent insulating ability helps maintain stable indoor temperatures, reducing the reliance on artificial heating and cooling systems, and leading to energy savings (Table 2).

Table 2 Characteristics of building material made from agricultural waste

Property/characteristic	Description
Lightweight	Agricultural waste-based construction materials are lightweight, which simplifies their handling during building projects and reduces transportation expenses and the requirement for heavy machinery (Peng et al. 2023)
Energy efficiency	Their lightweight nature promotes energy efficiency in buildings, reducing energy consumption for heating and cooling and fostering a more environmentally friendly built environment (Mlhem et al. 2022)
Thermal insulation	These materials exhibit superior thermal characteristics, naturally insulating buildings (Mlhem et al. 2022)
Fire resistance	Due to their fibrous composition and low thermal conductivity, agricultural waste-based materials exhibit outstanding fire resistance, thereby enhancing structural safety (Abu-Jdayil et al. 2021)
Sound absorption	Their porous and fibrous composition enables effective sound absorption, which is advantageous for mitigating urban noise pollution (Fediuk et al. 2021)
Biodegradability	Agricultural waste-derived materials are biodegradable, breaking down naturally without causing harm to the environment, which supports sustainable building methods (Maraveas 2020)

Furthermore, agricultural waste-based building materials exhibit remarkable fire resistance due to their fibrous composition and low heat conductivity. This fire-retardant property enhances the safety of structures and reduces the risk of fire-related accidents, making them particularly suitable for construction in fire-prone regions (Abu-Jdayil et al. 2019). Additionally, these materials have excellent sound absorption properties, thanks to their porous and fibrous structure. This attribute is advantageous in urban areas, where reducing noise pollution is essential for creating comfortable living and working environments. The biodegradability of agricultural waste-based building materials is a significant advantage. Being inherently biodegradable, these materials can decompose naturally without harming the environment at the end of their lifespan. This aligns with the principles of sustainable building practices and promotes a more circular approach to construction. Agricultural waste-based building materials present a promising solution to the environmental challenges faced by the construction industry. Their lightweight, thermal insulation, fire resistance, sound absorption, and biodegradability properties make them attractive alternatives to conventional materials, contributing to a more sustainable and greener future in the building sector.

4 Environmental Benefits of Using Agricultural Waste-Based Building Materials

Using agricultural waste-based building materials presents several environmental advantages, making it a sustainable and eco-friendly option in comparison to conventional construction materials. The primary benefit lies in its capacity to minimize waste and optimize resource utilization. Agricultural waste materials like rice straws, corn husks, coconut shells, and bagasse, which are often discarded or burned, contribute to air and soil pollution (Gupta et al. 2022). According to Awoyera and Adesina (2020), by repurposing these materials into building components such as bricks, panels, and insulation, we can divert waste from landfills and reduce greenhouse gas emissions, leading to a more sustainable waste management approach. Another significant environmental benefit of employing agricultural waste-based building materials is preserving natural resources. Unlike traditional building materials like concrete and steel, which require extensive mining and extraction processes leading to habitat destruction and resource depletion, agricultural waste materials are renewable and abundant, lessening the need for resource-intensive mining activities and reducing pressure on ecosystems (Akenji et al. 2016).

Furthermore, by incorporating agricultural waste-based materials in construction helps mitigate carbon dioxide emissions. The production of conventional building materials like cement is energy-intensive and emits a considerable amount of CO_2 (Sangmesh et al. 2023). Conversely, agricultural waste materials have a lower carbon footprint due to their less energy-intensive processing. Choosing these sustainable alternatives allows builders and homeowners to contribute to reduced carbon emissions and take positive action against climate change. Utilizing agricultural waste-based building materials also aligns with circular economy principles, promoting a more sustainable economy (Guillard et al. 2018). Instead of being discarded, waste from one industry becomes a valuable resource for another, fostering local, green industries and reducing the need for excessive resource extraction and disposal. Moreover, agricultural waste-based building materials play a crucial role in enhancing overall energy efficiency in buildings.

According to Wang et al. (2016), the use of agricultural waste-based building materials encourages sustainable land use practices. In many agricultural regions, crop residues are often burned to clear fields, leading to harmful pollutants and air quality degradation. Additionally, due to the high cost and restricted availability of the primary component of concrete, particularly in developing nations, agricultural wastes used as supplementary cementitious material (SCMs) in concrete manufacturing can improve the global economy and environmental friendliness of structures (Aprianti et al. 2015). By continuing to use agricultural waste, carbon emissions may be decreased, lowering the ecological impact of building development (Battampara et al. 2021). This could open up a pathway for the development of several experimental investigations to reveal a variety of agricultural by-products and recycled materials, which are unquestionably suitable for use as acoustic, rooftops and thermal panels (Domke and Mude 2015).

Furthermore, converting these waste materials into building resources discourages open burning and promotes sustainable land management techniques like mulching or composting, which enrich the soil and improve crop yields. Agricultural waste-based building materials offer numerous environmental benefits, including waste reduction, resource conservation, carbon emission mitigation, improved energy efficiency, and support for sustainable land use practices. Embracing these alternatives is essential for creating a more sustainable and resilient built environment for future generations. Governments, industries, and individuals should recognize the potential of agricultural waste-based building materials and support their adoption through policy incentives, research funding, and public awareness campaigns.

5 Economic Feasibility of Agricultural Waste-Based Building Materials

The viability of utilizing agricultural waste as a foundation for building materials has attracted significant attention due to the growing emphasis on sustainable and environmentally friendly solutions in the building sector. The construction industry has long been associated with excessive resource consumption and the generation of substantial waste, prompting a search for alternative materials that not only reduce environmental impact but also offer economic advantages. Agricultural waste, including materials like rice husks, straw, and bagasse, has emerged as a promising resource for crafting building materials such as particleboards, insulation, and even structural components (Karimah et al.). The economic feasibility of these materials arises from their abundance, low production costs, and potential to reduce dependence on traditional resource-intensive building materials (Adebanjo et al. 2023).

A primary driver of the economic viability of incorporating agricultural waste into construction lies in its readily available supply. Agricultural processes produce a substantial amount of waste, which is often either burned, resulting in air pollution, or left to decompose, contributing to emissions (Yrjälä et al. 2022). Repurposing this waste into construction materials not only addresses environmental concerns but also taps into a consistently available raw material source. For instance, rice husks, which are typically discarded after rice milling, can be transformed into particleboards or insulation materials using various techniques. Capitalizing on waste materials that are already abundant enables the construction industry to decrease reliance on scarce natural resources while also tackling waste management challenges. Kazmi et al. (2016) reported that the incorporation of rice husk ash (RHA) and sugarcane bagasse ash (SBA) in clay brick production, with a 5% proportion of RHA and SBA to clay weight, resulted in bricks meeting Pakistan building codes and ATSM standards for modulus of rupture and compressive strength. Similarly, Taurino et al., (2019) illustrated the creation of lightweight bricks using wine wastes (WWs) such as grape seeds, wine lees, and stalks combined with clay. These studies collectively highlighted the potential of agricultural waste in crafting building materials that align with prescribed construction benchmarks.

Likewise, researchers have explored the efficacy of agrowaste in generating economical and effective construction materials, particularly evident in green concrete applications. Akram et al. (2009) conducted an in-depth study showcasing how sugarcane bagasse ash (SCBA) could function as a viscosity-modifying agent in self-compacting concrete. This innovation bears notable economic implications, negating the need for vibrators during the compaction process. Such initiatives underscore the ability of agricultural waste to enrich the development of construction materials that offer economic benefits while upholding their performance. Another significant factor bolstering the economic feasibility of agricultural waste-based building materials is their cost-effectiveness. Traditional construction materials like cement, steel, and wood can be costly, and subject to market fluctuations. Conversely, agricultural waste materials are often considered surplus and can be procured at a fraction of the cost of conventional materials. According to Fico et al. (2022), the manufacturing processes required to convert these waste materials into usable construction components are generally less complex than those used for traditional materials. This not only reduces production expenses but also cuts down on the energy consumed during manufacturing. As a result, integrating agricultural waste materials into construction can lead to cost savings in both raw material acquisition and production processes, enhancing the economic viability of such materials.

Furthermore, the economic feasibility of utilizing agricultural waste-based building materials aligns with the principles of a circular economy. The concept of a circular economy revolves around minimizing waste and optimizing existing resources. By repurposing agricultural waste in construction, a cyclical process is established wherein waste materials are transformed into valuable products (Narisetty et al. 2022). This approach resonates with sustainability goals and lessens the need for continuous extraction and processing of new raw materials. As the demand for sustainable construction practices rises, businesses have an opportunity to position themselves as environmentally conscious entities while also benefiting from reduced costs associated with utilizing agricultural waste. The economic feasibility of construction materials derived from agricultural waste presents a compelling case for their integration into the construction sector. Their widespread availability, low production costs, and potential to foster a circular economy position them as a compelling alternative to conventional materials. As societies prioritize sustainability and environmental consciousness, these materials offer a means to reduce the ecological impact of the construction industry while potentially yielding economic benefits.

6 Challenges and Limitations of Agricultural Waste-Based Building Materials

Agricultural waste-based building materials represent an avenue towards sustainable construction practices, addressing environmental concerns tied to conventional construction materials. However, integrating these materials is accompanied by a series of challenges and limitations that necessitate thorough examination before widespread adoption. A central challenge arises from the inherent variability in raw materials derived from agricultural waste sources. These materials, procured from diverse agricultural by-products such as rice husks, straw, and bagasse, exhibit significant physical and chemical properties disparities. For instance, Xu et al. (2018) argued that bricks developed by using rice husk ash (RHA), and sugarcane bagasse ash (SBA) and clay could only be used in applications where lower structural loads were required in buildings. Consequently, ensuring consistency and predictability in the quality and performance of resulting building materials becomes challenging, impeding uniformity in construction projects.

Adding to this challenge is the absence of standardized manufacturing processes and established quality control procedures. In contrast to conventional materials benefiting from well-established production methods, agricultural waste-based materials lack standardized techniques for conversion and treatment (Awasthi et al. 2022). As a result, discrepancies in strength, durability, and other critical attributes of end products can emerge, eroding confidence in the suitability of these materials for construction purposes. A notable concern pertains to the durability of agricultural waste-based materials. Many of these materials are innately susceptible to biodegradation, insect infestations, and moisture-related deterioration (Kolawole et al. 2022). Unless subject to adequate treatment and protection, the longevity of constructed buildings might be compromised, necessitating frequent maintenance and repairs. Addressing this challenge involves developing effective treatment methodologies that bolster durability while upholding the sustainability of the materials.

Furthermore, attaining the requisite structural strength for load-bearing applications poses another hurdle. Agricultural waste-based materials often lack the inherent robustness found in conventional materials like concrete and steel (Kolawole et al. 2022). Considering this, engineers and material scientists are compelled to explore inventive approaches such as composites and advanced reinforcement techniques to ensure the materials can withstand the demanding structural requirements inherent in contemporary construction. Considerations regarding fire resistance and thermal performance are paramount due to their implications for building safety and energy efficiency. However, certain agricultural waste-based materials might exhibit inadequate fire resistance and subpar thermal insulation properties (Sonu et al. 2023).

Striving to adhere to stringent building code mandates for fire safety and energy efficiency proves challenging without introducing additives or treatments that might potentially compromise the environmentally friendly traits of the materials. Striking this balance is pivotal to determining the viability of these materials in construction contexts. Beyond technical complexities, constraints linked to availability and economic feasibility underscore the challenges. The dependable supply of agricultural waste hinges on factors such as geographic location, seasonal crop cycles, and localized agricultural practices (Yevich and Logan 2003). Such irregular supply chains can hinder large-scale construction projects that necessitate a stable material influx. Additionally, the conversion of agricultural waste into usable building materials often mandates specialized equipment and techniques. This can result in elevated energy consumption and augmented production costs compared to traditional materials, potentially affecting the cost-effectiveness of projects.

Furthermore, limited market familiarity and regulatory impediments represent barriers to comprehensive adoption. The construction sector typically leans on tried-and-true materials, necessitating comprehensive awareness campaigns to showcase the advantages and potential applications of agricultural waste-based materials. Regulatory frameworks might not yet be equipped to accommodate these materials, engendering uncertainty in project approvals. Collaboration with regulatory bodies and active advocacy for updated codes could facilitate a smoother assimilation process. The constraint of aesthetic limitations also warrants consideration. Agricultural waste-based materials might not offer the broad spectrum of colours, textures, and finishes characteristic of conventional materials. Architects and designers may need to exercise creativity to incorporate these materials into designs that align with their visual limitations. In conclusion, while agricultural waste-based building materials present an eco-conscious alternative, they are accompanied by a spectrum of challenges and limitations. Addressing these obstacles mandates the convergence of expertise from researchers, engineers, manufacturers, architects, policymakers, and the construction industry. Pioneering advancements in processing techniques, material treatments, quality control protocols, and regulatory frameworks will play a pivotal role in unlocking the potential of these materials and their contribution to sustainable construction practices.

7 Future Directions for Agricultural Waste-Based Building Materials in Construction and Conclusion

Looking ahead, the incorporation of agricultural waste-based building materials into the construction sector shows promise and potential for innovative and sustainable advancements. These materials not only bring about environmental benefits but also have the capacity to reshape the landscape of construction practices. There are several key avenues that are anticipated to shape the future trajectory of agricultural waste-based building materials in the construction domain. Firstly, the realm of material science research and innovation is expected to be a driving force. Progress in processing techniques, material compositions, and engineering methodologies is likely to enhance the overall properties and performance of agricultural waste-based building materials. Strengthening their structural integrity, fire resilience, and durability has the potential to make these materials more adaptable across a wide spectrum of construction applications. Secondly, establishing standardized guidelines and certifications tailored to these materials will be a critical step. The definition of clear benchmarks for quality, safety, and performance will instil confidence among architects, engineers, and construction practitioners to seamlessly incorporate agricultural waste-based materials into their projects. Additionally, obtaining certifications will streamline regulatory approvals and stimulate widespread adoption, creating a conducive environment for their integration.

Fostering collaboration across diverse sectors including agriculture, construction, and academia will be pivotal for driving progress. Building partnerships that bridge the gap between researchers, material producers, farmers, and construction experts can yield innovative solutions, efficient supply chains, and heightened awareness about the untapped potential of agricultural waste-based building materials. Market dynamics are also poised to exert substantial influence. The growing consumer demand for environmentally conscious and sustainable buildings is projected to drive the utilization of these materials. As individuals increasingly prioritize eco-friendly construction solutions, the market for such alternatives is expected to expand. This evolving consumer sentiment can incentivize manufacturers and builders to invest in refining and effectively utilizing agricultural waste-based materials. Moreover, government policies and incentives are expected to shape the trajectory of these materials in the construction sector. Favourable policies, including incentives like tax benefits for the adoption of sustainable materials and regulations aimed at waste reduction, can significantly accelerate the incorporation of agricultural waste-based building materials.

Government-driven initiatives can further fund research and development endeavours focused on optimizing these materials for construction-related applications. In conclusion, the outlook for agricultural waste-based building materials in the construction sector holds promise and diversity. Ongoing research endeavours, adherence to standardized norms, collaborative efforts that transcend sectors, changing market dynamics, and government backing are all pivotal factors that will guide their trajectory. As these materials continue to evolve and gain prominence, they stand to usher in a transformative era for the construction industry. These materials offer sustainable, economically viable, and ecologically conscious alternatives to conventional building materials. The integration of agricultural waste-based building materials has the potential to chart a sustainable and environmentally friendly course for the future of the built environment.

References

Abdel-Shafy HI, Mansour MSM (2018) Solid waste issue: sources, composition, disposal, recycling, and valorization. Egypt J Petrol 27(4). https://doi.org/10.1016/j.ejpe.2018.07.003

Abu-Jdayil B, Mourad AH, Hittini W, Hassan M, Hameedi S (2019) Traditional, state-of-the-art and renewable thermal building insulation materials: an overview. Constr Build Mater 214. https://doi.org/10.1016/j.conbuildmat.2019.04.102

Abu-Jdayil B, Barkhad MS, Mourad A-HI, Iqbal MZ (2021) Date palm wood waste-based composites for green thermal insulation boards. J Build Eng 43:103224

Adebanjo A, Kareem M, Olawuyi O, Ishola K, Odefemi A (2022) Effects of waste steel fibres on the mechanical properties of modified self compacting concrete. J Eng Stud Res 28(2), 7–16. https://doi.org/10.29081/jesr.v28i2.001

Adebanjo A, Oyewole K, Kumar V, Razak SNA, Emmanuel E, Singh P, Adebisi A (2023) Technoeconomic assessment of iron filings blocks, pp 485–494. https://doi.org/10.1007/978-981-99-1886-7_41

Adebanjo AU, Dahunsi BIO, Labiran JO (2021) Effects of concrete grades on strength characteristics of metakaolin modified recycled aggregate concrete. Nigerian J Technol Dev 18(3):184–193. https://doi.org/10.4314/njtd.v18i3.3

Aigbomian EP, Fan M (2013) Development of Wood-Crete building materials from sawdust and waste paper. Constr Build Mater 40. https://doi.org/10.1016/j.conbuildmat.2012.11.018

Akadiri PO (2015) Understanding barriers affecting the selection of sustainable materials in building projects. J Build Eng 4. https://doi.org/10.1016/j.jobe.2015.08.006

Akenji L, Bengtsson M, Bleischwitz R, Tukker A, Schandl H (2016) Ossified materialism: introduction to the special volume on absolute reductions in materials throughput and emissions. J Clean Prod 132. https://doi.org/10.1016/j.jclepro.2016.03.071

Akram T, Memon SA, Obaid H (2009) Production of low cost self compacting concrete using bagasse ash. Constr Build Mater 23(2). https://doi.org/10.1016/j.conbuildmat.2008.02.012

Aprianti E, Shafigh P, Bahri S, Farahani JN (2015) Supplementary cementitious materials origin from agricultural wastes—a review. Constr Build Mater 74. https://doi.org/10.1016/j.conbuildmat.2014.10.010

Awasthi MK, Sindhu R, Sirohi R, Kumar V, Ahluwalia V, Binod P, Juneja A, Kumar D, Yan B, Sarsaiya S, Zhang Z, Pandey A, Taherzadeh MJ (2022) Agricultural waste biorefinery development towards circular bioeconomy. Renew Sustain Energy Rev 158. https://doi.org/10.1016/j.rser.2022.112122

Awoyera PO, Adesina A (2020) Plastic wastes to construction products: status, limitations and future perspective. Case Stud Constr Mater 12. https://doi.org/10.1016/j.cscm.2020.e00330

Balda S, Sharma A, Capalash N, Sharma P (2021) Banana fibre: a natural and sustainable bioresource for eco-friendly applications. Clean Technol Environ Policy 23(5). https://doi.org/10.1007/s10098-021-02041-y

Battampara P, Ingale D, Guna V, Pradhan UU, Reddy N (2021) Green energy from discarded wool and fish scales. Waste Biomass Valoriz 12(12). https://doi.org/10.1007/s12649-021-01475-1

Benhelal E, Zahedi G, Shamsaei E, Bahadori A (2013) Global strategies and potentials to curb CO_2 emissions in cement industry. J Clean Prod 51. https://doi.org/10.1016/j.jclepro.2012.10.049

Bumanis G, Vitola L, Pundiene I, Sinka M, Bajare D (2020) Gypsum, geopolymers, and starch-alternative binders for bio-based building materials: a review and life-cycle assessment. Sustainability (Switzerland) 12(14). https://doi.org/10.3390/su12145666

Canavan S, Richardson DM, Visser V, Le Roux JJ, Vorontsova MS, Wilson JRU (2017) The global distribution of bamboos: Assessing correlates of introduction and invasion. AoB PLANTS 9(1). https://doi.org/10.1093/aobpla/plw078

Cardoen D, Joshi P, Diels L, Sarma PM, Pant D (2015) Agriculture biomass in India: Part 1. Estimation and characterization. Resour Conserv Recycl 102. https://doi.org/10.1016/j.resconrec.2015.06.003

Dassanayaka D, Hedigalla D, Weerasinghe D, Wanasekara N, Gunasekera U, Mohotti D (2023) Developing a compression-moulded composite partitioning panel from banana fibres and polylactic acid (PLA). Electron J Struct Eng 23(2). https://doi.org/10.56748/EJSE.233832

Ding GKC (2004) The development of a multi-criteria approach for the measurement of sustainable performance for built projects and facilities. A thesis submitted in fulfilment of the requirements for the award of Doctor of Philosophy

Ding GKC (2014) Life cycle assessment (LCA) of sustainable building materials: an overview—eco-efficient construction and building materials—3. In: Eco-efficient construction and building materials: life cycle assessment
Domke PV, Mude VD (2015) Natural fiber reinforced building materials. IOSR J Mech Civ Eng Ver II 12(3)
Fahim M, Haris M, Khan W, Zaman S (2022) Bamboo as a construction material: prospects and challenges. Adv Sci Technol Res J 16(3). https://doi.org/10.12913/22998624/149737
Fediuk R, Amran M, Vatin N, Vasilev Y, Lesovik V, Ozbakkaloglu T (2021) Acoustic properties of innovative concretes: a review. Materials 14(2):398
Fico D, Rizzo D, Casciaro R, Corcione CE (2022) A review of polymer-based materials for fused filament fabrication (FFF): focus on sustainability and recycled materials. Polymers 14(3). https://doi.org/10.3390/polym14030465
Gejdoš M, Potkány M (2022) The impact of global crisis accidents on the prices development of selected raw-wood assortments in the middle Europe region. In: 15th international scientific conference WoodEMA 2022—crisis management and safety foresight in forest-based sector and SMES operating in the global environment
Guillard V, Gaucel S, Fornaciari C, Angellier-Coussy H, Buche P, Gontard N (2018) The next generation of sustainable food packaging to preserve our environment in a circular economy context. Front Nutr 5. https://doi.org/10.3389/fnut.2018.00121
Guna V, Ilangovan M, Adithya K, Akshay AK, Srinivas CV, Yogesh S, Nagananda GS, Venkatesh K, Reddy N (2019) Biofibers and biocomposites from sabai grass: a unique renewable resource. Carbohyd Polym 218. https://doi.org/10.1016/j.carbpol.2019.04.085
Guna V, Ilangovan M, Reddy N, Radhakrishna PG, Maharaddi VH, Jambunath A, Rao AP (2021) Biobased insulating panels from mulberry stems. J Thermoplast Compos Mater. https://doi.org/10.1177/08927057211010884
Gupta N, Mahur BK, Izrayeel AMD, Ahuja A, Rastogi VK (2022) Biomass conversion of agricultural waste residues for different applications: a comprehensive review. Environ Sci Pollut Res 29(49). https://doi.org/10.1007/s11356-022-22802-6
Hegger M, Drexler H, Zeumer M (2017) Basics materials. Birkhäuser
Ilangovan M, Navada AP, Guna V, Touchaleaume F, Saulnier B, Grohens Y, Reddy N (2022) Hybrid biocomposites with high thermal and noise insulation from discarded wool, poultry feathers, and their blends. Constr Build Mater 345. https://doi.org/10.1016/j.conbuildmat.2022.128324
Jadhav T, Jadhav AD, Attar AC, Verma S (2020) Evaluation of silkworm rearing houses from Western Maharashtra. Indian J Tradit Knowl 19(4). https://doi.org/10.56042/ijtk.v19i4.44586
Kareem MA, Akintonde BB, Adesoye JS, Ishola K, Tijani IA, Adebanjo AU, Thanni MO, Oduoye WO, Adeyanju DO, Bello AA (2023) Influence of cashew leaf ash as partial replacement for cement on the properties of fresh and hardened concrete. Clean Waste Syst 4:100063. https://doi.org/10.1016/j.clwas.2022.100063
Kazmi SMS, Abbas S, Munir MJ, Khitab A (2016) Exploratory study on the effect of waste rice husk and sugarcane bagasse ashes in burnt clay bricks. J Build Eng 7. https://doi.org/10.1016/j.jobe.2016.08.001
Kolawole FO, Kolawole ID, Udebhulu OD, Kolawole SK, Aba MM, Shamaki PB (2022) An overview of the sources, structure, applications, and biodegradability of agricultural wastes. Hybrid Polym Nanocomposites Agric Waste. https://doi.org/10.1201/9781003170549-3
Korjenic A, Petránek V, Zach J, Hroudová J (2011) Development and performance evaluation of natural thermal-insulation materials composed of renewable resources. Energy Build 43(9). https://doi.org/10.1016/j.enbuild.2011.06.012
Koul B, Yakoob M, Shah MP (2022) Agricultural waste management strategies for environmental sustainability. Environ Res 206. https://doi.org/10.1016/j.envres.2021.112285
Kumar Sharma N (2020) Sustainable building material for green building construction, conservation and refurbishing. Int J Adv Sci Technol 29(10S)

Kunchariyakun K, Asavapisit S, Sombatsompop K (2015) Properties of autoclaved aerated concrete incorporating rice husk ash as partial replacement for fine aggregate. Cem Concr Compos 55. https://doi.org/10.1016/j.cemconcomp.2014.07.021

Laborel-Préneron A, Aubert J-E, Magniont C, Maillard P, Poirier C (2017) Effect of plant aggregates on mechanical properties of earth bricks. J Mater Civ Eng 29(12). https://doi.org/10.1061/(asce)mt.1943-5533.0002096

Lawson L, Degenstein LM, Bates B, Chute W, King D, Dolez PI (2022) Cellulose textiles from hemp biomass: opportunities and challenges. Sustainability (Switzerland) 14(22). https://doi.org/10.3390/su142215337

Maraveas C (2020) Production of sustainable and biodegradable polymers from agricultural waste. Polymers 12(5):1127

Mezher T (2011) Building future sustainable cities: the need for a new mindset. Constr Innov 11(2). https://doi.org/10.1108/14714171111124121

Mlhem A, Abu-Jdayil B, Tong-Earn T, Iqbal M (2022) Sustainable heat insulation composites from date palm fibre reinforced poly (β-hydroxybutyrate). J Build Eng 54:104617

Mokhlesian S, Holmén M (2012) Business model changes and green construction processes. Constr Manag Econ 30(9). https://doi.org/10.1080/01446193.2012.694457

Molaei Raisi E, Vaseghi Amiri J, Davoodi MR (2018) Mechanical performance of self-compacting concrete incorporating rice husk ash. Constr Build Mater 177. https://doi.org/10.1016/j.conbuildmat.2018.05.053

Montakarntiwong K, Chusilp N, Tangchirapat W, Jaturapitakkul C (2013) Strength and heat evolution of concretes containing bagasse ash from thermal power plants in sugar industry. Mater Des 49. https://doi.org/10.1016/j.matdes.2013.01.031

Mortier N, Velghe F, Verstichel S (2016) Organic recycling of agricultural waste today: composting and anaerobic digestion. In: Biotransformation of agricultural waste and by-products: the food, feed, fibre, fuel (4F) economy. https://doi.org/10.1016/B978-0-12-803622-8.00004-5

Nagrale SD, Hajare H, Modak PR (2012) Utilization of rice husk ash. Int J Eng Res Appl (IJERA) 2(4)

Narisetty V, Adlakha N, Kumar Singh N, Dalei SK, Prabhu AA, Nagarajan S, Naresh Kumar A, Amruthraj Nagoth J, Kumar G, Singh V, Kumar V (2022) Integrated biorefineries for repurposing of food wastes into value-added products. Bioresour Technol 363. https://doi.org/10.1016/j.biortech.2022.127856

Olonade KA, Adebanjo AU (2021) Carbon dioxide sequestration: potentials and opportunities for cement industry in Nigeria

Pansera M, Sarkar S (2016) Crafting sustainable development solutions: frugal innovations of grassroots entrepreneurs. Sustainability (Switzerland) 8(1). https://doi.org/10.3390/su8010051

Panyakaew S, Fotios S (2008) Agricultural waste materials as thermal insulation for dwellings in Thailand: preliminary results. In: PLEA 2008—towards zero energy building: 25th PLEA international conference on passive and low energy architecture, conference proceedings

Pappu A, Saxena M, Asolekar SR (2007) Solid wastes generation in India and their recycling potential in building materials. Build Environ 42(6). https://doi.org/10.1016/j.buildenv.2006.04.015

Parisi F, Asprone D, Fenu L, Prota A (2015) Experimental characterization of Italian composite adobe bricks reinforced with straw fibers. Compos Struct 122. https://doi.org/10.1016/j.compstruct.2014.11.060

Peng X, Jiang Y, Chen Z, Osman AI, Farghali M, Rooney DW, Yap P-S (2023) Recycling municipal, agricultural and industrial waste into energy, fertilizers, food and construction materials, and economic feasibility: a review. Environ Chem Lett 21(2):765–801

Pinto J, Paiva A, Varum H, Costa A, Cruz D, Pereira S, Fernandes L, Tavares P, Agarwal J (2011) Corn's cob as a potential ecological thermal insulation material. Energy Build 43(8). https://doi.org/10.1016/j.enbuild.2011.04.004

Prusty JK, Patro SK, Basarkar SS (2016) Concrete using agro-waste as fine aggregate for sustainable built environment—a review. Int J Sustain Built Environ 5(2). https://doi.org/10.1016/j.ijsbe.2016.06.003

Rajput D, Bhagade SS, Raut SP, Ralegaonkar RV, Mandavgane SA (2012) Reuse of cotton and recycle paper mill waste as building material. Constr Build Mater 34. https://doi.org/10.1016/j.conbuildmat.2012.02.035

Ramezani N, Sain M (2019) Non-catalytic green solvent lignin isolation process from wheat straw and the structural analysis. Renew Energy 140. https://doi.org/10.1016/j.renene.2019.03.026

Raut SP, Ralegaonkar RV, Mandavgane SA (2011) Development of sustainable construction material using industrial and agricultural solid waste: a review of waste-create bricks. Constr Build Mater 25(10). https://doi.org/10.1016/j.conbuildmat.2011.04.038

Rojas C, Cea M, Iriarte A, Valdés G, Navia R, Cárdenas-R, JP (2019) Thermal insulation materials based on agricultural residual wheat straw and corn husk biomass, for application in sustainable buildings. Sustain Mater Technol 20. https://doi.org/10.1016/j.susmat.2019.e00102

Sáez-de-Guinoa A, Zambrana-Vasquez D, Fernández V, Bartolomé C (2022) Circular economy in the European construction sector: a review of strategies for implementation in building renovation. Energies 15(13). https://doi.org/10.3390/en15134747

Sangmesh B, Patil N, Jaiswal KK, Gowrishankar TP, Selvakumar KK, Jyothi MS, Jyothilakshmi R, Kumar S (2023) Development of sustainable alternative materials for the construction of green buildings using agricultural residues: a review. Constr Build Mater 368. https://doi.org/10.1016/j.conbuildmat.2023.130457

Sathiparan N (2021) Utilization prospects of eggshell powder in sustainable construction material—a review. Constr Build Mater 293. https://doi.org/10.1016/j.conbuildmat.2021.123465

Satpathy AR, Sahu UN (2010) The role of Sabai grass industry in the economic development of Mayurbhanj District of Orissa (India). Int J Appl Agric Res 5

Shen L, Yang J, Zhang R, Shao C, Song X (2019) The benefits and barriers for promoting bamboo as a green building material in China—an integrative analysis. Sustainability (Switzerland) 11(9). https://doi.org/10.3390/su11092493

Shyam Kumar R, Balasundar P, Al-Dhabi NA, Prithivirajan R, Ramkumar T, Bhat KS, Senthil S, Narayanasamy P (2021) A new natural cellulosic pigeon pea (*Cajanus cajan*) pod fiber characterization for bio-degradable polymeric composites. J Nat Fibers 18(9). https://doi.org/10.1080/15440478.2019.1689887

Sonu Rani GM, Pathania D, Abhimanyu Umapathi R, Rustagi S, Huh YS, Gupta VK, Kaushik A, Chaudhary V (2023) Agro-waste to sustainable energy: a green strategy of converting agricultural waste to nano-enabled energy applications. Sci Total Environ 875. https://doi.org/10.1016/j.scitotenv.2023.162667

Sun JX, Sun XF, Sun RC, Su YQ (2004) Fractional extraction and structural characterization of sugarcane bagasse hemicelluloses. Carbohydr Polym 56(2). https://doi.org/10.1016/j.carbpol.2004.02.002

Tahmasebinia F, Ma Y, Joshua K, Sepasgozar SME, Yu Y, Li J, Sepasgozar S, Marroquin FA (2021) Sustainable architecture creating arches using a bamboo grid shell structure: numerical analysis and design. Sustainability (Switzerland) 13(5). https://doi.org/10.3390/su13052598

Taurino R, Ferretti D, Cattani L, Bozzoli F, Bondioli F (2019) Lightweight clay bricks manufactured by using locally available wine industry waste. J Build Eng 26. https://doi.org/10.1016/j.jobe.2019.100892

Van Dam JEG, Van Den Oever MJA, Keijsers ERP (2004) Production process for high density high performance binderless boards from whole coconut husk. Ind Crops Prod 20(1). https://doi.org/10.1016/j.indcrop.2003.12.017

Vitali F, Parmigiani S, Vaccari M, Collivignarelli C (2013) Agricultural waste as household fuel: techno-economic assessment of a new rice-husk cookstove for developing countries. Waste Manag 33(12). https://doi.org/10.1016/j.wasman.2013.08.026

Wang B, Dong F, Chen M, Zhu J, Tan J, Fu X, Wang Y, Chen S (2016). Advances in recycling and utilization of agricultural wastes in China: based on environmental risk, crucial pathways,

influencing factors, policy mechanism. Procedia Environ Sci 31. https://doi.org/10.1016/j.proenv.2016.02.002
Xu Q, Ji T, Gao S-J, Yang Z, Wu N (2018) Characteristics and applications of sugar cane bagasse ash waste in cementitious materials. Materials 12(1):39
Yevich R, Logan JA (2003) An assessment of biofuel use and burning of agricultural waste in the developing world. Global Biogeochem Cycl 17(4). https://doi.org/10.1029/2002gb001952
Yor Maikol SM, Maksimovich SV, Ivanovna BG, Ainur S (2020) Bamboo structures for modern sustainable architecture. ISVS E-Journal 7(3)
Yrjälä K, Ramakrishnan M, Salo E (2022) Agricultural waste streams as resource in circular economy for biochar production towards carbon neutrality. Curr Opin Environ Sci Health 26. https://doi.org/10.1016/j.coesh.2022.100339
Zhou Xy, Zheng F, Li Hg, Lu Cl (2010) An environment-friendly thermal insulation material from cotton stalk fibers. Energy Build 42(7). https://doi.org/10.1016/j.enbuild.2010.01.020

Innovative Green Infrastructure Initiatives for Development in the Construction Industry

Abiola U. Adebanjo, Nasir Shafiq, Vicky Kumar, Oladele J. Olatoyan, Siti N. Abd Razak, and Nur Fadilah Yusof

Abstract The current chapter presents a general overview of green infrastructure and sustainable materials as a viable solution for increasing the longevity and ecological soundness of the built environment. Environmental issues, resource efficacy, urban area quality, prospective solutions for green infrastructure and adoption of sustainable materials were all assessed. The study used a qualitative research approach, using data from various sources, including literature and case studies. Major findings highlight the multiple benefits of green infrastructure in terms of the environment, society, and economic effects. To thoroughly implement green infrastructure initiatives (GRINIT), it is critical to prioritise the use of environmentally sustainable features such as reused materials, renewable energy sources, environmentally friendly construction processes, and energy-saving components. Evaluation of case studies reveals the potential of GRINIT and sustainable materials to foster transformative change in decreasing environmental impacts, enhancing community well-being and developing more sustainable urban environments. Lessons from these cases suggest that using sustainable materials in GRINIT holds the key to a successful implementation and a more efficient approach to addressing some of the environmental issues associated with conventional methods. Adopting this innovative approach can potentially improve resource efficiency, and reduced environmental

A. U. Adebanjo (✉) · N. Shafiq · V. Kumar · S. N. Abd Razak · N. F. Yusof
Universiti Teknologi PETRONAS, Seri Iskandar, Malaysia
e-mail: abiola_22007022@utp.edu.com

N. Shafiq
e-mail: nasirshafiq@utp.edu.my

V. Kumar
e-mail: vicky_19000193@utp.edu.my

S. N. Abd Razak
e-mail: siti_0008995@utp.edu.my

N. F. Yusof
e-mail: nurfadilah.yusof@utp.edu.my

O. J. Olatoyan
Redeemer's University, Ede, Nigeria
e-mail: olatoyanola@run.edu.ng

E. Shaffie et al. (eds.), *Sustainable Green Infrastructure*,
https://doi.org/10.1007/978-981-96-1486-8_7

impact and rate of development (3-Rs). In conclusion, exemplary case studies can be a significant resource for the future implementation of comparable green infrastructure initiatives. Notwithstanding, the government also needs to support these initiatives by incentivising stakeholders in the built sector by implementing policies that encourage the adoption of GRINIT to develop resilient and sustainable structures.

Keywords Built environment · Circular economy · Green infrastructure · Sustainable materials · Urban sustainability

1 Introduction to Green Infrastructure

Sustainable development, circular economy and intelligent growth have become key concepts in contemporary environmental discourse due to the conflict and competition between human-driven economic growth and the natural ecosystem (Ying et al. 2022). Implementing environmentally friendly facilities is essential to achieving equitable growth by orchestrating environmental, social and economic progress (Bibri and Krogstie 2020). Green infrastructure (GI) is fundamental to achieving an equitable relationship between humanity and the natural world. This concept has evolved through extensive ideation and build-up of knowledge (Mukherjee and Shaw 2021). GI greatly aids in mitigating the impacts of climate change (Parker and Simpson 2020), improving the heat island effect (Shao and Kim 2022) and reducing environmental pollution (Sun et al. 2022). The potential benefits of GI on climate change have been extensively studied in research for several decades, not limited to its effects on urban areas as well as consequences and vulnerabilities. Studies have revealed that urban areas are increasingly acknowledged as important partakers in addressing climate change through the implementation of (Hunt et al. 2007).

The popularity of green infrastructure is growing as a strategy to enhance the sustainability and environmental friendliness of urban areas (Kim and Song 2019). As a result of this, there has been an increasing emphasis on the integration of natural elements and sustainable materials within both urban and rural landscapes, with the goal of addressing environmental issues and mitigating resource consumption (Li et al. 2017). Urban green space can be enhanced by implementing green infrastructure (GI) and tree planting. This approach offers various advantages, including the reduction of urban heat, prevention of floods, mitigation of air pollution, provision of recreational opportunities, promotion of health benefits, carbon sequestration, enhancing biodiversity and an aesthetically pleasing environment (Seddon et al. 2021). As shown in Fig. 1, incorporating sustainable practices into infrastructure projects requires consideration of key factors, including the use of recycled and renewable materials, renewable energy sources, implementation of environmentally friendly construction processes and integration of energy-saving components (Hoang et al. 2021). Prioritising the adoption of these technologies can further reduce the negative impact of the built environment on the natural world and consequently

Fig. 1 Key factors for consideration in the implementation of green initiatives

promote the realisation of the Sustainable Development Goals (SDGs) as depicted in Fig. 2 (Balali et al. 2020).

Recent technological advancements have facilitated the implementation of green infrastructure. Innovative solutions, including permeable pavements, rainwater harvesting systems and smart grid technologies, have effectively promoted sustainable urban development (Santhanam and Majumdar 2020). These innovations help to move closer to achieving the net-zero energy goals by improving the efficiency and effectiveness of green infrastructure (Raihan and Tuspekova 2023). Furthermore, green infrastructure has many advantages that go far beyond the realm of the environment, such as improved community well-being, increased value of properties, employment opportunities and reduced infrastructure maintenance costs (Liberalesso et al. 2020). Integrating sustainable materials and technologies in green infrastructure projects can lead to transformative change, fostering more sustainable and resilient urban environments (Vargas-Hernández and Zdunek-Wielgołaska 2021). This chapter aims to explore the various sustainable materials and technologies that could be utilised for green infrastructure. Subsequently, inferences are drawn from case studies of completed projects to demonstrate the practicality of green infrastructure. The challenges associated with green infrastructure, such as cost, limited public awareness, and regulatory obstacles, are carefully addressed. This chapter seeks to enhance knowledge of green infrastructure and the role of sustainable materials in attaining development objectives through an in-depth exploration of these areas of study.

Fig. 2 Benefits of adopting green building materials and their perceived impacts on SDGs. Adapted from Omer and Noguchi (2020)

2 Sustainable Materials for Green Infrastructure Initiatives

The utilisation of sustainable materials in green infrastructure projects is crucial for promoting environmental protection, resource efficiency and long-term sustainability (Hossain et al. 2020). Previous research has shown that by 2050 six billion people will be living in urban areas (Huang et al. 2019). Due to the increasing urban population and the resulting necessity to mitigate the adverse impacts of urbanisation on the natural environment, the need to address sustainability requirements for developing "green infrastructure" has never been more important than it is now (Chapman and Hall 2022). Implementing Green Infrastructure Initiatives (GRINIT) that utilise a wide range of recycled materials and eco-friendly resources has been considered a viable attempt towards achieving sustainability in the construction industry. The practicality of this concept has led to the emergence of a new range of resources and the reusability of by-products of industrial processes, which were previously categorised as waste materials (Sangmesh et al. 2023).

The applicability of incorporating sustainable materials, including recycled aggregates, reclaimed wood, bio-based composites and eco-friendly concrete mixes, into the construction process facilitates the advancement of circular economy principles and reduces the demand for fresh raw materials in the industry (Kurniawan et al. 2021). The concept of GRINIT provides insight into waste valorisation techniques, resource conservation and repurposing of materials such as timber, concrete, and soil, by understanding the dynamic nature of these materials across their life cycle (Hamid and Shamsudin 2022).

2.1 Recycled Aggregate Materials

Crushed concrete and reclaimed asphalt pavement are commonly studied recycled aggregates for potential utilisation in green infrastructure initiatives. Research has demonstrated the practicality and effectiveness of these materials in various applications, such as road building, permeable pavements, and green roofs. These alternative substitutes for traditional aggregates serve the dual purpose of reducing construction and demolition waste from landfills while exhibiting comparable or superior mechanical properties when used in concrete production (O'Connor et al. 2016). Using recycled aggregates in concrete can produce concrete with a compressive strength between the range of 20 and 40 MPa. Due to the higher absorption rate, they are often considered a better alternative for use as aggregates in pervious concrete.

2.2 Recycled Lumber Materials

The use of timber and reclaimed wood as boardwalks, bridges and park structures in green infrastructures has increased lately. This preference is primarily motivated by its renewable and biodegradable characteristics (Bowen and Lynch 2017; Ferrari et al. 2022). Using sustainably harvested wood in construction contributes to preserving forest habitats and mitigates pollution from timber mills. Furthermore, the use of alternative materials can help reduce the demand for newly harvested timber, consequently mitigating some of the challenges often attributed to deforestation. When appropriately harvested, recycled lumbers are an environmentally friendly renewable resource because they help reduce the disposal of usable materials in landfills and minimise pollution associated with producing new products (Abed et al. 2022). The fact that reclaimed wood primarily originates from old-growth trees makes it a better alternative in terms of density and durability when compared to timber sourced from newer forests, which presents it for better use in GIP. The lower environmental impact often associated with reclaimed wood gives it an additional advantage over newly harvested materials (Ramage et al. 2017). The use of reclaimed wood and sustainably sourced timber in infrastructure projects can also provide several environmental and economic benefits, including reduced demand for newly sourced lumber, unique look, lower impact manufacturing, and reduced carbon footprint (Sandberg and Kutnar 2016).

2.3 Bio-Based Materials

Bio-based composites, such as those obtained from bamboo, hemp, and kenaf, are gaining popularity as a sustainable building material. The strength-to-weight ratio, durability, and decay resistance of these materials make them useful in both structural and non-structural applications. Some of the implementation of bio-based materials in green infrastructure projects (GIP) include green roofs, facade systems and erosion control measures. These concepts have been credited with improving the long-term viability and operational efficiency of GIP (Farzana et al. 2022). Extensive research has been conducted on developing eco-friendly reinforced concrete by completely replacing the steel component with pretreated bamboo of various sizes. The outcome of such investigation has shown that the impressive tensile property of the treated bamboo compares favourably with those of the rebars and can be adopted when the applied moment does not exceed 50KNm in beams. The use of agricultural-based supplementary cementitious materials as pozzolans in concrete has been well documented by a lot of researchers (Arrigoni et al. 2020; Cassiani et al. 2021; Das et al. 2022; Kareem et al. 2023; Kumar and Prasad 2019). In some climes, based on the performance of these materials, there has been an implementation of green concrete policies which allows stakeholders in cement and concrete production to incorporate not more than 30% of these materials as a part of the concrete composites with a focus to reducing greenhouse gases as well as enhancing the durability and performance of GIP incorporating green concrete structures.

2.4 Recycled Plastic Materials

Plastics have become indispensable in many industries due to their versatile properties and broad compatibility with various demands (Maddah 2016). The production of plastic goods has experienced a 20-fold increase over the past five decades, and it is projected to reach 25 billion tonnes by 2050 (Cao and Wang 2021). The disposal of plastic waste in landfills, incinerators, and recycling facilities has become a growing concern for many countries because of the hazards of incineration and disposal methods. While incineration results in air pollution and energy consumption, landfilling, on the other hand, causes soil pollution and groundwater contamination (Joseph et al. 2021). Recent advancements in adopting plastic-based sustainable materials for green infrastructure have focused on developing innovative products such as recycled plastic composites, permeable geosynthetics and biodegradable erosion control materials. These materials offer unique properties and functionalities, contributing to the overall sustainability and effectiveness of GIP (Leão et al. 2012). Incorporating highly stable recycled plastic into the cement has proven to enhance the durability of mortar and concrete samples, providing an indirect approach to addressing issues of plastic waste management and resource scarcity (Chisholm et al. 2021; Ferreira et al. 2012). Similarly, adding recycled plastic fibre (RPF)

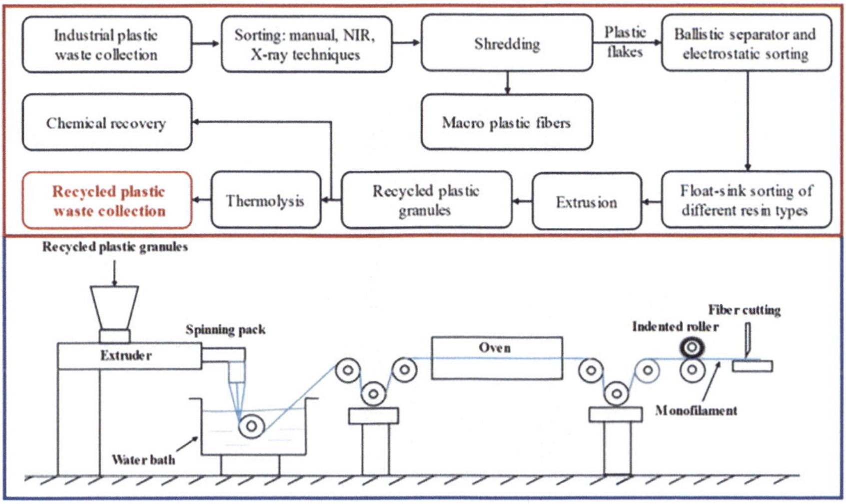

Fig. 3 Process of preparing the RPF derived from plastic wastes. Adapted from Duan et al. (2023)

to concrete has enhanced crack resistance, mechanical properties, durability, and corrosion resistance (Leão et al. 2012; Fig. 3).

2.5 Construction and Demolition Wastes (CDW)

One of the most pressing issues of the twenty-first century is waste management. Among the numerous wastes from different activities of life, construction and demolition waste (CDW) attracts so much attention primarily because it is considered the most voluminous waste stream produced worldwide (Papastamoulis et al. 2021). In the European Union zone, CDW waste accounts for about 900 million tonnes of total waste generated annually (Adebanjo et al. 2021; Balletto et al. 2013; Qiao et al. 2022; Xu et al. 2022). Figure 4 illustrates the waste volume produced within the European Union. CDW consists of various materials such as sand, gravel, concrete, metal, plastic, and glass. These materials do not decompose and are therefore categorised as inert and non-biodegradable. Excluding excavation soils, the inert component of waste constitutes 40–85% of the total volume (Dos Reis et al. 2021). Therefore, reusing these materials for GI applications is increasingly recognised as an effective measure for mitigating the environmental impact of construction activities (Colorado et al. 2022). For example, RCA derived from CDW has found considerable application in road-based materials, embankment support, and permeable pavements in rigid pavement construction. The utilisation of ceramic waste as a partial replacement for natural aggregates in concrete mixtures has also proven to enhance the

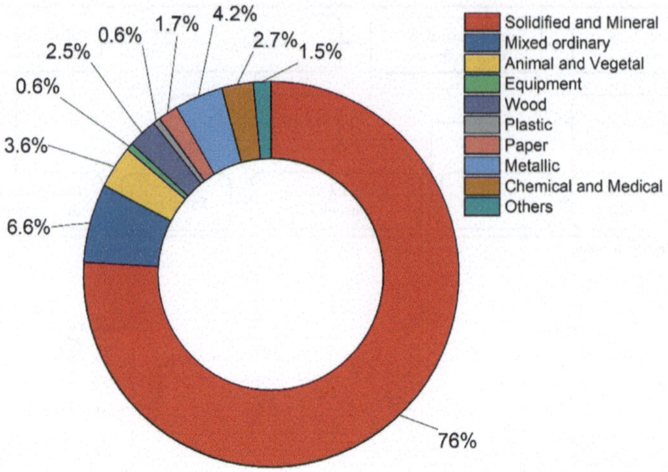

Fig. 4 Total EU waste produced by category. Adapted from Dos Reis et al. (2021)

mechanical properties of concrete while reducing the environmental impact associated with ceramic waste disposal (Dos Reis et al. 2021). Recycling and reusing CDW as sustainable materials for GRINIT conserve natural resources and reduce landfill congestion (Yu et al. 2022). Therefore, incorporating CDW aligns with the principles of the circular economy, promoting the efficient use of resources and minimising waste generation. CDW wastes are complex residual substances characterised by their diverse composition, encompassing various materials and potentially incorporating any constituent of a structure or infrastructure (Junaid et al. 2022). As shown in Fig. 5, the classification, quantity, as well as composition of CDW are affected by factors such as population growth, regional planning, legislation, soil characteristics, topography, construction materials and available technologies (Wang et al. 2020). In the near future, CDW is projected to undergo substantial annual growth on a global scale, thus making it a viable potential source of alternative materials for use in the full implementation of the GRINIT concept (Uche-Soria and Rodríguez-Monroy 2019). For large-scale implementation, there is a need for strategically organised approaches that are driven by innovation based on the classification, reuse, and recycling of CDW wastes (Dos Reis et al. 2021).

3 Technologies for Green Infrastructure

Technology is essential for the implementation of green infrastructure projects. Green infrastructure technologies can contribute to achieving Sustainable Development Goals (SDGs) such as stormwater management, energy efficiency, and climate resilience through innovative technology (Gordon et al. 2018). Smart grid technology

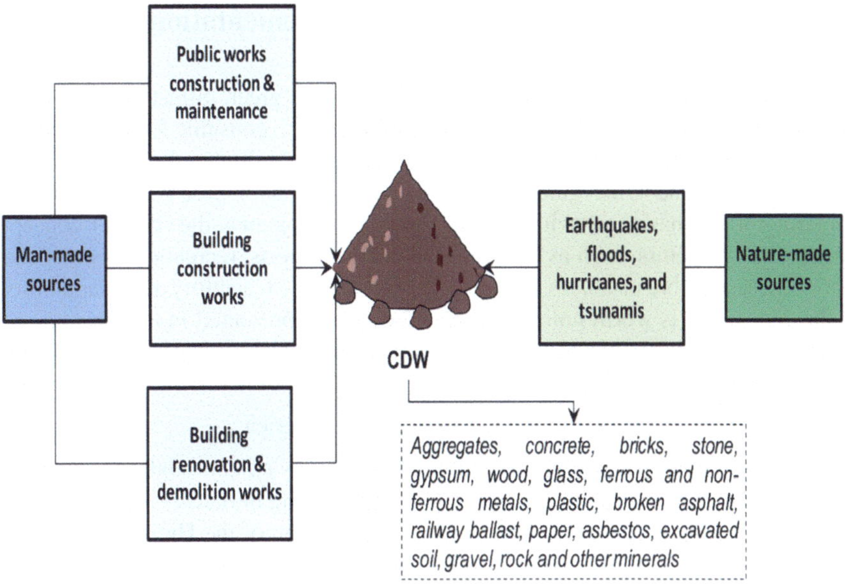

Fig. 5 CDW Classification according to the origin of source. Adapted from Dos Reis et al. (2021)

and advanced monitoring and control systems in green infrastructure initiatives facilitate real-time data collection and analysis. Data-driven optimisation enables energy consumption optimisation, system performance enhancement, and support for maintenance and operational decision-making (Ahmad et al. 2022). Rainwater harvesting, bioretention systems, and permeable pavements are methods for collecting, filtering, and reusing stormwater. However, porous pavements alleviate stormwater infrastructure stress by facilitating rainwater infiltration. Rainwater is stored in non-potable water systems. Bioretention systems utilise vegetation and engineered soils to absorb and filter runoff, recharging groundwater and removing contaminants (Sharma and Malaviya 2021).

Below are some key technologies commonly employed in green infrastructure:

- Green roofs mitigate runoff, reduce the effects of urban heat islands, and promote energy conservation. Shahmohammad et al. (2022) suggest that improved watering and modular green roof systems are viable alternatives.
- Permeable pavements contribute to reducing stormwater runoff and improving water quality. Water can be stored and infiltrated into porous materials such as concrete, porous asphalt, and interlocking pavers.
- Rainfall harvesting systems collect and store rainfall for various purposes, such as irrigation, toilet flushing, and laundry. Rainwater gathering techniques involve the utilisation of barrels, cisterns, and underground storage tanks.

4 Lessons from Case Studies of GIP Implementation

The literature highlights the significance of sustainable green infrastructure. Green buildings typically have a reduced environmental impact, consume fewer resources, exhibit aesthetic appeal, possess more outstanding durability, and necessitate less maintenance. Amidst the global surge in the adoption of green infrastructures, numerous efforts have been witnessed worldwide to implement the GRINIT concept across various sectors, such as buildings, railroads, airports, recreational parks, and more. These credible instances stand as a testament to the feasibility and adaptability of the concept. It is worth noting that the case studies presented in this chapter are provided solely to emphasise widespread adoption globally and do not reflect any bias from the authors.

A. High Line Park, New York City, USA—North America

The High Line Park, situated in the bustling and culturally rich borough of Manhattan in New York City, is a prominent illustration of pioneering initiatives in green infrastructure reshaping urban environments. The construction of the High Line initially took place during the 1930s, to integrate it into a freight rail network that catered to the industrial sectors of the city. However, due to the waning of the industrial sector and the transition towards road-based transport, the railway system experienced a decline in acceptability during the 1980s. Initially intended for demolition, the previously abandoned elevated railway line transformed remarkably into a distinctive urban sanctuary, providing a refuge from the bustling urban environment. The linear park, spanning 1.45 miles, is widely recognised and acclaimed for its exceptional design, integration of state-of-the-art technologies, and substantial contributions towards attaining sustainable development objectives.

The project incorporated several innovative GRINIT concepts, such as the implementation of sustainable drainage systems, the integration of natural and human-made structures, the utilisation of recycled and reclaimed materials, the adoption of alternative lighting techniques, and active community engagement. These concepts were instrumental in achieving the project's objectives.

Furthermore, the High Line has made noteworthy contributions towards several Sustainable Development Goals (SDGs), including SDG 11, which focuses on the development of sustainable cities and communities; SDG 13, which addresses climate action; SDG 15, which pertains to the preservation of life on land and SDG 17, which emphasises the importance of partnerships in achieving the goals. Despite being widely acknowledged as a promising model for sustainable infrastructure, the High Line, like any extensive undertaking, encounters specific socio-economic obstacles and aspects that require enhancement, which can offer valuable insights for forthcoming endeavours in implementing GIP. These challenges include inadequate accessibility for individuals with disabilities, a lack of equilibrium between natural preservation and commercialisation, the substantial costs associated with maintenance and the need for enhanced climate resilience. Addressing these areas of concern is crucial for successfully and effectively implementing future green infrastructure projects.

Source [The High Line Official Website].

B. The Bishan-Ang Mo Kio Park, Singapore, Asia

Bishan-Ang Mo Kio Park was established by rehabilitating the Kallang River, a watercourse significantly affected by urban development and environmental degradation. The objective of the project was to revitalise the river and its adjacent landscape, transforming a deteriorated setting into a flourishing green area that serves both recreational and ecological functions. The process of revitalising the park commenced in 2006 and was executed in multiple stages, ultimately culminating in the park's formal inauguration in 2012. The Bishan-Ang Mo Kio Park in Singapore exemplifies sustainable urban greenery, effectively redefining the conventional notion of a city park. In this case, implementing innovative technologies involves utilising natural materials to reshape and restructure the riverbanks, thereby facilitating the naturalisation process of the host river. Additional initiatives encompass utilising wetlands and bio-ponds as ecological filtration systems, implementing rain gardens to manage stormwater effectively, mitigating carbon emissions by embracing solar energy for illumination and establishing novel habitats to enhance biodiversity.

This project demonstrates adherence to various Sustainable Development Goals, including clean water, sustainable cities, climate action, life on land, and partnerships, making it a model for other cities seeking to enhance their urban green spaces and promote sustainable development. This project has established a commendable benchmark for urban green spaces. However, several areas could be improved upon to ensure that future projects can build upon its success. These areas include enhancing pedestrian connectivity, integrating climate-resilient facilities to cope with extreme temperatures and emphasising inclusion by raising awareness among residents and visitors.

Source [National Parks Board Singapore, 2023].

C. Copenhagen Sustainable Bicycle Mobility Infrastructure, Denmark, Europe

Sustainable urban mobility that attempts to reduce the dependency on fossil fuel-based transportation has been at the forefront of green infrastructure initiatives. In this regard, Copenhagen, the capital city of Denmark, has garnered global recognition as a prominent cycling metropolis, distinguished by its advanced and inventive bicycle infrastructure. Over time, the dedication to advancing cycling as an environmentally friendly means of transportation has resulted in significant changes to the urban environment and has served as a model for cities worldwide. The bicycle infrastructure in Copenhagen is a prominent example of sustainable urban mobility, illustrating the significant impact of prioritising cycling as a transportation mode. This emphasis on cycling has resulted in a noteworthy decrease in the carbon footprint associated with transportation methods reliant solely on fossil fuels.

The urban area has made notable advancements in promoting cycling and mitigating car dependency by implementing a comprehensive infrastructure of dedicated cycling lanes, innovative traffic signal systems and bike-sharing initiatives. The implementation of bicycle infrastructure has not only made significant contributions towards the achievement of Sustainable Development Goals, including the promotion of sustainable cities, improvement of public health, and mitigation of climate change but has also played a pivotal role in cultivating a sense of community and enhancing the overall well-being of local inhabitants. Copenhagen's bicycle infrastructure can serve as an exemplary model for other cities seeking to establish sustainable, healthy, and livable urban environments by focusing on various aspects such as safety enhancements, expansion into suburban areas, integration of user data and public awareness to enhance its achievements further.

D. The African Urban Oasis—Millennium Park, Abuja, Nigeria, Africa

In the context of the rapid urbanisation and environmental degradation experienced by numerous African cities, Millennium Park, located in Abuja, Nigeria, exemplifies a green infrastructure initiative that embodies the principles of sustainable development within the African continent. The park was created as a direct response to the urgent demand for green spaces and stormwater management within the urban environment. The significant benefits of adopting innovative technologies and design strategies align with the Sustainable Development Goals, specifically in sustainable cities, clean water, climate action, life on land and partnerships. The park integrates various sustainable practices, such as rainwater harvesting systems, previous concrete pavement for groundwater recharge, renewable energy for lighting, and natural wetlands to filtrate stormwater runoff.

By enhancing the integration of green transport, promoting inclusivity for individuals with diverse abilities and implementing circular water management practices, areas of concern can be improved for future implementation across the continent. In light of the urbanisation challenges faced by African cities, Millennium Park is a noteworthy case study in the effective incorporation of green spaces, sustainable water management practices and community engagement initiatives. These elements will collectively create a more sustainable and habitable urban environment. Millennium Park has the potential to serve as a valuable model and case study for future green infrastructure initiatives in Africa. Doing so can contribute to developing a more sustainable and resilient urban future for the entire continent.

E. Masdar City, Abu Dhabi, United Arab Emirates, Middle East

Situated in Abu Dhabi, United Arab Emirates, Masdar City is an innovative green infrastructure initiative in the Middle East that aspires to establish itself as one of the globe's foremost sustainable urban developments. Commenced in 2006, the present endeavour attempts to validate the practicability of urban living devoid of carbon emissions and waste, concurrently highlighting cutting-edge technologies and sustainable methodologies.

The city of Masdar in Abu Dhabi serves as a prime illustration of the Middle East's endeavours to execute environmentally friendly infrastructure initiatives and establish sustainable urban developments. Masdar City has emerged as a paradigmatic model for other cities in the region and beyond, owing to its adoption of cutting-edge technologies such as integrating renewable energy, implementing passive cooling techniques, autonomous electric vehicle systems, green building designs and establishing efficient waste management systems. The potential of the subject in addressing global challenges is exemplified by its contributions to Sustainable Development Goals, specifically in the areas of clean energy, sustainable cities, responsible consumption, climate action, and life on land. To sustain its status as a prominent sustainable urban development in the Middle East, Masdar City must prioritise identifying and resolving areas requiring improvement while actively adopting novel sustainable practices as it evolves and expands.

5 Challenges and Limitations of Green Infrastructure

While green infrastructure offers immense potential, it is essential to acknowledge the challenges and limitations that can hinder its practical implementation (Keeley et al. 2013). One major challenge of implementing green infrastructure is the economic aspect. While there are long-term benefits, initial investment costs are often higher than conventional infrastructure. For instance, constructing rain gardens or green roofs may require additional expenses. Additionally, the maintenance and monitoring of green infrastructure can be expensive (Foster et al. 2011). To address this challenge, it is crucial to develop cost-effective strategies and incentivise the adoption of green infrastructure through financial support, regulatory measures, recognition programmes, education, research funding, green public procurement, and community engagement to promote a circular economy and sustainable practices.

Urban areas face limited land availability, making it challenging to integrate green infrastructure effectively. The scarcity of available space poses a significant limitation when attempting to implement larger-scale green initiatives, such as urban forests or green corridors (Benton-Short et al. 2021). To overcome this challenge, urban planners need to adopt innovative techniques such as vertical greening, urban farming, and utilising unused spaces to maximise the potential of green infrastructure. Another critical limitation is the need to monitor and maintain green infrastructure regularly. Unlike traditional infrastructure, which may require less maintenance, green infrastructure necessitates ongoing care to ensure its effectiveness (Vogel et al. 2015). Monitoring vegetation growth, managing stormwater runoff, and controlling invasive species requires dedicated efforts and financial resources. Enhancing community involvement through citizen science programmes and partnerships can help monitor and maintain green infrastructure.

The role of green infrastructure in climate change adaptation is paramount as it effectively addresses various challenges, such as mitigating the urban heat island effect, managing flood risks, and promoting biodiversity. However, its limitations in addressing more extreme weather events like hurricanes, droughts, or heavy snowfall need to be acknowledged (Anand and Seetharam 2010).

While green infrastructure can be a part of the solution, it should be coupled with other adaptation strategies to ensure comprehensive resilience. Implementing green infrastructure requires specialised knowledge and expertise. A lack of understanding and training among professionals, policymakers, and community members can pose challenges during the planning, design, and installation stages. Overcoming this limitation will require increased education and training programmes, interdisciplinary collaboration, and knowledge-sharing platforms to bridge the gap between diverse stakeholders. Despite the numerous challenges and constraints, green infrastructure remains an indispensable tool for promoting sustainable development and addressing environmental issues (Fay 2012; Leal Filho et al. 2019).

By addressing economic considerations, integrating green infrastructure within urban planning, emphasising monitoring and maintenance, acknowledging its limitations in extreme weather events, and ensuring adequate knowledge and expertise, we can overcome these challenges and unlock the full potential of green infrastructure. A holistic approach involving collaboration between various stakeholders and sectors will be crucial in harnessing the long-term benefits of green infrastructure for a healthier and more sustainable future.

6 Conclusion

Green infrastructure has become increasingly important in addressing contemporary and future environmental challenges. The concept encompasses a diverse array of interconnected natural and constructed systems that contribute to preserving ecosystem services while concurrently fostering sustainable development. In light of the dynamic nature of our contemporary society, it is of utmost importance to delve into the potential future opportunities presented by green infrastructure.

The future of green infrastructure is poised to witness many exciting trends that will shape its trajectory. The first notable trend is increasing technology integration within green infrastructure systems. Advancements in sensor technology, big data analytics, and artificial intelligence will enable more efficient and adaptive management of green infrastructure networks. These developments will enhance decision-making, monitoring, and maintenance, resulting in improved resilience and effectiveness. Furthermore, incorporating nature-based solutions in urban planning will continue gaining momentum. These solutions emphasise using natural elements like vegetation, wetlands, and sustainable drainage systems to mitigate the impacts of climate change, enhance biodiversity, and improve air and water quality. This approach provides ecological benefits and enhances the well-being and quality of life for urban residents. Another emerging trend is recognising green infrastructure's role in mitigating heat island effects and combatting urban heat stress. The increasing temperatures resulting from climate change pose a significant challenge to urban areas, as it leads to heightened heat levels that pose risks to public health and contribute to increased energy requirements. Incorporating green roofs, vertical gardens and street trees into urban environments is becoming more prevalent to mitigate rising temperatures and deliver cooling benefits. Future cities will probably witness a heightened level of integration of these interventions.

The prospects for green infrastructure are optimistic and present a multitude of possibilities. The favourable depiction is shaped by the dynamic shifts in trends, the emergence of groundbreaking technologies, and the increasing comprehension of the manifold advantages it presents. Nevertheless, achieving this prospective outcome is contingent upon collaborative endeavours from policymakers, urban planners, researchers, and communities. Also, to promote the adoption of GRINIT, the government needs to incentivise using these technologies by providing financial and regulatory support for companies that use green materials in their construction projects. Integrating green infrastructure concepts into urban planning through collaborative approaches and government incentives is of utmost importance as it tackles societal needs and environmental challenges. By adopting a sustainable ideology and leveraging the capabilities of environmentally friendly infrastructure, we can lay the foundation for a future characterised by enhanced ecological well-being, improved public health, and increased capacity to withstand and recover from challenges.

Acknowledgements The authors would like to recognise Petroleum Technology Development Fund (PTDF), Nigeria for the award of a scholarship to Abiola Usman Adebanjo to study for a PhD in Civil Engineering at the Univiersiti Teknologi PETRONAS. All authors whose literature have formed the basis for which this chapter was conceived are duly acknowledged.

References

Abed J, Rayburg S, Rodwell J, Neave M (2022) A review of the performance and benefits of mass timber as an alternative to concrete and steel for improving the sustainability of structures. Sustainability (Switzerland) 14(9). https://doi.org/10.3390/su14095570

Adebanjo AU, Dahunsi BIO, Labiran JO (2021) Effects of concrete grades on strength characteristics of metakaolin modified recycled aggregate concrete. Nigerian J Technol Dev 18(3). https://doi.org/10.4314/njtd.v18i3.3

Ahmad T, Madonski R, Zhang D, Huang C, Mujeeb A (2022) Data-driven probabilistic machine learning in sustainable smart energy/smart energy systems: key developments, challenges, and future research opportunities in the context of smart grid paradigm. Renew Sustain Energy Rev 160. https://doi.org/10.1016/j.rser.2022.112128

Anand P, Seetharam K (2010) Climate change and living cities: Global problems with local solutions. In: Climate change and sustainable urban development in Africa and Asia. Springer, pp 21–35

Arrigoni A, Panesar DK, Duhamel M, Opher T, Saxe S, Posen ID, MacLean HL (2020) Life cycle greenhouse gas emissions of concrete containing supplementary cementitious materials: cut-off vs. substitution. J Clean Prod 263. https://doi.org/10.1016/j.jclepro.2020.121465

Balali A, Valipour A, Zavadskas EK, Turskis Z (2020) Multi-criteria ranking of green materials according to the goals of sustainable development. Sustainability (Switzerland) 12(22). https://doi.org/10.3390/su12229482

Balletto G, Naitza S, Mei G, Furcas C, Pani L, Francesconi L, Trulli N (2013) Challenges in the Cdw recovering in an Island Region—the Case of Sardinia. Sardinia 2013, …, 2007 (October 2013)

Benton-Short L, Keeley M, Rowland J (2021) Green infrastructure, green space, and sustainable urbanism: geography's important role. In: Geographic perspectives on urban sustainability. Routledge, pp 64–85

Bibri SE, Krogstie J (2020) Smart eco-city strategies and solutions for sustainability: the cases of Royal Seaport, Stockholm, and Western Harbor, Malmö, Sweden. Urban Sci 4(1). https://doi.org/10.3390/urbansci4010011

Bowen KJ, Lynch Y (2017) The public health benefits of green infrastructure: the potential of economic framing for enhanced decision-making. Curr Opin Environ Sustain 25:90–95
Cao H, Wang X (2021) Carbon dioxide copolymers: emerging sustainable materials for versatile applications. SusMat 1(1). https://doi.org/10.1002/sus2.2
Cassiani J, Martinez-Arguelles G, Penabaena-Niebles R, Keßler S, Dugarte M (2021) Sustainable concrete formulations to mitigate Alkali-Silica reaction in recycled concrete aggregates (RCA) for concrete infrastructure. Constr Build Mater 307:124919
Chapman C, Hall JW (2022) Designing green infrastructure and sustainable drainage systems in urban development to achieve multiple ecosystem benefits. Sustain Cities Soc 85. https://doi.org/10.1016/j.scs.2022.104078
Chisholm JM, Zamani R, Negm AM, Said N, Abdel daiem MM, Dibaj M, Akrami M (2021) Sustainable waste management of medical waste in African developing countries: a narrative review. Waste Manag Res 39(9). https://doi.org/10.1177/0734242X211029175
Colorado HA, Muñoz A, Neves Monteiro S (2022) Circular economy of construction and demolition waste: a case study of Colombia. Sustainability 14(12):7225
Das S, Saha P, Prajna Jena S, Panda P (2022) Geopolymer concrete: sustainable green concrete for reduced greenhouse gas emission—a review. Mater Today: Proc 60. https://doi.org/10.1016/j.matpr.2021.11.588
Dos Reis GS, Quattrone M, Ambrós WM, Cazacliu BG, Sampaio CH (2021) Current applications of recycled aggregates from construction and demolition: a review. Materials 14(7). https://doi.org/10.3390/ma14071700
Duan Z, Deng Q, Liang C, Ma Z, Wu H (2023) Upcycling of recycled plastic fiber for sustainable cementitious composites: a critical review and new perspective. Cem Concr Compos 105192
Farzana M, Maraz KM, Sonali SN, Hossain MM, Alom MZ, Khan RA (2022) Properties and application of jute fiber reinforced polymer-based composites. GSC Adv Res Rev 11(1):84–94
Fay M (2012) Inclusive green growth: the pathway to sustainable development. World Bank Publications
Ferrari F, Striani R, Fico D, Alam MM, Greco A, Esposito Corcione C (2022) An overview on wood waste valorization as biopolymers and biocomposites: definition, classification, production, properties and applications. Polymers 14(24). https://doi.org/10.3390/polym14245519
Ferreira L, De Brito J, Saikia N (2012) Influence of curing conditions on the mechanical performance of concrete containing recycled plastic aggregate. Constr Build Mater 36:196–204
Foster J, Lowe A, Winkelman S (2011) The value of green infrastructure for urban climate adaptation. Center Clean Air Policy 750(1):1–52
Gordon BL, Quesnel KJ, Abs R, Ajami NK (2018) A case-study based framework for assessing the multi-sector performance of green infrastructure. J Environ Manag 223. https://doi.org/10.1016/j.jenvman.2018.06.029
Hamid NH, Shamsudin MMH (2022) Recycled plastic bottles as sustainable materials. Green infrastructure: materials and applications, pp 1–20
Maddah HA (2016) Polypropylene as a promising plastic: a review. Am J Polym Sci. https://doi.org/10.5923/j.ajps.20160601.01
Hoang AT, Pham VV, Nguyen XP (2021) Integrating renewable sources into energy system for smart city as a sagacious strategy towards clean and sustainable process. J Clean Prod 305. https://doi.org/10.1016/j.jclepro.2021.127161
Hossain MU, Ng ST, Antwi-Afari P, Amor B (2020) Circular economy and the construction industry: existing trends, challenges and prospective framework for sustainable construction. Renew Sustain Energy Rev 130:109948
Huang K, Li X, Liu X, Seto KC (2019) Projecting global urban land expansion and heat island intensification through 2050. Environ Res Lett 14(11). https://doi.org/10.1088/1748-9326/ab4b71
Hunt JCR, Maslin M, Killeen T, Backlund P, Schellnhuber HJ (2007) Introduction. Climate change and urban areas: research dialogue in a policy framework. Philos Trans Royal Soc A: Math Phys Eng Sci 365(1860). https://doi.org/10.1098/rsta.2007.2089

Joseph B, James J, Kalarikkal N, Thomas S (2021) Recycling of medical plastics. Adv Ind Eng Polym Res 4(3):199–208

Junaid MF, ur Rehman Z, Kuruc M, Medveď I, Bačinskas D, Čurpek J, Čekon M, Ijaz N, Ansari WS (2022) Lightweight concrete from a perspective of sustainable reuse of waste byproducts. Constr Build Mater 319:126061

Kareem MA, Akintonde BB, Adesoye JS, Ishola K, Tijani IA, Adebanjo AU, Thanni MO, Oduoye WO, Adeyanju DO, Bello AA (2023). Influence of cashew leaf ash as partial replacement for cement on the properties of fresh and hardened concrete. Clean Waste Syst 4. https://doi.org/10.1016/j.clwas.2022.100063

Keeley M, Koburger A, Dolowitz DP, Medearis D, Nickel D, Shuster W (2013) Perspectives on the use of green infrastructure for stormwater management in Cleveland and Milwaukee. Environ Manag 51(6). https://doi.org/10.1007/s00267-013-0032-x

Kim D, Song SK (2019) The multifunctional benefits of green infrastructure in community development: an analytical review based on 447 cases. Sustainability (Switzerland) 11(14). https://doi.org/10.3390/su11143917

Kumar VVP, Prasad DR (2019) Influence of supplementary cementitious materials on strength and durability characteristics of concrete. Adv Concr Constr 7(2):75

Kurniawan TA, Lo W, Singh D, Othman MHD, Avtar R, Hwang GH, Albadarin AB, Kern AO, Shirazian S (2021) A societal transition of MSW management in Xiamen (China) toward a circular economy through integrated waste recycling and technological digitization. Environ Pollut 277. https://doi.org/10.1016/j.envpol.2021.116741

Leal Filho W, Tripathi SK, Andrade Guerra JBSOD, Giné-Garriga R, Orlovic Lovren V, Willats J (2019) Using the sustainable development goals towards a better understanding of sustainability challenges. Int J Sustain Dev World Ecol 26(2). https://doi.org/10.1080/13504509.2018.1505674

Leão AL, Cherian BM, De Souza SF, Kozłowski RM, Thomas S, Kottaisamy M (2012). Natural fibres for geotextiles. In Handbook of natural fibres. Elsevier, pp 280–311

Li F, Liu X, Zhang X, Zhao D, Liu H, Zhou C, Wang R (2017) Urban ecological infrastructure: an integrated network for ecosystem services and sustainable urban systems. J Clean Prod 163. https://doi.org/10.1016/j.jclepro.2016.02.079

Liberalesso T, Oliveira Cruz C, Matos Silva C, Manso M (2020) Green infrastructure and public policies: an international review of green roofs and green walls incentives. Land Use Policy 96. https://doi.org/10.1016/j.landusepol.2020.104693

Mukherjee M, Shaw R (2021) Uncertainties in urbanizing world and nature-based resilience building. Ecosystem-based disaster and climate resilience: integration of blue-green infrastructure in sustainable development, pp 1–25

O'Connor MP, Zimmerman JB, Anastas PT, Plata DL (2016) A strategy for material supply chain sustainability: enabling a circular economy in the electronics industry through green engineering. ACS Sustain Chem Eng 4(11). https://doi.org/10.1021/acssuschemeng.6b01954

Omer MAB, Noguchi T (2020) A conceptual framework for understanding the contribution of building materials in the achievement of Sustainable Development Goals (SDGs). Sustain Cities Soc 52:101869

Papastamoulis V, London K, Feng Y, Zhang P, Crocker R, Patias P (2021) Conceptualising the circular economy potential of construction and demolition waste: an integrative literature review. Recycling 6(3):61

Parker J, Simpson GD (2020) A theoretical framework for bolstering human-nature connections and urban resilience via green infrastructure. Land 9(8). https://doi.org/10.3390/LAND9080252

Qiao L, Tang Y, Li Y, Liu M, Yuan X, Wang Q, Ma Q (2022) Life cycle assessment of three typical recycled products from construction and demolition waste. J Clean Prod 376:134139

Raihan A, Tuspekova A (2023) Towards net zero emissions by 2050: the role of renewable energy, technological innovations, and forests in New Zealand. J Environ Sci Econ 2(1):1–16

Ramage MH, Burridge H, Busse-Wicher M, Fereday G, Reynolds T, Shah DU, Wu G, Yu L, Fleming P, Densley-Tingley D (2017) The wood from the trees: the use of timber in construction. Renew Sustain Energy Rev 68:333–359
Sandberg D, Kutnar A (2016) Thermally modified timber: recent developments in Europe and North America. Wood Fiber Sci 48(1):28–39
Sangmesh B, Patil N, Jaiswal KK, Gowrishankar TP, Selvakumar KK, Jyothi MS, Jyothilakshmi R, Kumar S (2023) Development of sustainable alternative materials for the construction of green buildings using agricultural residues: a review. Constr Build Mater 368:130457
Santhanam H, Majumdar R (2020) Permeable pavements as sustainable nature-based solutions for the management of Urban Lake Ecosystems. Nature-based solutions for resilient ecosystems and societies, pp 329–345
Seddon N, Smith A, Smith P, Key I, Chausson A, Girardin C, House J, Srivastava S, Turner B (2021) Getting the message right on nature-based solutions to climate change. Glob Change Biol 27(8). https://doi.org/10.1111/gcb.15513
Shahmohammad M, Hosseinzadeh M, Dvorak B, Bordbar F, Shahmohammadmirab H, Aghamohammadi N (2022) Sustainable green roofs: a comprehensive review of influential factors. Environ Sci Pollut Res 29(52):78228–78254
Shao H, Kim G (2022) A comprehensive review of different types of green infrastructure to mitigate urban heat islands: progress, functions, and benefits. Land 11(10). https://doi.org/10.3390/land11101792
Sharma R, Malaviya P (2021) Management of stormwater pollution using green infrastructure: the role of rain gardens. Wiley Interdiscip Rev Water 8(2). https://doi.org/10.1002/wat2.1507
Sun Y, Anwar A, Razzaq A, Liang X, Siddique M (2022) Asymmetric role of renewable energy, green innovation, and globalization in deriving environmental sustainability: evidence from top-10 polluted countries. Renew Energy 185. https://doi.org/10.1016/j.renene.2021.12.038
Uche-Soria M, Rodríguez-Monroy C (2019) An efficient waste-to-energy model in isolated environments. Case study: La Gomera (Canary Islands). Sustainability 11(11):3198
Vargas-Hernández JG, Zdunek-Wielgołaska J (2021) Urban green infrastructure as a tool for controlling the resilience of urban sprawl. Environ Dev Sustain 23(2). https://doi.org/10.1007/s10668-020-00623-2
Vogel JR, Moore TL, Coffman RR, Rodie SN, Hutchinson SL, McDonough KR, McLemore AJ, McMaine JT (2015) Critical review of technical questions facing low impact development and green infrastructure: a perspective from the Great Plains. Water Environ Res 87(9). https://doi.org/10.2175/106143015x14362865226392
Wang Z, Li H, Yang X (2020) Vision-based robotic system for on-site construction and demolition waste sorting and recycling. J Build Eng 32. https://doi.org/10.1016/j.jobe.2020.101769
Xu J, Liu Y, Simi A, Zhang J (2022) Recycling and reuse of construction and demolition waste: from the perspective of national natural science foundation-supported research and research-driven application. Case Stud Constr Mater 16:e00876
Ying J, Zhang X, Zhang Y, Bilan S (2022) Green infrastructure: systematic literature review. Econ Res-Ekonomska Istrazivanja 35(1). https://doi.org/10.1080/1331677X.2021.1893202
Yu Y, Yazan DM, Junjan V, Iacob ME (2022) Circular economy in the construction industry: a review of decision support tools based on Information & Communication Technologies. J Clean Prod 349. https://doi.org/10.1016/j.jclepro.2022.131335

Water Availability and Demand Analysis for Kedah River Basin Using Water Evaluation and Planning (WEAP) Model

Wei-Koon Lee, Sulaiman Kpakra Faba, and Noor Safwan Muhamad

Abstract Rapid population growth and industrialization have increased water stress for Kedah River basin. Kedah state is popularly known as the rice bowl of Malaysia. A study on the future water resource availability is important to ensure the sustainability of the basin development. A water allocation model was set up using Water Evaluation and Planning (WEAP) software. The model was calibrated using data from year 2002 to 2016, taking into consideration system demand, system losses, return flow, reservoirs inflow, and outflow. Validation based on the residual flow derived from the basin water balance study shows good agreement of historical water levels at the reservoirs. Using the projected future climate and water demand, model simulations were performed up to year 2060. The findings shows that wet and dry extremities attributed to climate change effect are likely to impact the basin significantly. Unmet water demand increases from year to year leading to potential water supply deficit in the basin. Kedah River basin is thus in pressing need for a comprehensive water resource study and planning in order to be well-prepared for these future scenarios.

Keywords Climate change · Kedah River basin · Water availability · Water demand and allocation · WEAP model

W.-K. Lee (✉) · S. K. Faba
School of Civil Engineering, College of Engineering, Universiti Teknologi MARA, Selangor, Malaysia
e-mail: leewei994@uitm.edu.my

S. K. Faba
e-mail: sfaba@nma.gov.sl

S. K. Faba
National Minerals Agency, New England Ville, Freetown, Sierra Leone

N. S. Muhamad
Civil Engineering Studies, College of Engineering, Universiti Teknologi MARA, Cawangan Pahang, Pahang, Malaysia
e-mail: safwanmuhamad@uitm.edu.my

E. Shaffie et al. (eds.), *Sustainable Green Infrastructure*,
https://doi.org/10.1007/978-981-96-1486-8_8

1 Introduction

Water cycles through the earth's structure as precipitation becomes usable as surface water and groundwater (Dayton 2019) to sustain livelihoods, human welfare, socioeconomic development, and ecosystem services. Globally, 2 billion people (26% of the population) do not have safe drinking water and 3.6 billion (46%) lack access to safely managed sanitation, according to the report published by UNESCO on behalf of UN-Water (UNESCO 2023). The world's population will continue to grow, which will affect not only the demand for water but also its price. Due to population growth, economic development, and dwindling supplies, water, which is recognized as a necessity for life, may pose the greatest challenge in future years in terms of its availability.

The attainment of the Sustainable Development Goals (SDGs) of the United Nations is dependent to a significant degree on the availability of water. Drought problems and extreme climate change have dominated the news in the late twenty-first century, causing significant problems for thousands of people (Bell et al. 2018). Due to potential changes in future climate conditions, temperature increases and variations in precipitation are likely to occur in many regions of the world. These are likely to have a negative impact on the water balance in many regions on the local, national, and global dimensions, causing water supply issues in many cities. The issue has prompted governments in many countries, including Malaysia, to adopt a more holistic water resource management approach.

Kedah is known as Malaysia's rice bowl. Paddy cultivation is the primary water user and is responsible for more than fifty percent of Malaysia's domestic rice production. The Muda Agricultural Development Authority (MADA) manages both the Kedah River basin and the adjacent Muda River basin, which combined, are the primary water sources for paddy irrigation. Water supply is also crucial for domestic, industrial, and livestock water needs in the state (Ismail et al. 2015). Rapid population growth in urban areas and industrial sectors and changing lifestyles in the basin has increased the demand for water (Anang et al. 2019). Furthermore, water contamination has degraded the water quality, thereby posing a threat to the environment, burden on the treatment system, and safe water availability (Mohd et al. 2011). With planned development of the state (Hashem et al. 2021), there is thus an urgent need to analyze the basin future water supply and demand scenarios using appropriate tool for planning purpose to safeguard against potential water stress and water scarcity.

In this study, the water evaluation and planning (WEAP) model are used to set up a water allocation for Kedah River basin. The model considers the current and future demand at the abstraction points, reservoir storage changes, and climate change effect. The model is calibrated and validated for simulations up to the year 2060 to assess the basin water availability. The study is significant to provide vital information for the effective planning of water works and to ensure equitable access of water for the sectorial users to establish a sustainable water future in Kedah.

2 Materials and Method

2.1 Study Area

Kedah River is one of the main rivers in Kedah state. It originates in the northern and northeastern mountainous regions bordering Perlis and Thailand and stretches over 100 km in length. The river travels through a hilly landscape before reaching a broad coastal plain before discharging into the Strait of Malacca. Kedah River basin is approximately 60 km wide and 80 km long, with a total area of 2,920 km^2. The basin extends from a height of 400 m to the coastal plains. Figure 1 depicts the river network of the basin, the basin boundary, and the administrative districts.

The land use in Kedah River basin comprises four main categories: 62% agriculture, 28% forest cover, 7% urban areas, and 3% bodies of water (Kamarudzaman et al. 2019) (Fig. 2). Most of the land is devoted to paddy crops. Non-paddy commodities include oil palm, fruits, vegetables, cash crops, herbs and seasonings, flowers, pepper, and industrial crops such as coconut, mushroom, nipa palm, areca nut, coffee, and roselle. Additionally, livestock production such as cattle, sheep, goats, poultry, and swine predominate in the areas.

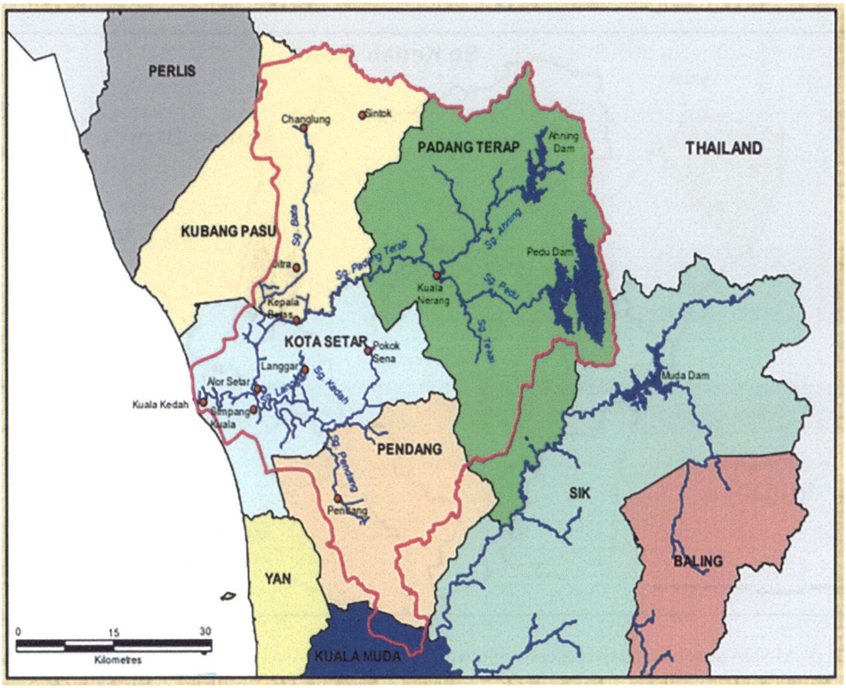

Fig. 1 Kedah River basin river network and administrative districts

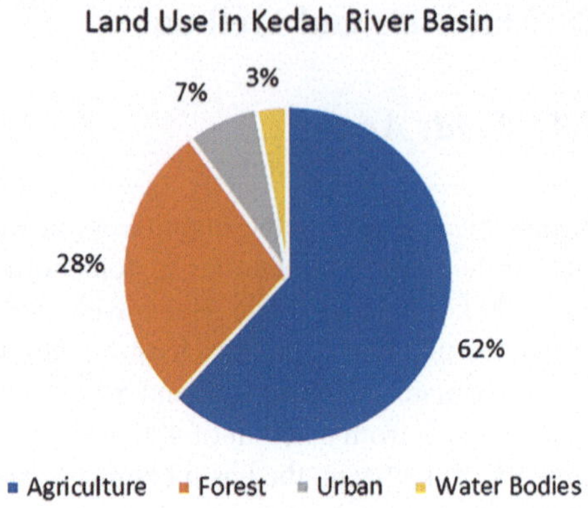

Fig. 2 Kedah River basin land use

MADA irrigation scheme in Kedah is the largest designated granary area of Malaysia (Fig. 3). It was introduced in the late 1960s, to transform the single-crop paddy field into a double-crop area. Today, the Muda irrigation scheme lies in the North–South region, about 20 km-wide and 65 km-long parallel to the coast of Kedah and Perlis states. Parcel I is in Perlis state, whereas Parcels II, III, and IV are in Kedah state.

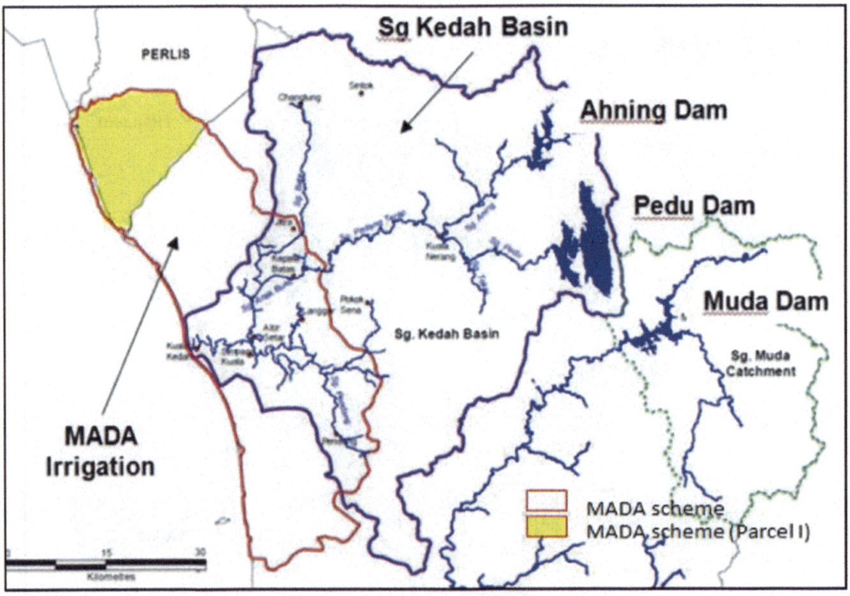

Fig. 3 MADA scheme and the reservoir network

2.2 *Basin Water Resources*

For the purpose of the study, the rainfall, evaporation are obtained from JPS and MADA. These data provide the background information of the climatic condition of the study area. In addition, the existing reservoir and inter-basin transfer are integral parts of the basin water resource and are elaborated.

The climate of Kedah River basin is generally humid and tropical. February is the warmest month of the year, whereas September is the coldest. It has a variable rainfall pattern, characterized by dry season and flooding during the rainy season (Tan 2018). Figure 4 shows the annual rainfall in the basin. According to MADA, the average annual rainfall is 2100 mm. Peninsular Malaysia's precipitation indicates that the southwest monsoon (SWM) and northeast monsoon (NEM) coincide with a primary seasonal cycle. Over the northwestern coast of the Peninsular, where Kedah River basin is located, the annual rainfall cycle exhibits extraordinary bimodality, with two peaks, one in April/May and another higher peak in October (Toriman et al. 2009). This is evident from the average monthly rainfall as shown in Fig. 5.

Figure 6 shows the annual total evaporation with an average of 1850 mm. From year 2003 to 2006 and 2014 to 2017, the higher evaporation rate indicates warm period. Figure 7 shows the monthly volumetric evaporation rate for Ahning and Pedu reservoirs in the basin. The large difference is attributed to the reservoir surface area (Table 1).

Pedu Reservoir is the primary impoundment storage which also stores water transferred via a 6.8 km-long Saiong Tunnel from Muda Dam with a design flow of 42.5 m^3/s. Note that Muda Reservoir has a much large catchment area of 984 km^2 but a relatively small storage capacity of 154×10^6 m^3. Figures 8 and 9 show the annual and average monthly transfer via Saiong Tunnel.

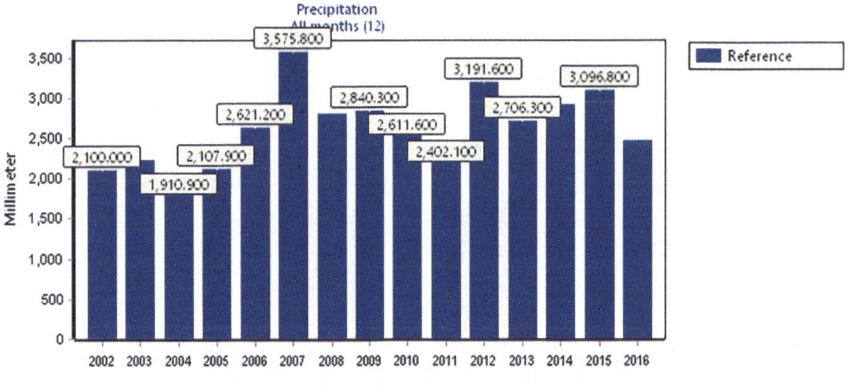

Fig. 4 Kedah River basin annual rainfall (2002–2016)

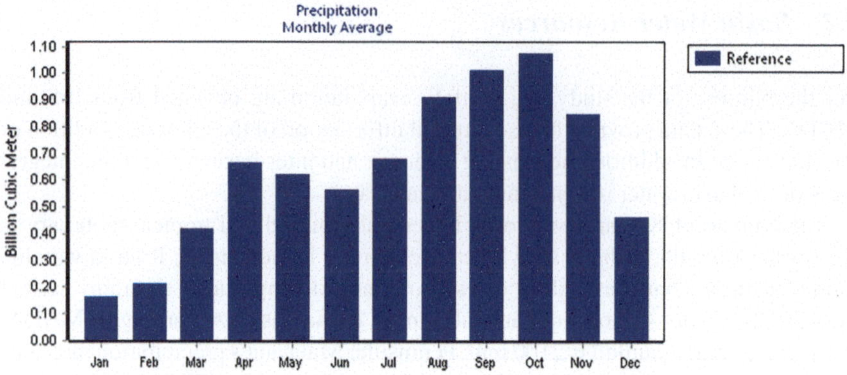

Fig. 5 Kedah River basin average monthly rainfall

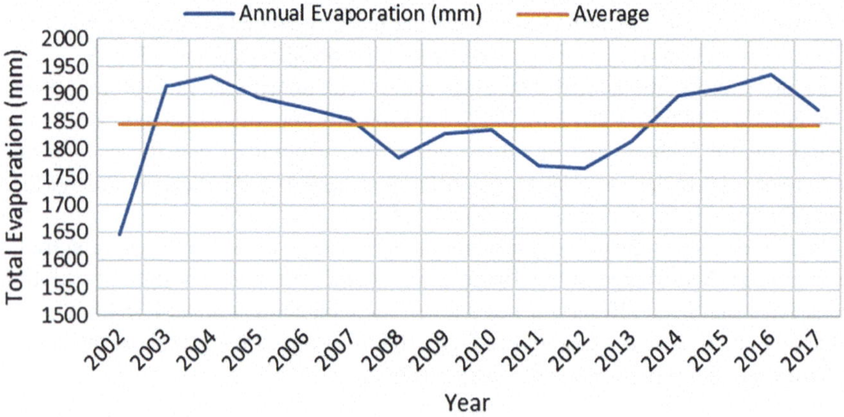

Fig. 6 Kedah River basin annual evaporation (2002–2017)

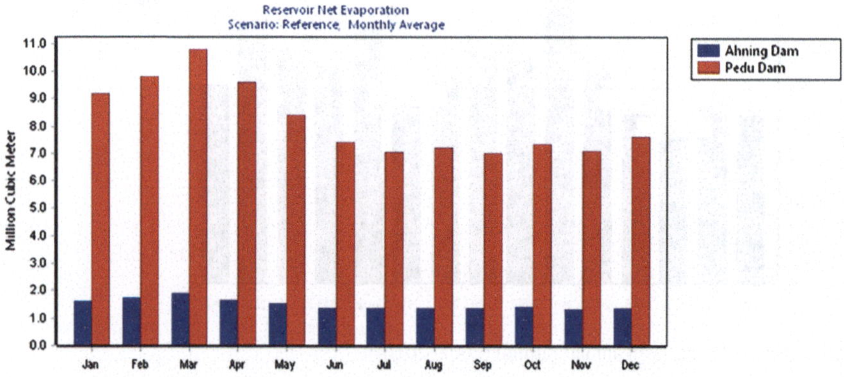

Fig. 7 Monthly net evaporation at Ahning and Pedu Reservoir

Table 1 Salient features of Pedu and Ahning Reservoir

Reservoir	Active storage (MCM)	Catchment area (km^2)	Reservoir area (km^2)	Useful drawdown level (m)
Pedu	1080	171	52	220 ft–320 ft MSL
Ahning	275	122	12	112.9–73 m MSL
Total	1355	293	64	

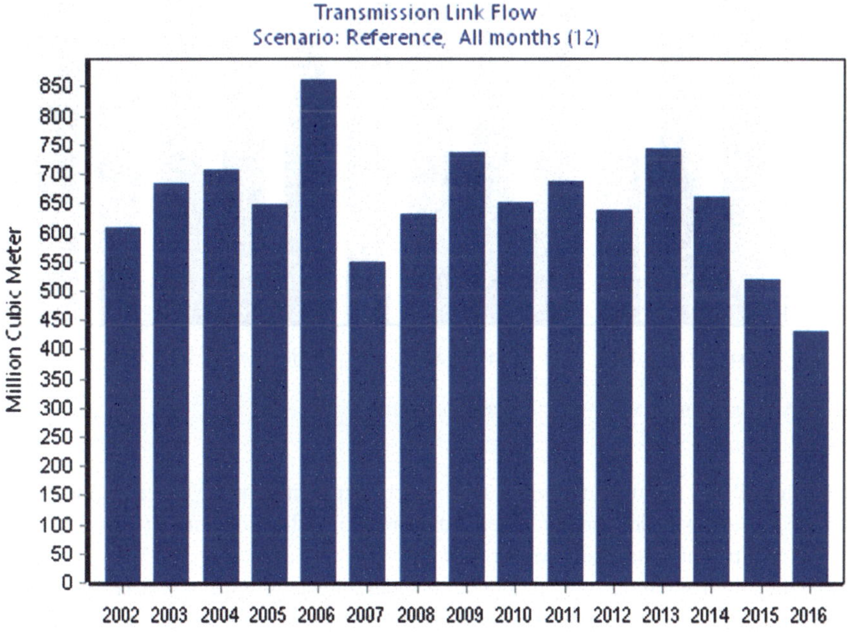

Fig. 8 Annual water transfer via Saiong Tunnel

2.3 WEAP Modeling

Water Evaluation and Planning System (WEAP) is established by the Boston Centre, Stockholm Environment Institute (SEI) at the Tellus Institute to promote participatory processes toward holistic, transparent, and robust decision-making. It enables policymakers and stakeholders to consider how water allocation management plans will affect cities, industries, and farms, as well as wildlife habitat and vulnerable communities.

Figure 10 summarizes the modeling framework using WEAP. It is one of several user-friendly water resources planning software, with a framework for connecting supply and demand needs. WEAP works on the fundamental principle of water balance for each node. The system's connection is subjected to demand priorities, supply references, and other constraints. The water accounting is computed such

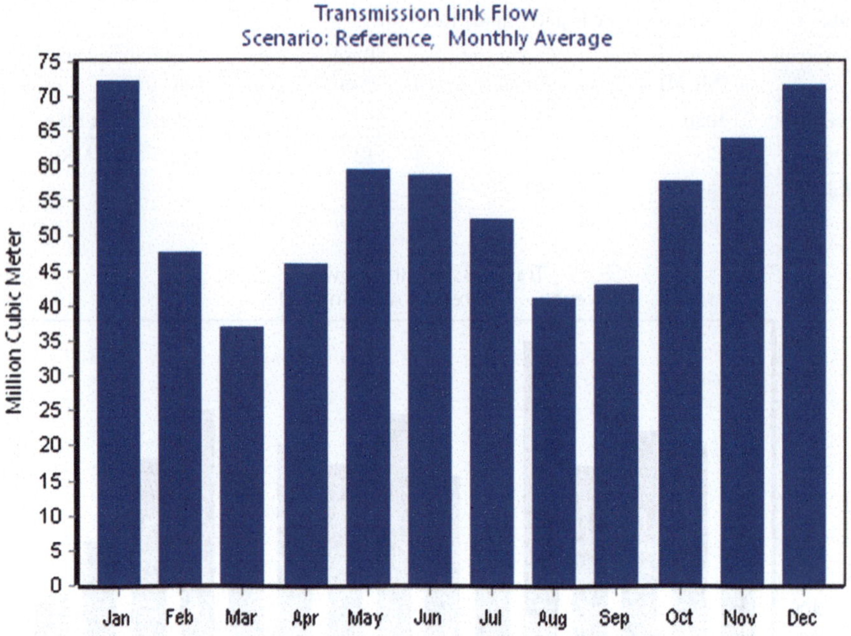

Fig. 9 Average monthly water transfer via Saiong Tunnel

that total inflows minus outflows equal the net change in storage based on the mass balance equations. Additionally, other equations are applied to simulate the processes such as infiltration losses and return flow.

Besides hydrometeorological data, additional data acquired for WEAP modeling include sectorial water demand, streamflow, reservoir stage-storage curve and records of reservoir water level, release, and spill. Water demand is derived based on the population data and projection, irrigation water requirement, and livestock population.

Figure 11 shows the model screenshot for the basin. The abstraction points correspond to the water treatment plants (WTPs) in the basin (green arrow). The demand centers (red nodes) include the irrigation schemes, livestock demand, and the urban centers. For the purpose of the model, the irrigated area is sub-divided into the northern scheme (12,239 ha), the southern scheme (104,831 ha), and the non-paddy scheme (49,160 ha). Urban water demand comprises domestic, commercial, industrial, and institutional sectors. Table 2 summarizes the current and future sectorial water demands. As shown in Fig. 11, a number of the demand centers located outside of the basin do not contribute to return flow (red arrow).

The reference scenario for the model is set to the year 2002. Calibration and validation are carried out using historical data up to the year 2016. Subsequently, simulation based on climate and demand projection is carried out up to year 2060.

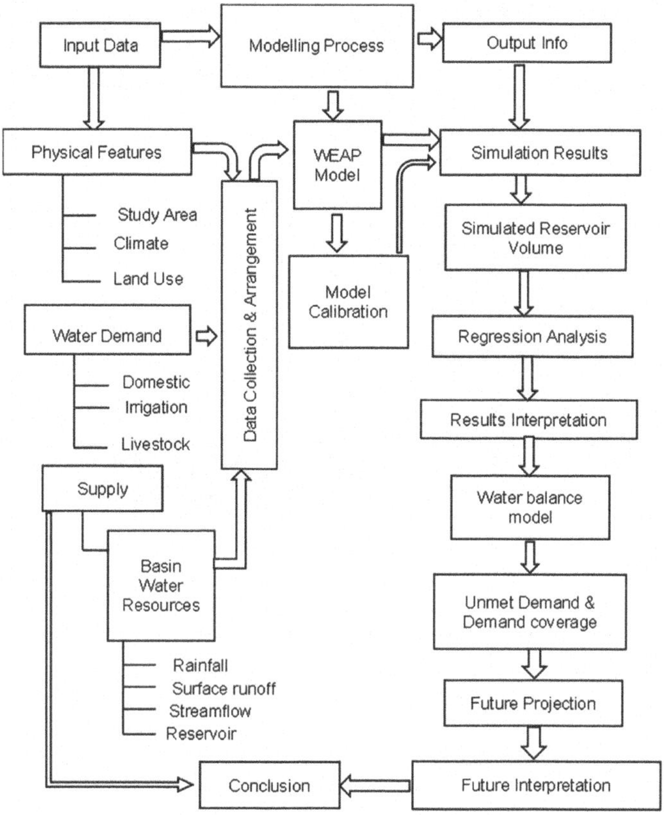

Fig. 10 WEAP modeling framework

Optimization is carried out by adjusting the reservoir inflow, initial reservoir water level, infiltration losses, and the fraction of return flow.

3 Results and Discussions

3.1 Calibration

Model calibration for the period from year 2002 to 2016 gives correlation coefficient (R^2) value of 0.8924 for Pedu Reservoir and 0.8865 for Ahning Reservoir. Figures 12 and 13 show the calibrated storage volumes are in good agreement with the observations.

This analysis reveals that the system experiences water loss through transmission to and return flow from the demand sites, and groundwater infiltration. The calibrated

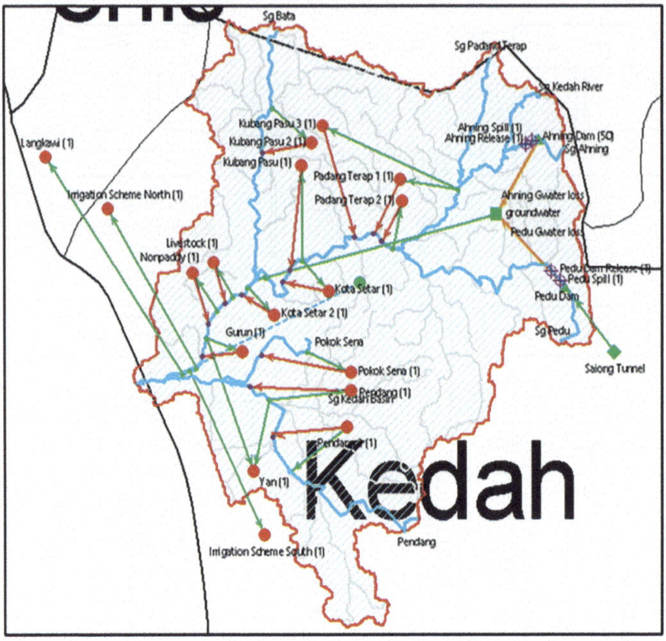

Fig. 11 WEAP model for Kedah River basin (green arrow indicates withdrawal, red circle indicates demand center, and red arrow indicates return flow)

Table 2 Sectorial current and future water demand (2002–2060)

Sector	Water demand (MCM)						
	2002	2016	2020	2030	2040	2050	2060
Potable	83	92.2	97.1	110.5	125.9	143.5	163.9
Irrigation	1024	1163.6	653	938.3	810.6	653	938.3
Livestock	10	15.6	17.6	23.6	31.7	42.6	57.3
Total	1117	1217.4	767.7	1072.4	968.2	839.1	1159.5

model predicts an average monthly return flow of 15.3 MCM from all demand centers. The non-paddy irrigation area has the largest monthly average return flow at approximately 6.3 MCM, accounting for 29% of overall return flows. The annual average transmission loss for all demand sites is approximately 14 MCM, while the annual average loss in return flow is 3.7 MCM, giving a total loss of 17.7 MCM. Annual groundwater loss from Ahning and Pedu is estimated at 6 MCM and 576 MCM, respectively. The loss from the Pedu Reservoir accounts for approximately 74% of the total water loss to groundwater over the entire basin.

Figures 14 and 15 show the calibrated inflows, outflows, and storage change for both Pedu and Ahning Reservoirs. The reservoir inflow is a function of the catchment

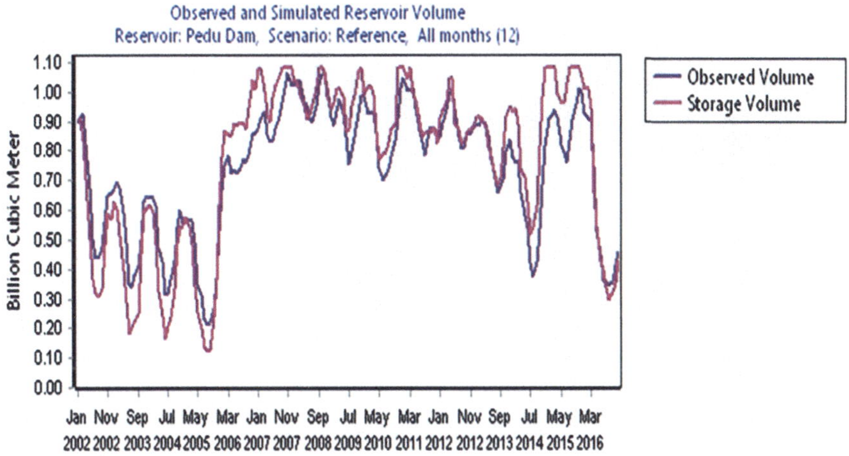

Fig. 12 Calibrated Pedu Reservoir storage volume

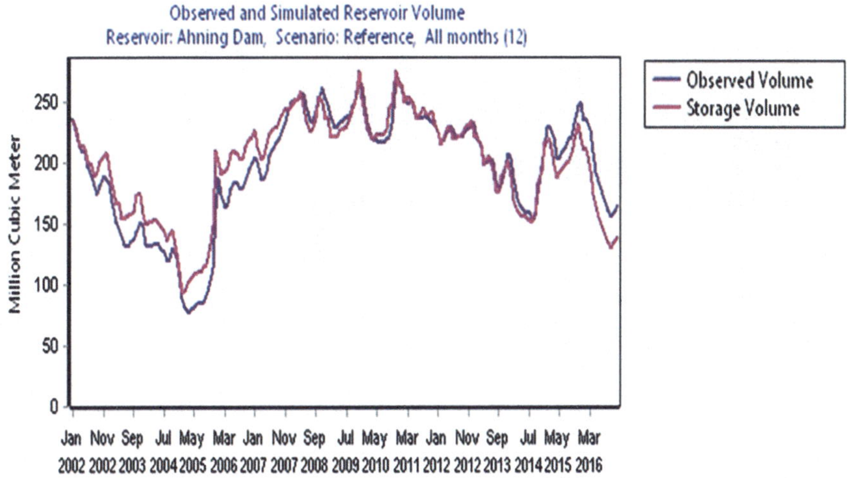

Fig. 13 Calibrated Ahning Reservoir storage volume

area and the rainfall depth, whereas the outflow comprises both release and spill, besides the evaporation and infiltration components.

3.2 Validation

The model calibration was validated by examining the basin residual flow. According to the report of national water resources study 2000–2050, the average Kedah River

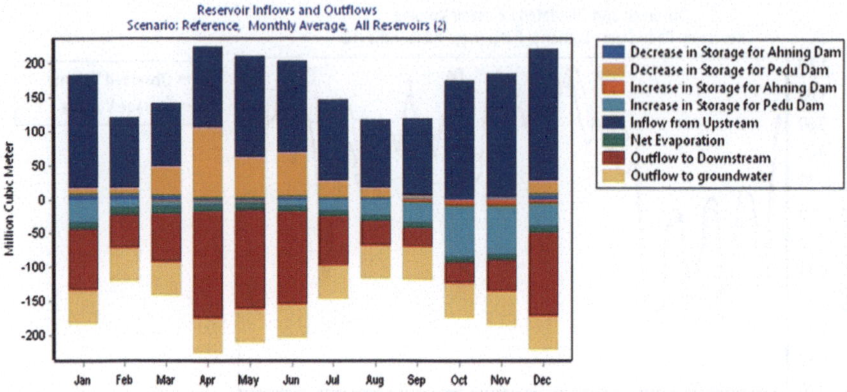

Fig. 14 Calibrated Pedu inflows, outflows, and storage change

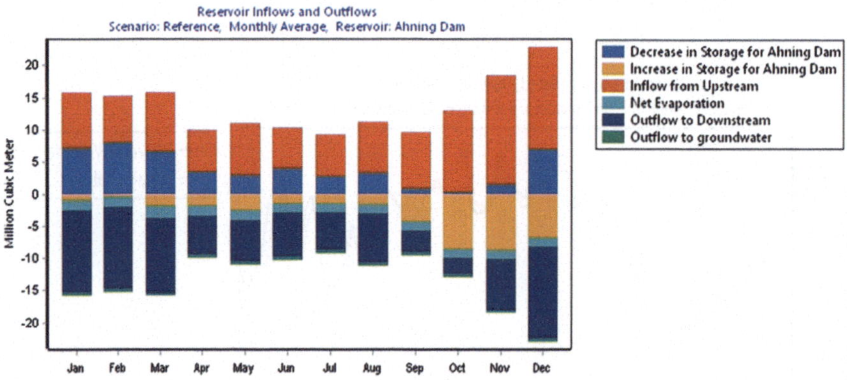

Fig. 15 Calibrated Ahning inflows, outflows, and storage change

streamflow that discharges into the sea is 135 m^3/s, equivalent to 4257 MCM per annum (DID 2011). Model simulation results (Table 3) for the year 2002 to 2016 gives average residual flow of 3890 MCM, which differs by only about 8.6%. This shows that the model provides good representation of the water inflow, outflow, and distribution in the Kedah system.

3.3 Impact of Climate Change on Water Availability

Climate change was assessed for its impact on Kedah River basin water resources. Table 4 shows the anticipated precipitation, evapotranspiration, and surface runoff for the basin from year 2020 through year 2060 at 10-year interval.

Table 3 Kedah River Basin water balance (2002–2016)

Water balance component	Annual volume (MCM)			
	Average 2002–2016	2014	2015	2016
Annual rainfall	7633	8459	8957	7136
Annual evapotranspiration	3695	3696	3820	3715
Inter-basin transfer (Saiong Tunnel)	651	662	518	432
Dam release and spill	983	1025	1006	1215
Annual runoff	3938	4763	5137	3421
Total catchment inflow	4921	5788	6143	4636
Water supply (WTP)	104	115	114	112
Irrigation	927	626	938	1164
Total utilization	1031	741	1052	1276
Residual flow	3890	5047	5091	3360

Source MADA

Table 4 Projected annual water balance

Component	Annual water balance (MCM)				
	2020	2030	2040	2050	2060
Precipitation P	6440	8215	6066	8116	5918
Evapotranspiration ET	3670	3782	3183	3898	3090
Surface runoff R	2764	4433	2883	4218	2828
R/P	0.43	0.54	0.48	0.52	0.48

Based on the results in Table 4, it is evident that climate change effect has significant impact on the basin water availability. The fluctuation of precipitation and evaporation can be attributed to the long-term climatic effect. The resulting surface runoff is observed to range from the order of 2800 MCM to 4400 MCM, showing huge contrast between a dry and a wet year. The runoff coefficient varies from a low of 0.43 to a high of 0.54. Note, however, the value is derived based on water demand and allocation only and does not take into consideration land use and land cover changes.

3.4 *Unmet Demand*

Unmet demand is the deficit that cannot be fulfilled by rivers or reservoirs. WEAP distributes water to demand areas using linear programming. Based on historical data between the year 2002 to 2016, all demand centers are satisfied except for Pokok Sena in March 2002 with a shortfall of 0.029 MCM. This happened during the dry

Table 5 Annual unmet demand under high population growth scenario

Demand center	Annual unmet demand (MCM)				
	2020	2030	2040	2050	2060
Potable water	**0.41**	**7.4**	**18.8**	**32.3**	**48.5**
Gurun	0.41	2.3	4.4	6.8	9.4
Langkawi	–	3.2	8.5	15.1	23.2
Kubang Pasu	–	1.2	3.3	5.7	8.7
Padang Terap	–	0.7	2.6	4.7	7.2
Kota Setar	–	–	–	–	–
Pendang	–	–	–	–	–
Pokok Sena	–	–	–	–	–
Yan	–	–	–	–	–
Irrigation	–	–	–	–	–
Livestock	–	–	–	–	–

season and may have been caused by pollution, water treatment plant failure, and climate change-related drought.

For future projection, Table 5 shows the unmet demand for potable water increases annually under the high population growth scenario. The urban centers with the highest water deficit in decreasing order are Gurun, Langkawi, Kubang Pasu, and Padang Terap. The sub-urban and rural areas are not expected to experience unmet demand. Meanwhile, both the irrigation and livestock water demand are also expected to be fully met. This may be partly attributed to higher irrigation efficiency and no increase in paddy area. The livestock sector is also not expected to expand significantly.

The present finding is consistent with Miniandi et al. (2023), who used the Soil and Water Assessment Tool (SWAT) to analyze water balance and water yield at Ulu Muda, Kedah, and discovered that water quantity has become a serious issue in Kedah.

4 Conclusions

The purpose of this research was to investigate the surface water availability of Kedah River basin considering the future climatic and basin socioeconomic water demand. The WEAP model is applied to simulate the basin water allocation based on present condition and future changes up to year 2060. Results show significant fluctuations of highs and lows in precipitation, evapotranspiration, and surface runoff due to climate change. Unmet demand increases over the years especially in urban demand centers, especially in Gurun, Langkawi, Kubang Pasu, and Padang Terap. The basin's total unmet demand for 2030, 2040, 2050, and 2060 is (in MCM) 0.41, 7.4, 18.8, 32.3, and 48.5, respectively. The granary area is the largest water user in the basin, but the demand is expected to be fulfilled due to anticipated improvement in irrigation efficiency and no increase in planted area. There is also no anticipated unmet demand in the livestock sector.

From the study, it may be concluded that future water stress in Kedah River basin is largely due to population growth and thus the corresponding potable water demand from the domestic, commercial, industrial, and institutional users. In order to ensure sustainable development and equitable water access, there is thus an urgent need for Kedah River basin to develop additional water infrastructure and explore the feasibility of alternative water resources.

Acknowledgements The authors acknowledge Department of Irrigation and Drainage (DID), Malaysia and Muda Agricultural Development Authority (MADA) for the data used in the present study. The second author acknowledges the sponsor, Islamic Development Bank (IsDB) for the financial support.

References

Anang Z, Padli J, Rashid, NKA et al (2019) Factors affecting water demand: macro evidence in Malaysia. Jurnal Ekonomi Malaysia 53(1):17–25. https://doi.org/10.17576/JEM-2019-5301-2

Bell JE, Brown CL, Conlon K et al (2018) Changes in extreme events and the potential impacts on human health. J Air Waste Manag Assoc 68(4):265–287. https://doi.org/10.1080/10962247.2017.1401017

Dayton GH (2019) Water availability. Encyclopedia Ecol (Second Edn) 3:656–662. https://doi.org/10.1016/B978-0-12-409548-9.11134-0

Department of Irrigation and Drainage (DID) (2011) Review of the national water resources study (2000–2050) and formulation of national water resources policy. Ranhill Consulting Sdn Bhd, Malaysia. https://www.water.gov.my/jps/resources/PDF/Hydrology%20Publication/Vol1ExecutiveSummary.pdf. Accessed 1 Sept 2023

Hashem AOA, Ahmad WAAW, Yusuf SY (2021) Water quality status of Sungai Petani River, Kedah, Malaysia. IOP Conf Ser Earth Environ Sci 646:01028. https://doi.org/10.1088/1755-1315/646/1/012028

Ismail BS, Prayitno S, Tayeb MA (2015) Contamination of rice field water with sulfonylurea and phenoxy herbicides in the Muda Irrigation Scheme, Kedah, Malaysia. Environ Monit Assess 187:406. https://doi.org/10.1007/s10661-015-4600-9

Kamarudzaman AN, Nor Asri NA, Ab Jalil MF, Abd Aziz R (2019) A study of water quality for Sungai Kedah during high and low tides. J Eng Res Educ 11:25–34

Miniandi ND, Sulaiman MS, Sathiamurty E (2023) SWAT model for water balance and water yield study in Ulu Muda Forest Reserve, Kedah. AIP Conf Proc 2484:050008. https://doi.org/10.1063/5.0110441

Mohd I, Mansor M, Awaluddin MR et al (2011) Pattern recognition of Kedah River water quality data by implementation of principal component analysis. World Appl Sci J 14:66–72

Tan KC (2018) Trends of rainfall regime in Peninsular Malaysia during northeast and southwest monsoons. J Phys Conf Ser 995:012122. https://doi.org/10.1088/1742-6596/995/1/012122

Toriman ME, Pereira JJ, Gasim MB et al (2009) Issues of climate change and water resources in Peninsular Malaysia: the case of north Kedah. Arab World Geographer 12(1–2):87–94. https://doi.org/10.5555/arwg.12.1-2.n2251970371424t6

United Nations Educational, Scientific and Cultural Organization (UNESCO) (2023) The United Nations world water development report 2023: partnerships and cooperation for water. UNESCO, Paris. https://unesdoc.unesco.org/ark:/48223/pf0000384655. Accessed 1 Sept 2023

The Importance of Green De-Stress Relief Spaces in Malaysian Airports: Enhancing Traveller Well-Being and Environmental Sustainability

Azlan Ariff Ali Ariff, Muhammad Izzat Syafiq Mohd Shahril, Zalina Samadi, and Embong Mohamad

Abstract Airports can be stressful environments for travellers with long queues and busy terminals. Studies show that exposure to green spaces positively impacts mental health and reduces stress. By relating the issue of stress in airports and the psychological benefits of green spaces, this study aims to establish the importance of providing leisure and green spaces in reducing stress levels among passengers and users towards improving the airport experience. Positive airport experiences enhance passenger satisfaction and reputation, leading to more return visits. Incorporating green de-stress relief spaces in Malaysian airports is crucial for traveller well-being and environmental sustainability. By integrating natural elements like indoor gardens, airports can offer a tranquil oasis for passengers seeking respite. These spaces align with Malaysia's environmental initiatives, showcasing the importance of nature preservation. The study employs a qualitative approach using thematic analysis, systematic literature review and visual observation as the research methodology. The relationship between biophilic design and attention restorative theory supports the integration of green space programs in providing relaxation activities and stress management tools for travellers. Overall, green de-stress relief spaces in Malaysian airports are vital in enhancing traveller experiences and promoting a greener, sustainable future that aligns with several sustainable development goals.

A. A. A. Ariff (✉) · M. I. S. M. Shahril · Z. Samadi · E. Mohamad
Studies of Architecture, School of Architecture and Interior Architecture, College of Built Environment, Universiti Teknologi MARA (UiTM) Cawangan Selangor Kampus Puncak Alam, 42300 Bandar Puncak Alam, Selangor, Malaysia
e-mail: azlanariff@uitm.edu.my

Z. Samadi
e-mail: zalin628@uitm.edu.my

E. Mohamad
e-mail: embong5863@uitm.edu.my

A. A. A. Ariff · M. I. S. M. Shahril
College of Built Environment, Universiti Teknologi MARA, 40450 Shah Alam, Selangor, Malaysia

E. Shaffie et al. (eds.), *Sustainable Green Infrastructure*,
https://doi.org/10.1007/978-981-96-1486-8_9

Keywords Green spaces · Stress relief · Airport design · Traveller well-being · Sustainable environment

1 Introduction

Airports can be stressful due to flight delays, security lines, and other disruptions, affecting passengers' well-being (McLinton et al. 2020). A survey by Malaysia Kaki Terbang and Malaysian Airports (MAHB) revealed that many Malaysians experienced anxiety at the KLIA airport, with 42% admitting missed trips, often due to losing something or someone important (AskLegal.my 2019). Space programs can offer activities like aromatherapy, music therapy, and guided meditation to help passengers relax (Batouei et al. 2020). Reduced stress levels positively impact passenger health and happiness, benefiting individuals and society. A positive airport experience improves its reputation and distinguishes it from competitors (DKMA 2014). Acknowledging the importance of stress reduction, airports should invest in space initiatives to enhance the traveller experience (DKMA 2014) to make flying a more pleasant and rewarding experience for passengers.

1.1 Problem Statement

The pressing issue of elevated physical and mental stress among travellers in Malaysian airports is caused by the crowded, demanding airport environment characterised by long queues, bustling crowds, and complex procedures (AskLegal.my 2019). Travellers' anxiety, frustration, and negative well-being are worsened by excessive information, missed flights, and long layovers (Donnelly 2021). These stressors harm both the travel experience and physical health (Zhang et al. 2021). Therefore, there is a critical need to investigate and propose the implementation of green de-stress relief spaces within Malaysian airports to mitigate the negative impact of stress and enhance travellers' well-being while creating a more holistic and positive travel environment.

1.2 Research Objectives

This research aims to establish the importance of reducing stress levels among passengers and users, encompassing three objectives:

- To define the green relief space in the airport.
- To identify the aspects of green relief space and determine the potential of green relief spaces to be considered in designing these spaces in airports.

- To highlight viable green de-stress relief spaces based on successful precedent project implementation.

1.3 Significance of Study

Implementing de-stressing programs at airports has the potential to significantly enhance user experience and community well-being compared to scenarios where such programs are not introduced (Stankov et al. 2022). This hypothesis contends that the incorporation of stress-relieving elements in airport public spaces will lead to an enhancement in the overall well-being of airport users by the reduction in stress levels. Wide arrays of activities and space programs within the green vicinity of airport facilities would help by providing a relaxed environment in alleviating anxiety and fatigue from long flights. Furthermore, this perspective suggests that de-stressing public space programs will effectively achieve their intended goal of offering users a more soothing and tranquil environment, ultimately contributing to an elevated sense of well-being.

2 Literature Review

2.1 Definition of Green De-Stress Relief Space

Green spaces refer to areas covered with vegetation, plants, and trees (Ramaiah and Avtar 2019). In the context of green de-stress relief spaces, these are environments designed to incorporate a significant amount of natural vegetation and plant life to create a visually appealing, calming, and nature-oriented atmosphere (Koops 2019).

De-stress refers to reducing or alleviating psychological and emotional stress (Roe and McCay 2021; Kilroy 2021). In the context of green de-stress relief spaces, de-stressing provides an environment that helps individuals relax, unwind, and temporarily escape the pressures and anxieties (Butler et al. 2022) of their everyday lives through nature-based elements (Kilroy 2021).

Relief spaces are designated areas within an environment intended to provide solace, comfort, and relaxation (Huggins 2022; Berman et al. 2022). In the context of green de-stress relief spaces, relief spaces are designed to offer a retreat from stressful situations (Shenoy 2023; DeLauer et al. 2022) by incorporating elements that soothe the senses and create a serene atmosphere (Jönsson 2020).

Green De-stress Relief Spaces can be defined as operationally specific areas within a larger environment, such as airports or public places, intentionally designed to incorporate abundant greenery, natural elements, and features conducive to relaxation and stress reduction. These spaces aim to provide a temporary respite from daily pressures and create an atmosphere where individuals can experience emotional relief, mental rejuvenation, and a sense of well-being, contributing to a soothing environment that promotes calmness and reduces stress.

2.2 Airport Stress and Its Impact on Travellers

Physical and mental stress caused by the airport environment is characterised by long lines, bustling crowds, and time-sensitive procedures can lead to heightened levels of physical and psychological stress among travellers (Donnelly 2021; Pronk et al. 2021). The uncertainty of travel and the need to navigate through a complex and often overwhelming environment [Guerreiro] contribute to this stress (Brill 2020). Cognitive overload happens when airport travellers are bombarded with information, signage, announcements, and security procedures (Foster and Jewell 2022; Kim et al. 2020). This cognitive overload can lead to mental fatigue, making it challenging for travellers to process information efficiently and exacerbating their stress levels (Armougum et al. 2020). The fear of missing flights, uncertainties about travel arrangements, and the pressure to adhere to schedules can trigger anxiety and frustration (Zhang et al. 2021; Jiang et al. 2020). Travellers may feel rushed, disoriented, and overwhelmed, negatively impacting their emotional well-being (Donnelly 2021). Long flights, changing time zones, and extended layovers contribute to travel fatigue (Batra 2021). Navigating airport facilities can exacerbate physical exhaustion, leaving travellers drained upon reaching their destinations (Donnelly 2021; Lin 2023). Airport stress can overshadow the overall travel experience, resulting in a negative travel experience (Zhang et al. 2021). Unpleasant encounters with security personnel (Kim et al. 2020), delays, and disorganised processes can leave travellers with negative impressions of the entire journey (Herjanto et al. 2020; Kiliç and Çadirci 2022). Chronic stress resulting from frequent travel can adversely affect physical health, including weakened immune systems, disrupted sleep patterns, and increased vulnerability to illnesses (Cohen and Kantenbacher 2020). To address these challenges and promote traveller well-being, airports are considering the need to increasingly focus on enhancing the airport experience.

2.3 The Need for Stress Relief Spaces in Airports

Airports incorporating stress relief green spaces, relaxation lounges, and quiet zones allow travellers to unwind and find solace away from the busy terminals (Hasanzade and Oel 2022). These spaces enable travellers to alleviate stress and regain their mental composure. Airports offering wellness services such as spa facilities, meditation rooms, and wellness programs help to cater to travellers seeking relaxation and stress reduction (Ebejer 2022). Improved clear signage, user-friendly layouts, and efficient wayfinding systems help to minimise confusion and help reduce traveller anxiety when navigating in the airport (Lin 2023). The integration of digital assistance technologies, such as mobile apps and interactive kiosks, provides travellers with real-time information, reducing uncertainty (Chingchuang et al. 2021; Rajab 2020) and enhancing their control over the travel process (Batra 2021). Traveller-focused services recognise the importance of customer service and empathy (Bogicevic et al. 2017) in addressing traveller stress (Batouei et al. 2020). Friendly and accommodating staff can mitigate stress by offering assistance and guidance (Ramsey et al. 2011). Efficient, streamlined security protocols, procedures and efficient check-in processes contribute to reducing stress and wait times for travellers (Chingchuang et al. 2021; Rajab 2020; Bogicevic et al. 2017).

Airports are responsible for ensuring traveller well-being by recognising and addressing airport stress. Efforts to create a more traveller-friendly environment through stress relief spaces, wellness services, efficient operations, and improved services contribute to a more positive and less stressful travel experience.

2.4 Green Spaces and Their Impact on Well-Being

The benefits of these spaces extend beyond mere aesthetics and touch upon various aspects of mental health, encompassing parks, gardens, and natural landscapes, and have been shown to exert a profound positive impact on psychological well-being (Raman et al. 2021; Cuong and Khan 2023). Green spaces are associated with improved mood and emotional well-being (Zhang et al. 2020). People who regularly engage with nature, doing even simple activities such as gardening, tend to experience less negative emotions, such as anger and depression, and report higher overall happiness levels (Chalmin-Pui et al. 2021). Natural settings evoke pleasure, serenity, and connectedness (Cuong and Khan 2023).

Attention Restoration Theory suggests that exposure to nature allows the brain to rest (Burn and Passmore 2022) from the cognitive demands of urban environments (White and Shah 2019; Mason et al. 2021). Green spaces offer a soft fascination, capturing attention without requiring concentrated focus, which leads to mental rejuvenation and increased ability to concentrate on subsequent tasks (Soendergaard 2020). Green spaces provide a respite from mental fatigue and cognitive overload caused by constant stimuli in urban settings (Cuong and Khan 2023), allowing the mind to recover and regain cognitive vitality (Sullivan and Li 2021). Green spaces often serve as communal areas where individuals can gather for social interaction with a sense of community (Lampert et al. 2021). These spaces promote social cohesion, the formation of friendships, and a sense of belonging, all of which contribute to mental well-being (Koay and Dillon 2020).

2.5 Biophilia Hypothesis and Its Application in Airport Design

The theoretical framework for this study draws on the potential of the Biophilia Hypothesis and its application in the design of airport environments. The Biophilia Hypothesis, proposed by Wilson (Sachs 2022), posits that humans possess an innate and evolutionary-based connection with nature (Andreucci et al. 2021; Gong and Zoltán 2023). This connection suggests that people have an inherent need to interact with and be surrounded by natural elements and that exposure to nature can positively affect psychological well-being. In the context of airport design, the Biophilia Hypothesis offers a valuable lens to examine how incorporating natural elements and biophilic design principles can contribute to enhancing traveller well-being and the overall airport experience (Bennett 2021).

Biophilic design principles explore the fundamental principles of biophilic design, such as incorporating natural materials, providing opportunities for nature views, integrating natural light (Richardson and Butler 2022), creating biomorphic shapes (Browning and Ryan 2020), and incorporating nature-inspired patterns (Hutson and Hutson 2023). These principles evoke a sense of connection with nature within built environments (Söderlund 2019; Hadley 2020).

The Biophilia Hypothesis suggests that exposure to nature can lead to stress reduction and mental restoration (DeLauer et al. 2022; Jönsson 2020). This component investigates how integrating biophilic design elements in airport spaces can provide travellers with opportunities for relaxation, stress relief, and cognitive restoration amid busy terminal environments (Hadley 2020; Lehto and Lehto 2019).

Biophilic design can improve traveller well-being by promoting positive emotions, reducing anxiety, and enhancing mood (Abdel-Gayed et al. 2023). It delves into how nature-inspired design elements can create environments that positively impact passengers' emotional experiences.

The Biophilia Hypothesis relates to the Attention Restoration Theory, which suggests that natural settings facilitate the restoration of attention and cognitive resources (Berto et al. 2022). This correlation may explore how biophilic design can mitigate the cognitive overload experienced by travellers in airports, leading to improved cognitive functioning and attention restoration.

By embracing the Biophilia Hypothesis, airport design can strengthen the user experience and foster a more humane and enjoyable travel environment (Lehto and Lehto 2019; Ibenna et al. 2023). This component investigates how biophilic elements create positive and memorable travel experiences.

Integrating Biophilic Design Strategies in Airports helps examine practical strategies for incorporating biophilic design in airport spaces, including integrating indoor gardens, green walls, natural materials, water features (DeLauer et al. 2022; Jönsson 2020), and nature-inspired artwork. It also considers challenges and considerations unique to airport design, such as security regulations and space constraints.

2.6 *Potential Green Relief Spaces for De-Stress Airport Design*

Airports can combine green relief spaces to provide de-stressing, therapeutic, and healing environments (Day and Gwilliam 2019).

- Zen Gardens (Marquez and Marquez 2019): Quiet, contemplative spaces with sand, rocks, and minimalistic design to promote relaxation and mindfulness.
- Botanical Gardens (Alquezar and Macedo 2019; Trofimova 2020): Diverse collections of plants, flowers, and trees in botanical gardens offer a visually appealing and calming environment.
- Courtyards (Santa et al. 2020; Preston 2023): Open-air spaces within the terminal with seating, plants, and water features providing a tranquil oasis for travellers.
- Indoor Green Walls (Trofimova 2020; Hussain and Ramdan 2020; Han et al. 2020): Vertical gardens with living plants that improve air quality and create a refreshing visual display.
- Outdoor Plazas (Zuskáčová 2023): Open spaces outside the terminal with seating, landscaping, and natural elements to encourage leisure and relaxation.
- Therapeutic Pathways (Battisto and Wilhelm 2019): Walking paths with greenery, benches, and art installations designed to offer a calming and rejuvenating experience.
- Reflecting Pools (Hussain and Ramdan 2020; Canepa and Ab Ghafar 2020): Water features enhance the visual appeal and produce soothing sounds for stress relief.
- Atriums (Grazuleviciute-Vileniske et al. 2022; Janicka-Świerguła 2022): Large, open spaces with abundant natural light and vegetation create a sense of openness and connection to nature.
- Roof Gardens (Kireeva 2020): Utilising rooftop spaces for gardens or green areas, providing travellers with a serene view and a respite from the bustling environment.
- Meditation Rooms (Federici 2022): Quiet spaces designed for meditation, prayer, or simply a moment of solitude, often adorned with soothing décor and natural elements.
- Play Areas (Kefalidou et al. 2019) with Nature Themes: Spaces for children with natural play structures and elements, encouraging families to engage in nature-oriented activities (Anabitarte et al. 2021).
- Wellness Lounges (Stankov et al. 2022; Maitra and Joseph 2022): Dedicated areas for relaxation, offering comfortable seating, aromatherapy, yoga or stretching sessions.

By integrating these green spaces, airports can create environments that help reduce stress, enhance well-being, and offer travellers a sense of calm amidst the busy travel experience.

3 Research Methodology

A combination of thematic analysis and systematic literature review methodologies is adapted when researching to obtain the outputs related to integrating green spaces in airports for stress reduction, therapeutic benefits, and healing.

3.1 Thematic Analysis

Thematic analysis is a qualitative research method used to identify and analyse patterns or themes within qualitative data, which involves the following steps.

- Data Collection: Relevant articles related to airport design and green spaces are collected.
- Generating Initial Codes: Data coding is initiated by identifying segments that relate to the benefits of green spaces in airports (de-stress, mental health, sustainability). These codes are identified in single words or phrases.
- Searching for Themes: Similar codes are grouped together to form potential themes. The codes related to "mindfulness," "relaxation," and "stress reduction" could develop a theme of "traveller's well-being".
- Reviewing Themes: The identified themes are reviewed and refined to ensure accuracy in capturing the essence of the data.
- Defining and Naming Themes: Each theme is given a descriptive name and compiled in a list of sub-themes within each main theme.

3.2 Systematic Literature Review

A systematic literature review involves systematically searching, selecting, and evaluating existing research studies to provide a comprehensive overview of airport green relief spaces.

- Research Hypothesis: Clear research hypothesis is defined based research aim and objective, such as the benefits of integrating green spaces in airports for stress reduction, therapeutic effects, and healing.
- Search Strategy: A comprehensive search strategy is developed using relevant themes, keywords and codings developed from thematic analysis. This strategy yields many articles about airport design, green spaces, and well-being.
- Study Selection: Inclusion and exclusion criteria are applied to the search results based on selected studies that directly address the research question and meet predefined quality standards.
- Data Extraction: Relevant information from the selected studies is extracted, including study characteristics, methodologies, key findings, and conclusions.

- Quality Assessment: The quality and validity of each selected study is evaluated to ensure that only reliable and rigorous research contributes to the review.
- Synthesis of Findings: The key findings from the selected studies are summarised and synthesised, which identify the common themes and patterns related to the benefits of green spaces in airports.

Combining thematic analysis and systematic literature review methodologies provides a better understanding of the benefits of integrating green spaces in airports for stress reduction, therapeutic effects, and healing. Thematic analysis helps capture insights from qualitative data, while a systematic literature review offers a broader perspective by analysing existing research studies.

3.3 Visual Observation

The research methodology involves obtaining permission to observe travellers' interaction in specific green spaces at Jewel Changi Airport. This selection of case study is based on the similarity of tropical climate setting, sun orientation, relative demography and culture, and comparable building size. Through visual observation, behaviours that suggest public engagement with green elements are recorded using photos and short videos. This data is then analysed to identify recurring patterns, assessing whether the travellers use these spaces for stress relief. The findings are compared with the outcome from thematic analysis and systematic literature review to draw insights. The observations highlight how travellers interact with green spaces for stress relief and suggest ways to enhance their well-being through improved design.

4 Results and Discussion

A combination of thematic analysis and systematic literature review methodologies is adapted when researching to obtain the outputs related to integrating green spaces in airports for stress reduction.

4.1 Thematic Analysis

From 77 journal articles reviewed, it is found that nine main themes constitute the overall discussion on green de-stress relief spaces in Malaysian airports towards enhancing traveller well-being and environmental sustainability, followed by 51 keywords in the form of coding that enriches the elaboration on the importance for green de-stress relief spaces in Malaysian airports towards enhancing traveller well-being and environmental sustainability.

- Traveller Well-being: Relaxation, Stress Reduction, Mental Health, Physical Comfort, Emotion Regulation, Mindfulness, Escape from Crowds, Social Interaction.
- Environmental Sustainability: Biodiversity, Carbon Sequestration, Energy Efficiency, Local Plant Species, Sustainable Materials, Eco-friendly Design, Climate Resilience, Water Conservation.
- Design and Infrastructure: Greenery Integration, Biophilic Design, Natural Light Optimisation, Sustainable Architecture, Efficient Resource Usage, Multi-functional Spaces, Aesthetic Appeal.
- Community Impact: Eco-tourism Promotion, Positive Local Engagement, Cultural Connection, Education on Sustainability, Health and Well-being Awareness.
- User Behaviour and Interaction: User Engagement with Green Spaces, Utilisation of Amenities, Interaction with Nature Elements, Emotional Responses, Positive Experiences.
- Stress Relief Strategies: Natural Soundscapes, Visual Aesthetics, Comfortable Seating, Interaction with Water Features, Multi-sensory Engagement, and Playful Elements.
- Health and Wellness Programs: Yoga/Meditation Sessions, Wellness Activities, Stress Relief Workshops, and Fitness Opportunities.
- Passenger Experience and Satisfaction: Positive Travel Experience, Enhanced Passenger Comfort, High Satisfaction Levels.
- Challenges and Opportunities: Space Limitations, Economic Benefits, Environmental Impact Assessment, Public Awareness.

A thematic analysis diagram is formed to understand the weight of each theme and keywords mentioned repeatedly in the discussion of the importance of green de-stress relief spaces in Malaysian airports towards enhancing traveller well-being and environmental sustainability, indicated by the font scale represented in the word cloud below. These themes provided an outline for developing the systematic literature review's aspects and classification (Fig. 1).

4.2 Systematic Literature Review

From 77 journal articles reviewed, 58 journals discussed the topic of green space and stress in the airport, and none addressed the topic specifically concerning the Malaysian context, which suggests the novelty of this research. It is found that five main areas that contributed to green de-stress relief spaces design discussion are airport stress and its impact on travellers, the need for stress relief spaces in airports, the impact of green spaces on well-being, Biophilia Hypothesis and potential green relief spaces. Each aspect is classified according to respective discussion points

Fig. 1 Thematic analysis diagram is represented as a word cloud. *Source* Authors

and mapped according to a specific problem with their corresponding solution to understand better how each issue and strategy, as represented in keywords, are related. These main criteria are further discussed in elaborated plans for developing green relief spaces. Based on these interdepending strategies, a conceptual framework is drawn to provide an overall view of green spaces as highly potential stress relief green space strategies in airports (Table 1).

4.3 Visual Observation

Jewel Changi Airport in Singapore is often praised for successfully integrating green spaces, creating a unique and inviting environment for travellers. The key elements that contributed to the success of implementing green spaces under one roof are described based on visual observations.

- Rain Vortex: The Rain Vortex, the world's tallest indoor waterfall, not only serves as a captivating visual spectacle but also helps regulate the indoor climate by generating a cooling effect through evaporative cooling (Fig. 2).
- Canopy Park: Canopy Park offers various attractions like walking trails, play areas, sky nets, and mazes, all surrounded by greenery. These recreational spaces provide opportunities for relaxation, play, and exploration.
- Diverse Flora: Jewel houses various plant species worldwide, showcasing rich biodiversity and contributing to the overall aesthetic and tranquillity.

Table 1 Systematic classification of literature review mapping with theoretical framework towards green de-stress relief spaces design

Literature review components			Theoretical framework	
Airport stress and its impact on travellers	The need for stress relief spaces in airports	Impact of Green Spaces On Well-being	Biophilia hypothesis	Potential green relief spaces
Complex and overwhelming environment (Guerreiro et al. 2019)	Stress relief spaces (Hasanzade and Oel 2022)	Enhanced mood (Raman et al. 2021; Cuong and Khan 2023; Zhang et al. 2020; Chalmin-Pui et al. 2021)	Biophilic design principles (Sachs 2022; Andreucci et al. 2021; Gong and Zoltán 2023)	• Outdoor Plazas (Zuskáčová 2023) • Roof Gardens (Kireeva 2020) • Courtyard (Santa et al. 2020; Preston 2023) • Play areas with nature themes (Kefalidou et al. 2019; Anabitarte et al. 2021)
Physical and mental stress (Donnelly 2021; Pronk et al. 2021)			Stress reduction and restoration (DeLauer et al. 2022; Jönsson 2020; Hadley 2020; Lehto and Lehto 2019)	• Reflecting pools (Hussain and Ramdan 2020; Canepa and Ab Ghafar 2020) • Botanical Gardens (Alquezar and Macedo 2019; Trofimova 2020)
Cognitive overload (Foster and Jewell 2022; Kim et al. 2020; Armougum et al. 2020)	Improved signage and navigation (Lin 2023)	Restoration and attention restoration (Burn and Passmore 2022; White and Shah 2019; Mason et al. 2021)	Attention restoration (Berto et al. 2022)	• Atrium (Grazuleviciute-Vileniske et al. 2022; Janicka-Świerguła 2022) • Therapeutic pathways (Battisto and Wilhelm 2019) • Indoor green walls (Trofimova 2020; Hussain and Ramdan 2020; Han et al. 2020)

(continued)

Table 1 (continued)

Literature review components			Theoretical framework	
Airport stress and its impact on travellers	The need for stress relief spaces in airports	Impact of Green Spaces On Well-being	Biophilia hypothesis	Potential green relief spaces
Effect on health (Cohen and Kantenbacher 2020)	Wellness service (Ebejer 2022)	Mindfulness and relaxation (Soendergaard 2020)	Enhanced well-being (Bennett 2021; Abdel-Gayed et al. 2023)	• Zen garden (Marquez and Marquez 2019) • Meditation rooms (Federici 2022) • Wellness lounges (Stankov et al. 2022; Maitra and Joseph 2022)
Travel fatigue (Donnelly 2021; Batra 2021; Lin 2023)	Traveller-focused services (Batouei et al. 2020; Bogicevic et al. 2017; Raman et al. 2021)	Reduced mental fatigue (Sullivan and Li 2021)	Strengthen user experience (Lehto and Lehto 2019; Ibenna et al. 2023)	
Negative travel experience (Zhang et al. 2021; Herjanto et al. 2020; Kiliç and Çadirci 2022)	Efficient security protocols (Chingchuang et al. 2021; Rajab 2020; Bogicevic et al. 2017)	Stress reduction (Chalmin-Pui et al. 2021; Sullivan and Li 2021)		
Anxiety and frustration (Jiang et al. 2020; Zhang et al. 2021)	Digital assistance (Batra 2021; Chingchuang et al. 2021; Rajab 2020)	–		

Source Authors

Fig. 2 View of a massive crowd around the rain vortex during the Disney 100 celebration. *Source* Authors

- Natural Light and Space: The architecture of Jewel maximises the use of natural light, reducing the reliance on artificial lighting. The expansive open spaces, glass facades, and garden views create a seamless connection between indoor and outdoor environments.
- Iconic Indoor Garden: The centrepiece of Jewel Changi Airport is the "Forest Valley," a lush indoor garden spanning five stories. It features thousands of trees, plants, and a large waterfall, creating a serene and refreshing atmosphere amidst the bustling airport environment (Fig. 3).
- Retail and Dining Integration: Green spaces are seamlessly integrated with retail and dining areas, allowing travellers to enjoy their meals or shop amidst lush surroundings creating a sense of leisure and relaxation (Fig. 4).
- Wellness and Events: Jewel offers wellness services like early morning yoga sessions and meditation classes, emphasising its focus on promoting well-being. It also hosts various events and performances in its Canopy Park, attracting travellers and locals (Fig. 5).

The success of Jewel Changi Airport in integrating green spaces lies in its ability to seamlessly merge nature with architecture, creating an environment that offers more than just functional travel services. Its combination of iconic attractions, diverse plant life, and attention to visitor comfort has set a precedent for other airports looking to enhance the traveller experience through similar design approaches, validating the viability of integrating stress relief green spaces in airports. The on-site photography above clearly demonstrated the correlation of the Biophilia Hypothesis and Attention Restoration Theory in Jewel Changi Airport of how integrating green spaces in airport

Fig. 3 Multi-level lush green terraces of Jewel Changi Indoor Garden. *Source* Authors

Fig. 4 Concept of dining in the garden at the top floor of Jewel Changi airport. *Source* Authors

design contributes to stress reduction among the visitors through participation in vibrant activities.

Fig. 5 Visitors enjoying the relaxed setting of water vortex and entertained by the light show at Jewel Changi Rain Vortex, the world's tallest indoor waterfall. *Source* Authors

5 Recommendation for Future Implementation

Recently, there has been a growing interest in adapting airports as relaxing and enjoyable public areas. The design of airports of the future provides significant promise for the effective integration of public space components and architectural methods that reduce stress. Airports have the potential to become much more than merely a waiting area for travellers with the implementation of new concepts and technology.

A detailed study on de-stress programs and strategies can be carried out to create individualised plans for relieving stress based on the individual's preferences and the factors that cause stress. In addition to this, comprehensive literature research can be carried out to find existing programs proven for relieving stress used in various contexts, including healthcare institutions, workplaces, and public areas, that are applicable to improve airport functions. Understanding user preferences, interests, and future requirements of airport passengers concerning leisure and relaxation should be a priority for any study in designing future airports. Gaining significant insights from this understanding can influence the creation of focused design strategies by conducting in-depth surveys, interviews, and focus groups with travellers of many ages and backgrounds. Incorporating biophilic elements into airport areas can be explored by integrating interior vegetation, natural lighting, and water features to encourage relaxation and a sense of connection with the surrounding natural world.

6 Conclusion

Integrating green spaces and stress reduction strategies in airport design is pivotal for improving traveller well-being and sustainability in Malaysia. The hypothesis drawn from the aim and objectives of the research is tested using a triangulation of three research methodologies; thematic analysis, systematic literature review and visual observation. The thematic analysis identifies nine themes: travel well-being, environmental sustainability, design and infrastructure, community impact, user behaviour, stress relief strategies, health and wellness programs, passenger experience and satisfaction. These themes provide an outline of the systematic literature review. Through systematic literature review, the application of stress relief space, improved signage and navigation, traveller-focused and wellness services, efficient security protocols, digital assistance passenger mood, mindfulness and relaxation, aids in attention restoration, reduced stress and mental fatigue. The correlation of the Biophilia Hypothesis and Attention Restoration Theory further justifies the importance of integrating green spaces and stress reduction strategies in airport design. Seamless connectivity and digital assistance improve wayfinding, boosting overall efficiency and comfort. Such approaches prioritise passenger well-being, enhance travel experiences, and transform airports into more than transit points. Visual observation at Jewel Changi Airport revealed how different green spaces are applied successfully in the building, proving the viability of green space integration within airport facilities. With future airport designs, promising integration of public spaces and stress-reducing architecture can redefine their purpose. The significance of green de-stress relief spaces aligns with Malaysia's sustainable goals, enhancing airport experiences and promoting eco-tourism for resilient urban development. By embracing new concepts and technology, airports have potential beyond waiting areas, offering travellers comfort, relaxation, and inspiration.

Acknowledgements Special thanks to the team of researchers comprised of students and supervising lecturers as research supervisors from Studies for Architecture, School of Architecture and Interior Architecture, Universiti Teknologi MARA for their dedication and commitment in making this research paper possible.

References

Abdel-Gayed AH, Hassan TH, Abdou AH et al (2023) Travelers' subjective well-being as an environmental practice: do airport buildings' eco-design, brand engagement, and brand experience matter? Int J Environ Res Public Health 20(2):938

Alquezar RD, Macedo RH (2019) Airport noise and wildlife conservation: what are we missing? Perspect Ecol Conserv 17(4):163–171

Anabitarte A, García-Baquero G, Andiarena A et al (2021) Is brief exposure to green space in school the best option to improve attention in children? Int J Environ Res Public Health 18(14):7484

Andreucci MB, Loder A, McGee B et al (2021) Exploring regenerative co-benefits of biophilic design for people and the environment. Urban services to ecosystems: green infrastructure benefits from the landscape to the urban scale. Springer International Publishing, Cham, pp 391–412

Armougum A, Gaston-Bellegarde A, Joie-La Marle C et al (2020) Physiological investigation of cognitive load in real-life train travelers during information processing. Appl Ergon 89:103180

AskLegal.my (2019) M'sians who are scared of flying get spot-checked more, and other weird airport stats. Available via AskLegal.my. https://asklegal.my/p/airport-klia-mahb-passenger-behavior-immigration-customs-survey. Accessed 18 May 2023

Batouei A, Iranmanesh M, Mustafa H et al (2020) Components of airport experience and their roles in eliciting passengers' satisfaction and behavioural intentions. Res Transp Bus Manag 37:100585

Batra A (2021) Beyond travel dreaming, planning, check-in, boarding… stress? Anatolia 32(1):46–58

Battisto D, Wilhelm JJ (eds) (2019) Architecture and health: guiding principles for practice. Routledge, New York

Bennett A (2021) [Re] discovering natureleco-tourism & natural resource conservation: the 'potential' for a sustainable symbiotic relationship

Berman N, Mehta D, Matsuo A (2022) The sticky campus in Japan: re-evaluating campus spaces. In: Globalisation, societies and education, pp 1–10

Berto R, Barbiero G, Nasar JL (2022) Biophilic design rationale: theory, methods, and applications. Front Psychol 13:978689

Bogicevic V, Bujisic M, Bilgihan A (2017) The impact of traveler-focused airport technology on traveler satisfaction. Technol Forecast Soc Chang 123:351–361

Brill T (2020) How does stress shape experience and perceptions in public space? Doctoral dissertation, The Ohio State University

Browning WD, Ryan CO (2020) Nature inside: a biophilic design guide. Routledge, London

Burn AS, Passmore J (2022) Outdoor coaching: the role of attention restoration theory as a framework for explaining the experience and benefit of eco-psychology coaching. Int Coach Psychol Rev 17(1):21–36

Butler G, Szili G, Hay I et al (2022) Searching for sanctuary during COVID-19: exploring regional South Australians' nature-based tourism experiences. Rural Soc 31(1):1–14

Canepa S, Ab Ghafar N (2020) Water in architecture, architecture of water. J Civ Eng Archit 14:249–262

Chalmin-Pui LS, Griffiths A, Roe J (2021) Why garden? Attitudes and the perceived health benefits of home gardening. Cities 112:103118

Chingchuang C, Ono K, Watanabe M (2021) The study on the information architecture for the future airport information system. A case study of Suvarnabhumi airport and don muang airport in Thailand. J Sci Des 5(1):87–96

Cohen SA, Kantenbacher J (2020) Flying less: personal health and environmental co-benefits. J Sustain Tour 28(2):361–376

Cuong VH, Khan WA (2023) Exploring the relationship between green spaces and mental health in urban environments. J Sustain Urban Futures 13(6):22–33

Day C, Gwilliam J (2019) Living architecture, living cities: soul-nourishing sustainability. Routledge, New York

DeLauer V, McGill-O'Rourke A, Hayes T et al (2022) The impact of natural environments and biophilic design as supportive and nurturing spaces on a residential college campus. Cogent Soc Sci 8(1):2000570

DKMA (2014) Why focus on improving the passenger experience? 9 great reasons to make customer service a top priority. DKMA. Available via DKMA http://www.dkma.com/en/images/downloads/customerservice/Why%20focus%20on%20the%20passenger%20experience.pdf. Accessed 18 May 2023

Donnelly A (2021) Improving the air travelers experience through airport design. Doctoral dissertation, State University of New York at Buffalo

Ebejer J (2022) The benefits of tourism and travel to the tourist's well-being: a conceptual discussion. In: International conference of the international association of cultural and digital tourism. Springer Nature, Switzerland, Cham, pp 607–618
Federici A (2022) Digital humanities in the study of multifaith prayer rooms in airports: theoretical aspects and a case study. Digital humanities in the study of multifaith prayer rooms in airports: theoretical aspects and a case study, pp 143–155
Foster MJ, Jewell ST (2022) Introduction to research syntheses and the PIECCESS framework. Piecing together systematic reviews and other evidence syntheses: a guide for librarians, p 3
Gong Y, Zoltán ES, János G (2023) Heathy dwelling: the perspective of biophilic design in the design of the living space. Buildings 13(8):2020
Grazuleviciute-Vileniske I, Daugelaite A, Viliunas G (2022) Classification of biophilic buildings as sustainable environments. Buildings 12(10):1542
Guerreiro J, Ahmetovic D, Sato D et al (2019) Airport accessibility and navigation assistance for people with visual impairments. In: Proceedings of the 2019 CHI conference on human factors in computing systems, pp 1–14
Hadley M (2020) Hurry up and wait: spatial strategies for urban stress relief. Master's thesis, University of Waterloo
Han H, Koo B, Chua BL et al (2020) Travelers' intentions for green behaviors at airports: exploring the effect of green physical surroundings using mixed methods. J Hosp Tour Manag 45:569–579
Hasanzade MP, van Oel CJ (2022) Passengers' preferences for architectural design characteristics in the design of airport terminals. Archit Eng Design Manage: 1–16
Herjanto H, Byrnes M, Rivas P et al (2020) How high can you fly? LCC passenger dissatisfaction. Asian J Bus Res 10(2):72–90
Huggins L (2022) Discovering restorative spaces on campus to reduce stress for first-year college students. Int J Soc Sci Educ Stud 9(1):308
Hussain AS, Ramdan AS (2020) Sustainability in the design of passenger terminals for airports. IOP Conf Ser Mater Sci Eng 745(1):012141
Hutson J, Hutson P (2023) Neuroinclusive workplaces and biophilic design: Strategies for promoting occupational health and sustainability in smart cities. Global Health Econ Sustain 1(1)
Ibenna PO, Obiadi BN, Onuorah IM et al (2023) The architecture of biophilic tourism, and the positive impact in urban development and human health. UNIZIK J Educ Res Policy Stud 16(1):90–116
Janicka-Świerguła K (2022) Biophilic design—a remedy for the challenges of the modern world? Wydawnictwo Politechniki Śląskiej Gliwice, 71
Jiang Y, Li S, Huang J et al (2020) Worry and anger from flight delay: antecedents and consequences. Int J Tour Res 22(3):289–302
Jönsson J (2020) Urban refuge-design for decreasing urban stress in a dense environment
Kefalidou G, D'Cruz M, Castro A et al (2019) Designing airport interiors with 3D visualisations. In: Extended abstracts of the 2019 CHI conference on human factors in computing systems, pp 1–8
Kiliç S, Çadirci TO (2022) An evaluation of airport service experience: an identification of service improvement opportunities based on topic modeling and sentiment analysis. Res Transp Bus Manag 43:100744
Kilroy MC (2021) Utilising community gardens to promote stress reduction and mental well-being. Doctoral dissertation, California State Polytechnic University
Kim C, Lee KC, Costello FJ (2020) The intention of passengers towards repeat use of biometric security for sustainable airport management. Sustainability 12(11):4528
Kireeva TV (2020) Green roofs in Russia: classification and typology. IOP Conf Ser Earth Environ Sci 459(5):052034)
Koay WI, Dillon D (2020) Community gardening: Stress, well-being, and resilience potentials. Int J Environ Res Public Health 17(18):6740
Koops S (2019) A green campus-urban ecology, nature's effects on well-being, and how to implement this into a practical design. Doctoral dissertation, University of Groningen

Lampert T, Costa J, Santos O et al (2021) Evidence on the contribution of community gardens to promote physical and mental health and well-being of non-institutionalised individuals: a systematic review. PLoS ONE 16(8):255621
Lehto XY, Lehto MR (2019) Vacation as a public health resource: toward a wellness-centered tourism design approach. J Hosp Tourism Res 43(7):935–960
Lin YS (2023) Architectural wayfinding improvement based on technologies—potential application to hospitals in Honolulu. Doctoral dissertation, University of Hawai'i at Manoa
Maitra R, Joseph MA (2022) New avenues of wellness tourism in untouched aspects of healing in India. Int J Res Tourism Hosp (IJRTH): 1–8
Marquez V, Marquez V (2019) "The liquid airport": security, permeability, and containment in airports. In: Landside|airside: why airports are the way they are, pp 203–243
Mason L, Ronconi A, Scrimin S et al (2021) Short-term exposure to nature and benefits for students' cognitive performance: a review. Educ Psychol Rev: 1–39
McLinton SS, Drury D, Masocha S et al (2020) "Air rage": a systematic review of research on disruptive airline passenger behaviour 1985–2020. J Airline Airport Manage 10(1):31–49
Preston L (2023) Tranquilitree: the potential of trees to mitigate aircraft noise pollution from Schiphol airport
Pronk D, Pronk B, Curtis T (2021) The resilience shield. Macmillan Publishers, Australia
Rajab V (2020) DesİQual: destination in motion. Emotional engagement as a determinant of service quality. Service design for a personalised travelling experience and well-being
Ramaiah M, Avtar R (2019) Urban green spaces and their need in cities of rapidly urbanising India: a review. Urban Sci 3(3):94
Raman TL, Abdul Aziz NA, Yaakob SSN (2021) The effects of different natural environment influences on health and psychological well-being of people: a case study in Selangor. Sustainability 13(15):8597
Ramsey JR, Nassif Leonel J, Zoccal Gomes G (2011) Cultural intelligence's influence on international business travelers' stress. Cross Cultural Manage Int J 18(1):21–37
Richardson M, Butler CW (2022) Nature connectedness and biophilic design. Build Res Inf 50(1–2):36–42
Roe J, McCay L (2021) Restorative cities: urban design for mental health and well-being. Bloomsbury Publishing, London
Sachs NA (2022) EO Wilson: a legacy of "biophilia". HERD Health Environ Res Design J 15(4):283–286
Santa SLB, Ribeiro JMP, Mazon G et al (2020) A green airport model: proposition based on social and environmental management systems. Sustain Cities Soc 59:102160
Shenoy RG (2023) A place of pause, architecture that can de-stress. Doctoral dissertation, Virginia Tech
Söderlund J (2019) The emergence of biophilic design. Springer Nature, Switzerland
Soendergaard S (2020) Effects of nature exposure on cognitive-perceptual-motor performance. The Florida State University
Stankov U, Gretzel U, Filimonau V (2022) Mindfulness and tourism. The mindful tourist: the power of presence in tourism. Emerald Publishing Limited, Bingley, pp 23–48
Sullivan WC, Li D (2021) Nature and attention. Nature and psychology: biological, cognitive, developmental, and social pathways to well-being, 7–30
Trofimova TE (2020) Greenspace design for airports. IOP Conf Ser Earth Environ Sci 459(5):052033
White H, Shah P (2019) Focus: attention science: attention in urban and natural environments. Yale J Biol Med 92(1):115
Zhang Y, Mavoa S, Zhao J, Raphael D et al (2020) The association between green space and adolescents' mental well-being: a systematic review. Int J Environ Res Public Health 17(18):6640
Zhang Y, Ramsey JR, Lorenz MP (2021) A conservation of resources schema for exploring the influential forces for air-travel stress. Tour Manage 83:104240
Zuskáčová V (2023) 'Where are you?': (auto) ethnography of elite passage and (non)-placeness at London Heathrow Airport. Mobilities, pp 1–16

Performance Evaluation of Sustainable Built-Up Composite Cold-Formed Steel Column Filled with Rubberwood Dust Pellet Concrete

Mohd Syahrul Hisyam Mohd Sani and Fadhluhartini Muftah

Abstract Cold-formed steel (CFS) is a steel-based material that is used in construction as a building material. CFS is made by rolling, pressing, and bending processes at ambient temperature which is classified as different process of production when compared with hot-rolled steel. CFS with a variety of shapes, grades, and cross-sections is studied to propose as a structural element in construction such as beams and columns due to its advantages for example fast production, lightweight, and anti-corrosion. However, CFS with a thin surface and open-shaped section is exposed to structural failures and problems which could affect the whole structure and lastly the structure will collapse. Structural failures that have always been discussed are local buckling, distortional buckling, global buckling, and torsion. Besides, rubberwood dust which is collected from the sanding process of rubberwood in furniture industry factories is used as artificial aggregate in concrete. Rubberwood is categorised as a waste material which normally disposed of in a landfill area and creates a lot of environmental problems. With the intention to solve the problems, the CFS channel section with 50 mm of flange element, 100 mm of web element, 12 mm of lipped element, and 1.55 mm of thickness is selected and proposed a built-up section by attaching two CFS channel sections in a face-to-face configuration. Rubberwood is moistened with tap water, shaped into pellets, and lastly sintered in an oven to form rubberwood dust pellet which replace the coarse aggregate in concrete. Then, the built-up face-to-face CFS section is filled with concrete with rubberwood dust pellets and forms the sustainable composite structure. The main objective of the study is to determine the mechanical performance of rubberwood dust pellet concrete-filled built-up CFS stub columns and observe the failure mode of the structure. There are three parts of the experimental activity including the material performance of CFS, the chemical and physical rubberwood dust pellet, and the mechanical performance of built-up CFS column filled with rubberwood dust pellet concrete. From the testing

M. S. H. M. Sani (✉) · F. Muftah
Civil Engineering Studies, College of Engineering, Universiti Teknologi MARA (UiTM) Cawangan Pahang, 26400 Bandar Jengka, Pahang, Malaysia
e-mail: msyahrul210@uitm.edu.my

F. Muftah
e-mail: fadhlu@uitm.edu.my

E. Shaffie et al. (eds.), *Sustainable Green Infrastructure*,
https://doi.org/10.1007/978-981-96-1486-8_10

and observation, the mechanical performance of the built-up CFS column filled with rubberwood dust pellet especially the ultimate load of the stub column increased by approximately 53.29% when compared with the stub column without concrete and decreased by about 23.31% when compared with the built-up CFS column filled with normal concrete.

Keywords Built-up cold-formed steel · Rubberwood dust pellet · Composite column · Concrete · Sustainable

1 Introduction

A column known as a compression member is classified as one of the structural elements in the building which sustains the action from the upper side in compression mode and also carries the self-weight of the above structure and itself. Normally, the column is utilised to link up and support the horizontal member, truss, and arches. Nowadays, the column is designed by producing varieties of shapes, dimensions, and heights, and divided into two categories, namely short and slender.

When a load categorised as exceeding the section's critical buckling load is applied, the cold-formed steel section loses its stability due to the value of the width-to-thickness ratio increasing (Liu et al. 2023). Therefore, many CFS sections are proposed to have the stiffener by adding the lipped on the flange element or by adding the stiffener on the web element in single or double conditions. However, when the CFS is suggested as a beam or column element under slender circumstances, it still exhibits local and distortional buckling. According to Dubina and Ungureanu (2023), it is challenging to distinguish between the local and distortional modes in CFS open sections, whether or not edge stiffeners are present, as any kind of open shaped or cross-section is still in the active mode. According to Teoh et al. (2023a), when compression load was applied, the CFS with open-section had a substantially lower torsional stiffness and resulted in distortional buckling.

Matsubara and Batista (2023) have discussed the CFS section as a column and beam element are extremely vulnerable to several structural instability phenomena either individual (local, distortional, and global buckling) or interaction (local-distortional, local–global, distortional-global, and local-distortional-global buckling) and in contrast to isolated modes, the buckling interaction of CFS section as structural element may cause strength loss, hence the structural design must take this into account. Figure 1 illustrates the CFS section of the individual and interaction buckling mode. Chen et al. (2023a) have stated that the CFS reacts as a compression member in the condition of a slender cross-section is usually borne to local buckling failure.

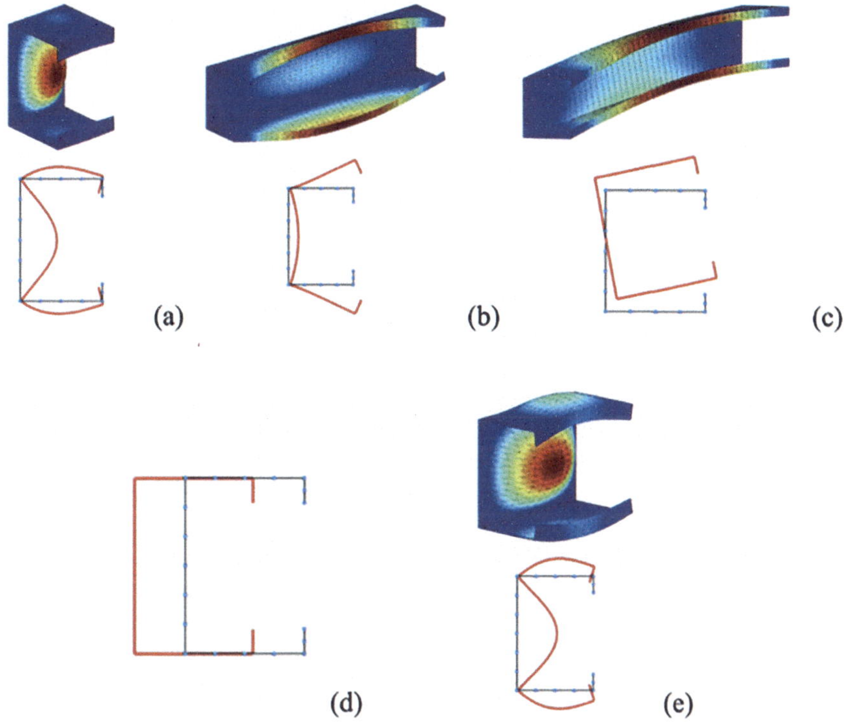

Fig. 1 Example of the individual and interaction of buckling mode of CFS section **a** local, **b** distortional, **c** global flexural–torsional, **d** global flexural, and **e** local-distortional (Matsubara and Batista 2023)

1.1 Cold-Formed Steel

CFS is produced by using techniques of rolling, bending, pressing, and curving the steel sheet to form various shapes and cross-sections and is sometimes recognised as a thin-walled structure. The production of the CFS is different when compared with hot-rolled steel which uses high temperature, high energy consumption, high technology machines, and many skilled workers. Wu et al. (2023) have discussed the benefits of CFS, which have led to its increasing popularity as a structural element in residential buildings. These benefits include being lightweight, resistant to corrosion, emitting less carbon dioxide, and environmentally friendly when compared to masonry or reinforced concrete structures. Due to its advantages in installation, production, and transportation, CFS has been advocated as a structural element and utilised in buildings across the world (Pham 2023). Additionally, CFS especially the channel section failed due to web crippling when utilised as primary girders which subjected to concentrated transverse action (Young et al. 2023). Young et al. (2023) have studied the web crippling analysis of the CFS channel section and reported that the CFS channel section reduced the effect of web crippling by producing the section with having stiffened flanges and stiffened web when classified as a slender web section.

1.2 Rubberwood Dust

Rubber is categorised as an important crop that is planted worldwide and especially in Southeast Asia such as Thailand, Malaysia, and Indonesia are classified as the main producers of latex and rubber tree products. The rubberwood tree is chopped and cut after it becomes too old to continue producing latex and processed to form a suitable product. More than 2.7 million cubic metres of rubberwood were produced in Thailand, according to Srivaro et al. (2021), and were used to make toys, furniture, culinary utensils, kitchen boards, and industrial boards. Normally, the process of producing furniture consists of measuring, cutting, sanding, polishing, painting, and assembly. The process of cutting, sanding, and polishing wood produces wood dust, ash, and other waste materials which would be dumped in landfill regions.

Rubberwood dust that is gathered from the furniture industry during the sanding process typically causes pollution and environmental issues at a factory or in a landfill region. Suliman et al. (2019) have reported that the waste wood dust in the landfill region is growing daily and having an impact on the environment that is deemed important for use in concrete production. Furthermore, the environmental impact that would happen in the landfill region is air pollution, groundwater contamination, soil pollution, and hazardous gas emissions. Finally, the researchers tried to study the significance of the rubberwood dust used in concrete technology as a fine aggregate replacement, coarse aggregate replacement, or additional material, and lastly supported the 3R (Reduce, Reuse, and Recycle) approach.

1.3 Aggregate Pellet in Concrete

Normal concrete is a composite mixing material that quickly hardens in 24 h and consists of cement, fine aggregate, coarse aggregate, and tap water in a fluid state. Owing to its versatility, uniformity, longevity, and efficiency, concrete is a widely utilised building material globally. A significant portion of the concrete is made up of coarse aggregate, which accounts for 60% to 80% of the total weight and indirectly influences the strength of the mixture. Nonetheless, research is being done to substitute the cement or aggregate in concrete, either partially or completely. Utilising waste material, both treated and untreated, is how to substitute material for aggregate is chosen in order to lower manufacturing costs, lower carbon dioxide emissions, and lessen the environmental effect. Conventional resources have been significantly depleted as a result of this intensive use, and hence, recycling waste products is becoming more and more important for sustainability in order to minimise the exploitation of natural raw materials and lessen the impact on the environment. Additionally, Dong et al. (2022) have documented that natural aggregate became scarce due to extensive construction activities, the development of urban areas and infrastructure, and lastly the introduction of alternative aggregate as a means of safeguarding environmental degradation and maintaining natural topography.

Nowadays, utilising industrial or agricultural waste to substitute coarse aggregate in concrete in pellet form is becoming more and more common. In general, the aggregate pellet is divided into two components, unsintered pellet and sintered pellet. The complete process of the unsintered sludge pellet according to a study by Zhu et al. (2023) is shown in Fig. 2. Paktiawal and Alam (2022) have reported that the aggregate pellet formed by compacting and through other processes of sawdust or organic waste, such as rice husk, straw, shell, crop, and dust. Besides, the study of using fresh unused concrete with the process of pelletising into a round shape is introduced and reported less physical and mechanical properties when compared with natural coarse aggregate (Bassani et al. 2019). The density, strength, and workability of concrete are often impacted by the aggregate pellet (Paktiawal and Alam 2022). The study of aggregate pellets in concrete by using waste material or raw material is considered too extensive, for example, industrial solid red mud wastes and calcium carbide slag (Zhao et al. 2023), fly ash (Dong et al. 2022), waste paper ash (Mohd Sani et al. 2019), industrial copper sludge (Tay et al. 2000), ground-granulated blast-furnace slag (Joy et al. 2021), and cut rubber tyre (Ganiron 2014). Figure 3 shows an example of the aggregate pellet.

1.4 Built-Up Cold-Formed Steel

The open-section CFS is used as a wall frame, roof truss system, and storage rack and is sold in the building material sector. Chen et al. (2023b) have shown that the CFS may function as secondary load-bearing components in building construction, for example, wall girts, cladding, side rails, and purlins. CFS sections are made in a range of shapes, cross-sections, and grades; however, because of their open-section and asymmetrical cross-section, they are vulnerable to unstable situations and compromise structural integrity when used as structural elements or main load-bearing components. As a result, the CFS is created to construct built-up sections by assembling and joining methods from basic shapes either plain sections or edge-stiffened sections. The various method and materials of the fastener are used for instance bolt and nut, welds, self-drilling screws, rivets, and self-tapping screws in the assembly process to produce CFS closed sections or symmetrical sections. Craveiro et al. (2022) have stated that the built-up CFS section is classified as a versatility section, possesses pointedly greater load-bearing capacity, and reduces global slenderness. Li et al. (2023) have found that the structural performance is largely dependent on the fastener spacing, especially the self-drilling screw used to assemble the CFS plain section into a built-up CFS section. From the production of the built-up CFS section, several experimental activities and numerical analysis

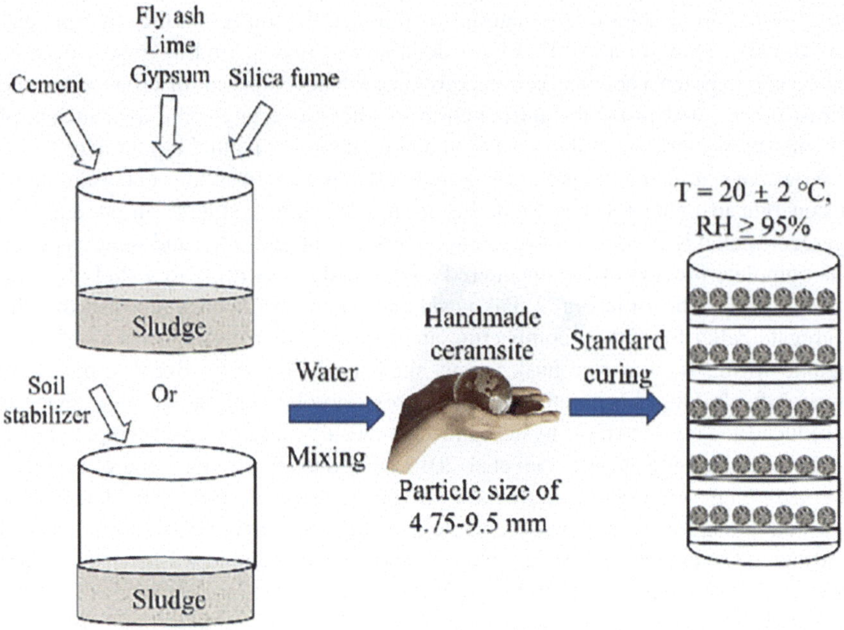

Fig. 2 The schematic diagram of the preparation and process of unsintered sludge pellet (Zhu et al. 2023)

Fig. 3 Example of aggregate pellet such as **a** burnt wood pellet (Paktiawal and Alam 2022), **b** unused concrete pellet (Bassani et al. 2019)

of buckling issues and structural performance with various parameters for example geometry properties, cross-section, fastener type, fastener spacing, and boundary end condition are studied (Abbasi et al. 2023).

Chen et al. (2023a) have reported that the built-up CFS section is shown having properties better than the initial open sections in terms of load resistance and structural stability. The example of a built-up CFS section employing plain sections is shown in Fig. 4. The non-prismatic character of the built-up CFS cross-sections makes it

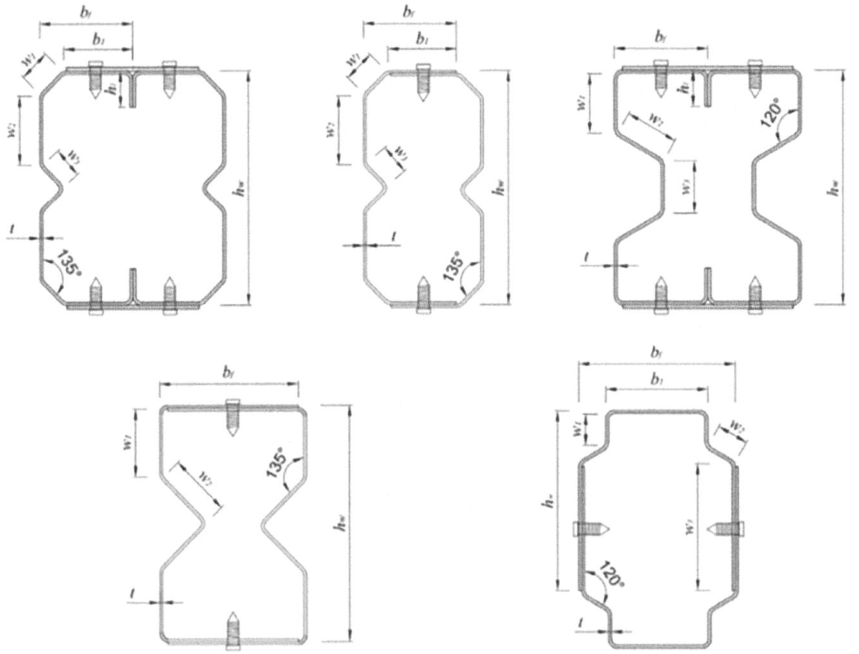

Fig. 4 An illustration of a built-up CFS section secured with a self-drilling screw (Chen et al. 2023a)

difficult to anticipate their structural performance, even while the manufacturing flexibility allows for the optimisation of the appropriate cross-section (Selvaraj and Madhavan 2023). The structural performance of the built-up CFS section is indicated as more strength and stable when subjected to load due to the double restraint effect of the section joining together. Teoh et al. (2023a) noted that the built-up CFS section with a two-axis symmetric cross-section may successfully eradicate the eccentricities between the gravity and shear centres, and boost the stability of the section. The length of the column in the practical and actual use of built-up columns normally varies from 2.7 to 3.7 m (Teoh et al. 2023a). Lastly, there is still no complete or limited information on the design code of the methodology for proposing a built-up CFS section.

1.5 Built-Up Cold-Formed Steel Filled with Normal Concrete and Special Concrete

The built-up CFS filled with normal concrete or special concrete is classified as a high-strength composite structure because the compression load is applied directly to concentric between these two materials. Huang et al. (2023) have reported that the built-up CFS with the filling of the concrete is able to spread the lateral confining pressure and reduce the internal local bending issue. There are a lot of advantages when built-up CFS is filled with concrete such as fast construction, cost-effectiveness, additional reinforcement elimination, and structural integrity reductions. Rahnavard et al. (2023) have reported that the built-up CFS is responsible as a permanent formwork, saves time on construction, and sustainability practices, more affordable, low-cost structure system, enhancing fire resistance by reducing temperature change and greater axial compression load-bearing capacity. The slender built-up CFS with concrete is able to reduce the buckling issue especially local and distortional buckling of the section. The built-up CFS with concrete infill was examined by Chen et al. (2023a), and it was determined that this was a suitable way to address the local buckling problem by offering internal restraint and ultimately increasing the structural stiffness. Figure 5 shows the example of the built-up CFS section for column elements filled with concrete.

The strengthened composite structure may be achieved by considering the steel tube's depth-to-thickness ratio and the maximum lateral confining pressure while constructing a built-up CFS section in a square cross-section filled with concrete

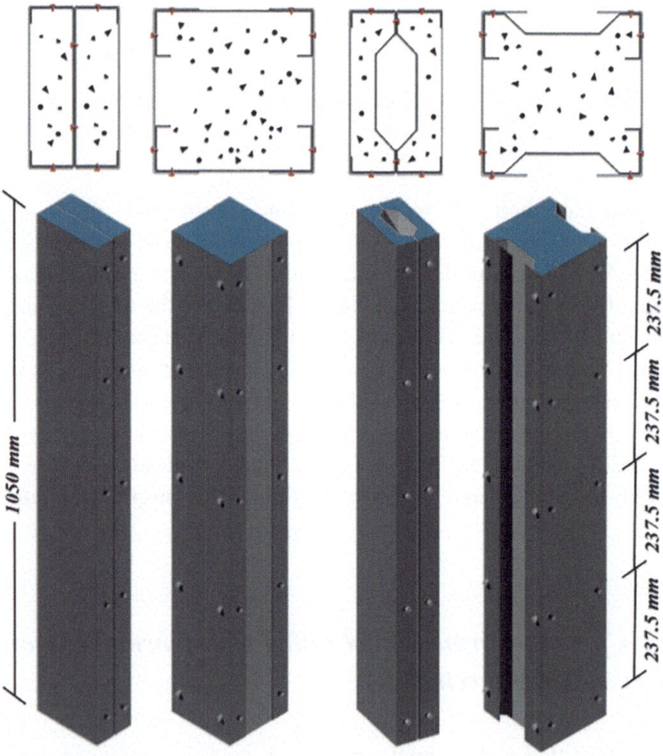

Fig. 5 The example of a built-up CFS column filled with concrete in a variety of cross-sections (unit in mm) (Rahnavard et al. 2022)

(Huang et al. 2023). The composite structure is suitable for use as tall buildings, offshore structures, tower structures, bridges, and foundation piers. Numerous studies have attempted to fill the built-up CFS section with special concrete rather than normal concrete for utilising the waste material as concrete ingredient replacement, including self-compacting lightweight concrete (Teoh et al. 2023a), lightweight aggregate concrete (Teoh et al. 2023b), rubberised concrete (Sabbagh et al. 2022), ultra-lightweight-concrete (Hegyi and Dunai 2016), and washed bottom ash concrete (Mohd Sani et al. 2022). Lastly, there is classified as an uncommon study, and limited research is done on the compression behaviour of built-up CFS sections filled with normal or special concrete.

1.6 Summary

From the observation and analysis of the previous study, the study on the plain CFS section to form a built-up section is classified as a popular research area to solve the buckling issue by reducing production costs. The study of combining the built-up CFS with other materials also becoming popular due to the high demand for alternative building materials as structural elements to reduce production and material costs in parallel with the development of a country. However, there is no information available about the design code that would describe the whole design flowchart and rules for constructing the built-up CFS section as a column element. Additionally, the study of the rubberwood dust collected from the sanding process of rubberwood furniture as aggregate pellet is limited and there is no information on rubberwood dust pellet concrete that filled in built-up CFS column or beam. The rubberwood dust utilised in concrete as additional material or replacement material is also not reported broadly in the area of concrete technology. Rubberwood dust without any 'reduce, reuse, and recycle' process would affect the environmental impact and increase the cleaning and waste disposal costs. The main objective of the study is to evaluate the mechanical performance of built-up cold-formed steel columns filled with rubberwood dust pellet concrete as a sustainable composite structure element.

2 Experimental Material and Method

This session is divided into two main parts including the experimental material and experimental method. The experimental material is covered in detail by the CFS, rubberwood dust pellet, and concrete. Furthermore, the experimental method is prepared for the material properties of CFS, physical and chemical properties of the rubberwood dust pellet by using inductively coupled plasma, and lastly the mechanical performance of built-up composite CFS column filled with rubberwood dust pellet concrete.

2.1 Experimental Material

CFS with channel shape and with dimensions of 100 mm × 50 mm × 12 mm × 1.55 mm is selected and used. The CFS channel section possesses the lipped element at both flanges and also contains a double web stiffener. The factory yield strength of the section is 450 MPa, the area is 329 mm^2, the second moment of area, I_{xx} is 0.53 × 10^6 mm^4, and the second moment of area, I_{yy} is 0.11 × 10^6 mm^4. The material properties of the CFS are obtained at ambient temperature by a series of coupon tensile tests on flat elements especially flange and web which are based on BS EN 10002-1:2001. Two CFS channel sections are combined and attached in a face-to-face configuration to produce a hollow part at the middle side of the built-up CFS section. The attachment of these two CFS channel sections is by using the three times of spot weld in three locations; top end, mid-height, and bottom end on two sides.

The rubberwood dust employed in this study is taken from a by-product of the sanding process and normally is used for creating cement boards. The rubberwood dust is stored and left in the open polyethylene tank for 1 month for the reason of reducing the sugar content and ensuring that the bond between the concrete ingredients does not dissolve when mixed. After that, the rubberwood dust is moistened and combined with water in a 2:1 ratio, and finally through the pelletising process at room temperature with a diameter size range of 10–15 mm to produce the rubberwood dust pellet as shown in Fig. 6. The rubberwood dust pellet is sintered in an oven at 200 °C for 1 day to obtain a hard and stiff pellet, and fully submerged into a shellac liquid for reducing the water absorption, avoiding the termite attack, and strengthening the pellet surface.

The built-up CFS column with a dimension of 100 mm × 100 mm × 250 mm which is categorised as a stub column is produced and filled with the conventional concrete ingredient as recognised as normal concrete and by replacing the coarse aggregate in full percentage with rubberwood dust pellet as known as rubberwood dust pellet concrete. Figure 7 shows an example of fresh normal concrete and fresh

Fig. 6 The shape and size of the rubberwood dust pellet

Fig. 7 The example of the fresh **a** normal concrete and **b** rubberwood dust pellet concrete

rubberwood dust pellet concrete. A concrete design mix is proposed by using a ratio with grade 30 and a water-cement ratio of 0.5. There is another specimen that formed without any concrete-filled in the built-up section.

2.2 *Experimental Method*

The material properties testing of CFS is started by using the universal testing machine with 100 kN of capacity and three specimens for each element (flange and web) are conducted to obtain the average value of the ultimate stress, yield stress, deformation at ultimate stress, and deformation at yield stress. The pace rate of the testing for material properties is 1 mm/min. Then, the rubberwood dust pellet is checked for their chemical and physical properties before being used in the concrete mix. The experimental activity concerning the performance evaluation particularly the compression behaviour of built-up composite CFS stub column filled with normal and rubberwood dust pellet concrete is investigated. The failure mode of all specimens is observed. The axial compression test at a compression machine with pinned end support condition and with 7 kN/s of pace rate is proposed. There are three specimens of the column as given in Table 1. The average data of the ultimate load of every specimen is recorded.

Table 1 The description of the column specimen

Description of the specimens	Specimen label
Built-up CFS column without concrete	BWC
Built-up CFS column filled with normal concrete	BFNC
Built-up CFS column filled with rubberwood dust pellet concrete	BFRDPC

3 Results and Discussion

The result and discussion of the experimental activity are divided into three sections; material properties of CFS, chemical and physical properties of rubberwood dust pellet, and mechanical performance of built-up CFS column filled with rubberwood dust pellet concrete. All result data and failure modes are determined and reported.

3.1 *Material Properties of Cold-Formed Steel*

From the testing, the ultimate load of the web element is 10.0 kN and the flange element is 10.6 kN. The web and flange elements are found to exhibit ultimate and yield stresses that ranged from 557 to 581 MPa and 498 MPa to 528 MPa, respectively. According to the results, it showed that the stress of the flange element is higher than the web element attributable to the process of bending and shaping of the element. Then, it is stated that the yield stress of the web element is 498 MPa and that of the flange element is 528 MPa, both of which are higher than the 450 MPa value of the factory yield stress. The value of the modulus of elasticity of both elements is determined to be 210 GPa. In the light of the fact that the experimental value exceeded the factory yield stress value, the CFS channel section is suitable for further action and designated as complying.

3.2 *Chemical and Physical Properties Rubberwood Dust Pellet*

The chemical properties of the rubberwood dust pellet are tested for determining the chemical and sugar composition or content before being used as a coarse replacement in the concrete mix and tabulated in Table 2. From the test, the highest value of chemical content of the rubberwood dust pellet is iron (Fe) and aluminium (Al). The highest value of the sugar content of the rubberwood dust pellet is arabinose and the sugar content is reported able to affect the properties of the concrete such as setting time, bonding, durability, and strength development. The amount of glucose and xylose in the rubberwood dust pellet is not particularly large, but a significant amount of glucose and xylose will cause concrete to cure more slowly.

Table 2 The chemical content and sugar content of the rubberwood dust pellet

Parameter	Rubberwood dust pellet
Chemical content	
Fe (ppm)	116.59
Si (ppm)	1.75
Al (ppm)	21.36
Ca (%)	0.22
Mg (%)	0.09
% Loss on ignition (LOI)	48.13
Sugar content	
Glucose (ppm/mg/L)	55.18
Xylose (ppm/mg/L)	67.07
Arabinose (ppm/mg/L)	167.84

Table 3 The physical properties of the rubberwood dust pellet

Parameter	Coarse aggregate	Rubberwood dust pellet
Mass (kg) (1 unit)	0.0100	0.0030
Density (kg/m^3)	1,322	373

Furthermore, the physical properties of the rubberwood dust pellet are tested and the result is shown in Table 3. The percentage difference of the mass for 1 unit between the conventional coarse aggregate and rubberwood dust pellet is 70%. 71.79% of the percentage difference of the density between the conventional coarse aggregate and rubberwood dust pellet is reported. The density of the aggregate will have an impact on the stiffness and strength of the concrete, as well as the density of the concrete. With low elastic modulus and strength, the aggregate with pores can be characterised as having low aggregate density.

3.3 Mechanical Performance of Built-Up Cold-Formed Steel Column Filled with Rubberwood Dust Pellet Concrete

The result data of the mechanical performance especially the ultimate load of the column is given in Table 4. 55.72% and 37.63% are reported when comparing the mass between BFNC with BWC and between BFRDPC with BWC, respectively. Furthermore, the value of the mass is decreased by approximately 29.01% when compared BFNC with BFRDPC. Therefore, the BFRDPC is classified as a lightweight column structure. The percentage difference of the average ultimate load between the BFNC and BWC, and between the BFRDPC and BWC is 53.29% and 23.31%, respectively. In addition, the percentage difference of the ultimate load between the BFNC and BFRDPC is 39.09%. Thus, the BFRDPC is suitable to be

Table 4 The mass and ultimate load of the column

Specimen	Mass (kg)	Average ultimate load (kN)	Failure mode
BWC	2.8000	196.1	Local buckling at the end of the column
BFNC	6.3240	419.8	Local buckling far from the end of the column
BFRDPC	4.4895	255.7	Local buckling far from the end of the column

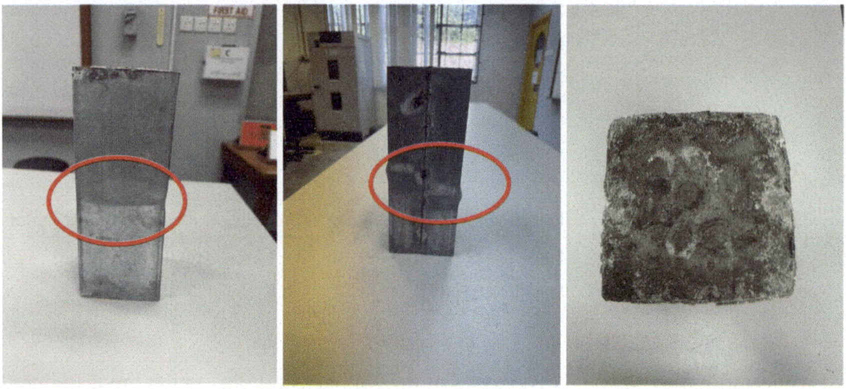

Fig. 8 The failure mode of the BFNC specimen

used as a structural element of the column because the percentage difference between BFNC and BFRDPC is less than 50%. If referring to failure mode observation, all specimens are illustrated to have the local buckling however the location of the failure is different. The column specimen without concrete normally failed in buckling mode at the end of the column while the column specimen with concrete either normal or rubberwood dust pellet concrete is far from the end of the column or situated at half or quarter of the height of the column. There is no failure mode reported on the top surface of the concrete. Figure 8 is the failure mode of the BFNC specimen, and Fig. 9 is the failure mode of the BFRDPC specimen. The outward buckling is also observed to occur, and the failure mode is similar to the study by Qian et al. (2024).

4 Conclusion and Recommendation

From the investigation and result, the following conclusion may be made:

(1) Rubberwood dust is a by-product of the sanding process at the furniture manufacturing factory and is classified as having the potential to replace a coarse aggregate utilised partially or entirely in concrete production. During mixing,

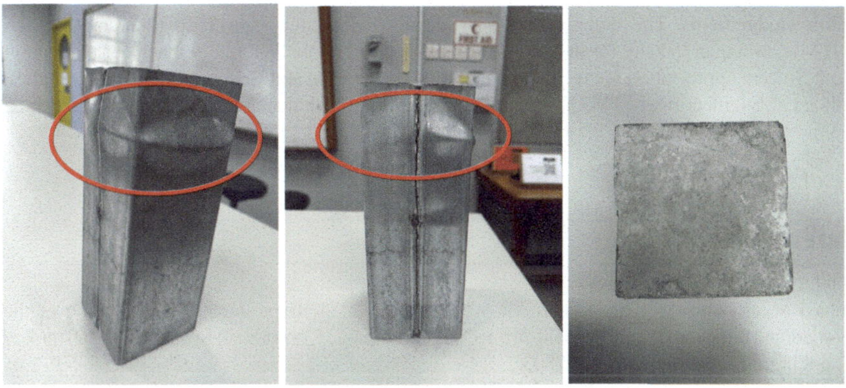

Fig. 9 The failure mode of the BFRDPC specimen

rubberwood dust pellet in concrete that require a particular mix proportion and handling approach must be evidently prepared. Rubberwood dust pellet is viewed as a replacement material in concrete to reduce the need for natural aggregate in the future.

(2) The chemical and physical properties of the rubberwood dust pellet are determined and shown that they are suitable to combine with other ingredients in the concrete. The rubberwood dust pellet is classified as lightweight coarse aggregate with a rounded shape when compared with conventional coarse aggregate.

(3) The mechanical performance especially the ultimate load of the built-up CFS column without concrete is increased by about 23.31% when added the rubberwood dust pellet aggregate and 53.29% when added the normal concrete. The comparison value of the built-up CFS column with normal concrete and the built-up CFS column with rubberwood dust pellet concrete is 39.09% which is classified as suitable and significant to use as a structural column element. The built-up CFS column with rubberwood dust pellet concrete is categorised as a lightweight column structure due to the percentage difference with the built-up CFS column filled with normal concrete approximately 29.01%. The built-up CFS column filled with rubberwood dust pellet concrete is considered to support the sustainable development goal.

Some recommendation for further activity of research is discussed such as comparing the result with the design code of practice or numerical analysis by utilising finite element modelling for predicting the value of the ultimate load and failure mode. In addition, the height of the column should be determined by selecting the slenderness ratio with categories of stub, intermediate, and slender, and should be studied the fastener spacing and configuration of the built-up CFS section before being proposed as a column.

Acknowledgements The authors gratefully acknowledged the Universiti Teknologi MARA (UiTM) Cawangan Pahang for their cooperation in providing the laboratory machines and equipment. Sincere appreciation is given to the faculty members and staff of Civil Engineering Studies, College of Engineering, UiTM Cawangan Pahang for their technical guidance.

References

Abbasi M, Rasmussen KJR, Khezri M, Schafer BW (2023) Experimental investigation of the sectional buckling of built-up cold-formed steel columns. J Constr Steel Res 203:107803. https://doi.org/10.1016/j.jcsr.2023.107803

Bassani M, Diaz Garcia JC, Meloni F, Volpatti G, Zampini D (2019) Recycled coarse aggregates from pelletized unused concrete for a more sustainable concrete production. J Clean Prod 219:424–432. https://doi.org/10.1016/j.jclepro.2019.01.338

Chen MT, Zhang T, Young B (2023a) Behavior of concrete-filled cold-formed steel built-up section stub columns. Thin-Wall Struct 187:110692. https://doi.org/10.1016/j.tws.2023.110692

Chen M, Shen K, Lu WB, Zhang XY, Shi Y, Yang CF, Wang HL (2023b) Compression tests of cold-formed steel built-up T-shaped columns. Structures 53:1172–1185. https://doi.org/10.1016/j.istruc.2023.04.125

Craveiro HD, Rahnavard R, Laim L, Simoes RA, Santiagi A (2022) Buckling behavior of closed built-up cold-formed steel columns under compression. Thin-Wall Struct 179:109493. https://doi.org/10.1016/j.tws.2022.109493

Dong B, Chen C, Wei G, Fang G, Wu K, Wang Y (2022) Fly ash-based artificial aggregates synthesized through alkali-activated cold-bonded pelletization technology. Constr Build Mater 344:128268. https://doi.org/10.1016/j.conbuildmat.2022.128268

Dubina D, Ungureanu V (2023) Local/distortional and overall interactive buckling of thin-walled cold-formed steel columns with open cross-section. Thin-Walled Struct 182:110172. https://doi.org/10.1016/j.tws.2022.110172

Ganiron TU (2014) Pelletized cut rubber: an alternative coarse aggregate for concrete mixture. Int J Adv Sci Technol 64:21–30. https://doi.org/10.14257/ijast.2014.64.03

Hegyi P, Dunai L (2016) Experimental investigation on ultra-lightweight-concrete encased cold-formed steel structures part II: stability behaviour of elements subjected to compression. Thin-Wall Struct 101:100–108. https://doi.org/10.1016/j.tws.2016.01.003

Huang WF, Shao YB, Hassanein MF (2023) Behaviour and confinement-based direct design of concrete-filled cold-formed stiffened steel tubular short columns. J Constr Steel Res 202:107773. https://doi.org/10.1016/j.jcsr.2022.107773

Joy A, Anilkumar A, Gopika MM, Panicker GS, Martin M (2021) Production of pelletized cold bonded GGBS aggregate. IOSR J Mech Civ Eng 18(3):1–8

Li Y, Li R, Zhou T, Han A, Xie Y (2023) Novel design method of screwed connections of cold-formed steel built-up plates. Structures 55:2433–2444. https://doi.org/10.1016/j.istruc.2023.07.056

Liu H, Wang W, Deng L, He Y (2023) Inelastic buckling analysis of cold-formed steel members with residual stresses based on CFSM. Structures 56:104871. https://doi.org/10.1016/j.istruc.2023.06.142

Matsubara GY, Batista EM (2023) Local-distortional buckling mode of steel cold-formed columns: generalized direct strength design approach. Thin-Wall Struct 183:110356. https://doi.org/10.1016/j.tws.2022.110356

Mohd Sani MSH, Ismail MI, Rahman MA, Muftah F (2019) Waste paper ash pellets as coarse aggregate replacement in concrete. Int J Eng Adv Technol 8(4):1112–1117

Mohd Sani MSH, Muftah F, Mohsan NM, Kado B (2022) Behavior of built-up cold-formed steel stub columns infilled with washed bottom ash concrete. Adv Technol Innov 7(2):92–104

Paktiawal A, Alam M (2022) Temperature effect on compressive strength and quality grading of standard grade concrete with partial replacement of natural coarse aggregate by burnt wood pellet aggregate. Mater Today Proc 56:574–580. https://doi.org/10.1016/j.matpr.2022.02.387
Pham NH (2023) Sectional capacities of cold-formed perforated steel channel columns. Mater Today Proc 85:113–117. https://doi.org/10.1016/j.matpr.2023.05.273
Qian Z, Song Q, Huang H, Wang W (2024) Axial compressive behavior of spiral stirrup-reinforced concrete-filled cold-formed steel built-up box stub columns. Eng Struct 308:118017. https://doi.org/10.1016/j.engstruct.2024.118017
Rahnavard R, Craveiro HD, Simoes RA, Laim L, Santiago A (2022) Fire resistance of concrete-filled cold-formed steel (CF-CFS) built-up short columns. J Build Eng 48:103854. https://doi.org/10.1016/j.jobe.2021.103854
Rahnavard R, Craveiro HD, Simoes RA, Santiago A (2023) Concrete-filled cold-formed steel (CF-CFS) built-up columns subjected to elevated temperatures: test and design. Thin-Wall Struct 188:110792. https://doi.org/10.1016/j.tws.2023.110792
Sabbagh AB, Jafarifar N, Deniz D, Torabian S (2022) Development of composite cold-formed steel-rubberised concrete semi-rigid moment-resisting connections. Structures 40:866–879. https://doi.org/10.1016/j.istruc.2022.04.069
Selvaraj S, Madhavan M (2023) Structural behaviour of cold-formed steel built-up closed cross-section columns-assessing the influence of parameters and design methods. Eng Struct 294:116600. https://doi.org/10.1016/j.engstruct.2023.116600
Srivaro S, Leelatanon S, Setkit M, Matan N, Khongtong S, Jantawee S, Tomad J (2021) Effects of manufacturing parameters on properties of rubberwood-cross laminated timber manufactured via hot pressing. J Build Eng 44:102703. https://doi.org/10.1016/j.jobe.2021.102703
Suliman NH, Abdul Razak AA, Mansor H, Alisibramulisi A, Mohd Amin N (2019) Concrete using sawdust as partial replacement of sand: is it strong and does not endanger health? MATEC Web Conf 258:01015. https://doi.org/10.1051/matecconf/201925801015
Tay JH, Hong SY, Show KY (2000) Reuse of industrial sludge as pelletized aggregate for concrete. J Environ Eng 126(3):279–287
Teoh KB, Chua YS, Pang SD, Kong SY (2023a) Experimental investigation of flexural buckling behaviour of self-compacting lightweight concrete-filled cold-formed built-up box section (CFBBS) columns. Thin-Wall Struct 187:110751. https://doi.org/10.1016/j.tws.2023.110751
Teoh KB, Chua YS, Pang SD, Kong SY (2023b) Experimental investigation of lightweight aggregate concrete-filled cold-formed built-up box section (CFBBS) stub columns under axial compression. Eng Struct 279:115630. https://doi.org/10.1016/j.engstruct.2023.115630
Wu H, Sui L, Liu X, Liang S, Zhou T (2023) Shear-slip constitutive model of screw-fastened connections and application in numerical analysis of built-up back-to-back cold-formed steel columns. Thin-Wall Struct 186:110710. https://doi.org/10.1016/j.tws.2023.110710
Young B, Ellobody E, He J (2023) Web crippling tests on cold-formed high strength steel channel sections having different stiffened flanges and stiffened web. Thin-Wall Struct 190:110995. https://doi.org/10.1016/j.tws.2023.110995
Zhao Q, Shi Y, Xue C, Jia Y, Guo W, Wang D, Wang S, Gao Y (2023) Investigation of various curing methods on the properties of red mud-calcium carbide slag-based artificial lightweight aggregate ceramsite fabricated through alkali-activated cold-bonded pelletization technology. Constr Build Mater 401:132956. https://doi.org/10.1016/j.conbuildmat.2023.132956
Zhu Y, Fu H, Wang P, Xu P, Ling Z, Wei D (2023) Pore structure characteristics, mechanical properties, and freeze-thaw resistance of vegetation-previous concrete with unsintered sludge pellets. Constr Build Mater 382:131342. https://doi.org/10.1016/j.conbuildmat.2023.131342

Analyzing the Obstacles and Process of Introducing Building Information Modeling (BIM) in the Design Stage of C&S Engineering Consulting Firm Within Malaysia

Mohd Rashid Bin Ya'acob, Che Maznah Mat Isa, Siti Hamidah Abdull Rahman, and Salmaliza Salleh

Abstract The twelfth Malaysia Plan (12th MP) outlined by the Malaysian government for 2021–2025 emphasizes the significance of digitalization, specifically through the implementation of Building Information Modeling (BIM) in the construction industry. Despite BIM's introduction in 2007 and the government's target of achieving an 80% implementation rate by 2025, its adoption among civil and structural (C&S) consultants remains limited. This chapter presents insights from a pilot survey involving 30 respondents, employing various statistical analyses such as validity, reliability tests, and factor analysis. The survey revealed significant obstacles to BIM adoption among C&S consultants, categorized into five themes: process, knowledge, policy, cost, and technology. To address these challenges, the chapter introduced the BIM Implementation Framework (BIMIF) for government building projects, adapted from the practices identified in Case Study A. This framework prioritizes a conventional contract approach, particularly during the design stage. Ultimately, this study aims to enhance the integration of BIM within Malaysian government projects through conventional contracts, thereby refining BIM-driven construction processes.

Keywords BIM · Civil and structural engineering consultants · Government building construction

M. R. B. Ya'acob · C. M. M. Isa (✉) · S. H. A. Rahman
School of Civil Engineering, College of Engineering, Universiti Teknologi MARA, Shah Alam, Selangor, Malaysia
e-mail: chema982@uitm.edu.my

S. Salleh
Universiti Teknologi Petronas, Seri Iskandar, Perak, Malaysia

E. Shaffie et al. (eds.), *Sustainable Green Infrastructure*,
https://doi.org/10.1007/978-981-96-1486-8_11

1 Background

In the evolving landscape of the construction industry, Building Information Modeling (BIM) has emerged as a transformative technology and methodology, revolutionizing how projects are conceptualized, executed, and managed. Rooted in the digital realm, BIM encompasses not only geometric representations but also a comprehensive data-rich framework that encapsulates every facet of a construction project's lifecycle. The adoption of BIM promises enhanced collaboration, streamlined processes, and improved project outcomes, all contributing to the overarching goal of modernizing the construction sector.

At its core, BIM transcends the boundaries of traditional design practices by introducing a dynamic and interconnected environment where stakeholders interact with a shared digital model. This model serves as an authoritative repository of project information, enabling real-time collaboration, visualization, and analysis. The significance of BIM lies in its ability to transcend the limitations of conventional two-dimensional drawings and fragmented information exchange. BIM redefines the process by fostering a holistic approach that encompasses planning, design, construction, operation, and the eventual transition toward refurbishment or demolition.

The integration of BIM into the Malaysian construction landscape marks a pivotal juncture in the country's pursuit of industry advancement. Spearheaded by the Construction Industry Development Board (CIDB), BIM adoption has gained traction as a catalyst for efficiency, sustainability, and improved project delivery. The CIDB Malaysia Strategic Plan (2015–2020) laid the foundation for BIM's widespread incorporation, mandating its application in construction projects by the year 2020, encompassing both public and private sectors. This strategic initiative aligns with the broader framework of the Industrial Revolution 4.0, where digitalization takes center stage in the transformation of various industries, including construction.

While the trajectory of BIM adoption in Malaysia showcases encouraging growth, it is not devoid of challenges that warrant meticulous consideration. Among the most salient of these challenges are the persisting gaps in awareness and knowledge about BIM technology. While strides have been made in raising awareness, there remains a pervasive lack of understanding about the transformative potential of BIM across diverse echelons of the construction industry. Financial considerations also loom large in the BIM adoption landscape. The initial costs associated with procuring software, providing training, and establishing the necessary infrastructure can be daunting, particularly for smaller entities. This financial burden can impede the entry of deserving stakeholders into the realm of BIM. The industry's historical reluctance to embrace change, compounded by the intricate nature of civil and structural engineering work, further compounds the challenges. The complexity of transitioning from established practices to BIM-powered methodologies requires overcoming resistance and fostering a culture of adaptability. Another critical challenge is the absence of standardized guidelines that provide a clear roadmap for BIM implementation. The lack of uniformity in processes, data exchange protocols, and Level of Detail (LOD) requirements introduces confusion and hampers effective collaboration among stakeholders. Moreover, the delayed implementation of the BIM mandate in Malaysia has necessitated a compressed adaptation timeline for organizations. This urgency to comply with the mandate could lead to hurried implementations and suboptimal results.

A study by Idris et al. (2021) indicates that Malaysia's progress in BIM implementation falls short due to factors like lack of awareness, high costs, slow adaptation, absence of clear guidelines for organizations and policymakers, and delayed mandating of BIM. Yasser Yahya et al. (2020) note that many Malaysian construction businesses are unaware of BIM technology, while Samad et al. (2018) highlight a lack of understanding and willingness to change in BIM adoption. Muñoz-La Rivera et al. (2019) report deficiencies in structural engineering companies (SECs) hindering their productivity and collaboration, with and structural engineering consultant services.

In light of these obstacles, this chapter seeks to address a critical gap in current understanding by focusing on BIM implementation obstacles specific to civil and structural engineering (C&S) consultants in Malaysia. By comprehensively examining these obstacles, the chapter aims to provide actionable insights that can inform policy decisions, industry practices, and strategies for smoother BIM integration within this specialized sector. In the following sections, this chapter will delve deeper into the intricacies of these challenges through a structured questionnaire survey, extracting invaluable perspectives from practitioners at the forefront of BIM adoption in Malaysia's civil and structural engineering domain. Through this exploration, a richer understanding of the impediments to BIM adoption will emerge, shedding light on potential pathways to surmount these obstacles and usher in a new era of construction innovation and excellence.

2 Literature Review

Strengthening the metamorphosis of the construction industry is a pivotal necessity for driving resilient economic growth. Spanning from 2016 to 2020, the Construction Industrial Transformation Program (CITP), under the oversight of CIDB, has been designated as Malaysia's premier agenda for reshaping the construction landscape toward enhanced economic viability and ecological sustainability. An integral facet of this transformation is the seamless integration of Building Information Modeling (BIM), an innovative technology, into the industry. To facilitate the widespread adoption of BIM across Malaysia's construction sector, CIDB launched MyBIM in 2017 as a central hub for training and resource provision. Through collaborative endeavors with the construction sector, CIDB seeks to amplify awareness and foster the smooth assimilation of BIM into professional and organizational practices. Projections suggest that BIM implementation could trigger a significant 3.6% upswing in labor productivity during RMK-12 (Unit Perancang Ekonomi 2021). Additionally, a statement from CIDB on March 21, 2021, underscores an impressive 60% surge in construction productivity during the CITP phase, with intentions to sustain this momentum through the Construction 4.0 Strategic Plan (CSP) spanning from 2021 to 2025 (CIDB Strategic Plan 2021–2025).

Amid the realm of the Construction 4.0 revolution, BIM emerges as a cornerstone technology, with the potential to heighten productivity, safety, and sustainability across the construction domain. BIM is one of the pivotal technologies among the dozen identified (Construction Industry Development Board Malaysia 2020). For the Public Works Department (JKR), tasked with overseeing public construction ventures, the integration of BIM since 2007 has been pivotal in enhancing project outcomes. This strategic trajectory continues with the JKR Strategic Plan 2021–2025, mandating BIM implementation in projects exceeding RM 10 million, with an envisaged annual increment of 10% in its application.

Within the domain of construction projects, BIM functions as a robust safeguard against variation orders (VO), often arising from conflicts among diverse project elements. Such orders inadvertently lead to escalated costs. BIM's integration accelerates tasks like drafting and quantity assessment through tools like Autodesk Revit, Civil 3D, and Cost X. Furthermore, BIM augments project visualization through its 3D approach, surpassing conventional 2D methods. Nonetheless, ongoing research underscores that BIM's assimilation is still in its infancy. Obstacles encompass technical adoption hurdles, behavioral constraints, implementation intricacies, management bottlenecks, and a deficiency of digital education, all impeding its seamless integration (Waqar et al. 2023). Additional constraints encompass inadequate financial and technical support, unclear standards and protocols, and the absence of a comprehensive national BIM framework (Porwal and Hewage 2013; Muñoz-La Rivera et al. 2019; Zahrizan et al. 2013).

Localized studies underscore the multitude of challenges faced in Malaysia's BIM journey, including limited awareness, high costs, gradual adaptation, and a scarcity of precise implementation guidelines (Zahrizan et al. 2013; Samad et al. 2018; Al-Ashmori et al. 2020; Idris et al. 2021; Manzoor et al. 2021). Alreshidi et al. (2017) categorize factors influencing BIM efficiency into five dimensions: ICT, socio-organizational, practitioner, BIM process, and financial/legal (Idris et al. 2021) which underscore the pivotal role of the BIM process in its implementation, uncovering a research gap concerning civil and structural engineering consultants in Malaysia. The following are the identified obstacles from these studies:

i. Inadequate Awareness and Limited Knowledge:

   Despite the positive growth in BIM adoption rates, a pervasive challenge remains the insufficient awareness and limited understanding of BIM technology within the Malaysian construction industry. Many construction businesses, especially smaller firms, remain unfamiliar with the transformative potential of BIM. The lack of awareness translates into missed opportunities for improved project efficiency, reduced costs, and enhanced collaboration. Addressing this challenge is crucial to ensuring that stakeholders are well-informed about the advantages of BIM, enabling them to make informed decisions.

ii. Financial Implications and Cost Concerns:

The financial burden associated with BIM implementation has proven to be a significant hurdle. The costs involved in acquiring appropriate BIM software, providing training to staff, and setting up the necessary hardware infrastructure can be perceived as substantial. This is particularly problematic for smaller firms operating on tight budgets. While BIM promises long-term gains in terms of efficiency and accuracy, the initial investment can be daunting. Strategies to mitigate these cost concerns, such as government subsidies or industry-wide collaborations, are vital to fostering broader adoption.

iii. Resistance to Change and Slow Adaptation:

The construction industry is known for its conservative approach to technological change. This resistance to adopt new practices is also observed in the realm of BIM implementation. Established methodologies and processes, deeply ingrained in the industry's culture, can make introducing BIM a challenging endeavor. Convincing stakeholders to shift from familiar practices to the novel BIM approach requires a concerted effort. Education campaigns, case studies showcasing successful implementations, and the endorsement of influential industry players can aid in overcoming this resistance and driving faster adaptation.

iv. Absence of Clear Guidelines and Standards:

The absence of comprehensive and standardized guidelines for BIM implementation has left practitioners grappling with uncertainty. The lack of a uniform framework for BIM processes, Level of Detail (LOD) requirements, data exchange, and collaboration protocols contributes to confusion and inefficiencies. Inconsistencies in understanding these aspects can lead to discrepancies in project execution, affecting overall project quality and timelines. The development and dissemination of clear and well-defined guidelines are pivotal to facilitating smoother BIM integration across projects.

v. Delayed Mandate Implementation:

While the Malaysia BIM Roadmap mandates the use of BIM technology in construction projects from 2020 onward, the relatively recent introduction of this requirement has caused challenges. Organizations, both public and private, might find themselves scrambling to meet the mandate within tight timelines. The delayed implementation of the mandate has limited the lead time available for adequate preparation, training, and adjustments to existing workflows. This situation highlights the importance of gradual and well-planned transitions to prevent rushed adoptions that could lead to suboptimal results.

vi. Complex Nature of Civil and Structural Engineering:

The challenges faced by civil and structural engineering consultants in adopting BIM are amplified by the complex nature of their work. Unlike architectural firms, these consultants must deal with intricate structural analysis, load calculations, and intricate design considerations. Translating these complex aspects into a digital BIM model requires specialized skills and tools, adding another layer of complexity to BIM implementation within this sector.

vii. Gaps in Consultant Service Agreement (CSA):

The Consultant Service Agreement (CSA), a foundational document for government projects, lacks explicit directives for civil and structural engineering consultants regarding BIM. The absence of clear guidelines for Level of Detail (LOD) requirements, data sharing, and the transition of BIM models between different project phases leaves consultants uncertain about best practices. This lack of guidance forces reliance on external modelers or third parties, leading to additional costs and potential loss of control over the project's digital representation.

In summary, the critical obstacles identified within the context of BIM implementation among civil and structural engineering consultants in Malaysia reveal a multifaceted landscape of challenges. Addressing these challenges requires a holistic approach involving awareness campaigns, financial support mechanisms, tailored training programs, industry collaboration to establish standards, and the revision of existing contractual frameworks. By systematically tackling these obstacles, the construction industry in Malaysia can unlock the full potential of BIM, enhancing project efficiency, collaboration, and overall project outcomes.

3 Methodology

This chapter employed a quantitative research design to explore the obstacles faced in implementing Building Information Modeling (BIM) among civil and structural (C&S) engineering consultants. The sampling method utilized was non-probability convenient sampling, involving the selection of respondents based on their accessibility. The participants were drawn from the list of registered C&S engineering consultants on the Board of Engineers Malaysia (BEM) website as of March 31, 2022. This choice of sample was driven by the mandatory registration of engineering consultants in Malaysia under the Registration of Engineers Act 1967, falling into categories such as Body Corporate, Multi-discipline, Sole Proprietor, or Partnership.

The chapter population encompassed 1905 registered engineering consulting organizations in the civil and structural discipline, according to the Board of Engineers Malaysia data. Following the guidelines outlined by Krejcie and Morgan (1970), a sample size of 321 respondents was deemed suitable. Saunders (2007) suggests a minimum sample size of 10 for a pilot study, while Roscoe (1975) recommends sample sizes larger than 30 and less than 500 for most research studies, with a minimum of 30 for each category in subsample studies, several times larger than the number of variables in multivariate research, and as small as 10–20 for simple experimental research with tight controls. Therefore, for the initial phase of this pilot study, a smaller subset of 30 respondents was randomly selected for analysis. The process of reaching out to these respondents involved various means such as email correspondence, phone calls, and an online survey constructed using Google Form.

The survey questionnaire was constructed using a 5-point Likert scale to assess the perceived significance of different viewpoints, with responses ranging from 1 (insignificant) to 5 (very significant). This questionnaire encompassed twenty (20) obstacles related to BIM adoption Table 1, identified through previous research as outlined in the literature review. Thorough statistical analysis was applied to the collected data. The preliminary steps included evaluating the validity and normality of the data, followed by reliability tests to ensure the questionnaire's consistency and stability.

To uncover the pivotal factors influencing the limited adoption of BIM among C&S engineering consultants in Malaysia, a factor analysis was conducted using IBM SPSS software. This involved scrutinizing the component matrix post-rotation. The findings and implications of each method are elaborated upon comprehensively in the results section, providing a clear understanding of their respective purposes and outcomes.

This chapter's focus on C&S engineering consultants engaged in conventional contracts for government projects arises from a notable observation. Among the twenty BIM projects falling under RMKe11, only five have been carried out through conventional contracts, whereas the remaining fifteen have opted for design and build (D&B) contracts. This preference for D&B contracts is driven by challenges

Table 1 BIM obstacle

No.	Factor
B1	Interoperability data exchange
B2	Current software not compatible
B3	High specification of hardware
B4	BIM software difficult
B5	High cost on CAPEX
B6	Financial constraint
B7	Unwillingness to invest
B8	Time barrier of BIM
B9	Lack of BIM standard
B10	Lack of BIM best practice
B11	No BIM workflow
B12	BIM coordination difficult
B13	Lack of BIM demand
B14	No legal contractual of BIM
B15	Lack of BIM incentive
B16	Lack of BIM clear policy
B17	Lack of BIM knowledge
B18	Lack of BIM experience
B19	Reluctance to change
B20	Lack of BIM training

stemming from unclear BIM processes and guidelines. In this chapter, Case Study A is one of five case studies that is a pioneer project that uses BIM implementation from design until the construction stage used as a sample project to obtain process BIM in the C&S design stage.

In the context of D&B contracts, the successful implementation of the BIM method is facilitated by the main contractor's comprehensive involvement in BIM processes, spanning from project inception to completion. This contractor assumes responsibility for managing the conceptual design at Level of Details (LOD) 100, progressing through to the publication of the as-built model at LOD 500. Notably, the D&B contractor possesses the necessary insights into BIM-related costs and can seamlessly integrate them into the overall project budget. In contrast, conventional contracts lack the provision for additional BIM-related expenses within the consultant fee structure.

Compounded by insufficient skills, limited experience, and an absence of well-defined BIM processes and frameworks, C&S engineering consultants often opt to engage external modelers instead of conducting BIM tasks internally. This approach, however, tends to diminish their consultant fees. Given this backdrop, the present research aims to construct a robust framework and systematic process that empowers C&S engineering consultants to effectively implement BIM within the context of conventional tenders for government projects.

This research endeavor will encompass the entire project lifecycle, commencing from the planning phase and extending through the project's handover to the client. It will specifically address the various stages outlined in the Consultancy Service Agreement (CSA) set forth by the Kementerian Kewangan Malaysia (KKM). These stages include the Preliminary, Design, Tender, Construction, and Defect Liability Period phases.

4 Survey Result

4.1 Reliability Assessment

To gage the internal consistency of the BIM obstacles studied, a reliability test using Cronbach's coefficient was performed. The outcomes of this test are presented in Table 2. The results from the modified factors obtained through the rotated component matrix indicate that the factors, as listed in Table 5, are not only acceptable but also reliable for inclusion in the paper of BIM obstacles. The Cronbach's Alpha value, at 0.91 (> 0.7), reflects the respondents' consensus on these factors.

Table 2 Reliability test of BIM obstacle factor for C&S engineering consultant

Cronbach's alpha	N of items
0.910	20

Table 3 Mean score ranking of obstacles factor

BIM Obstacle	Mean	Ranking	SD
B1	3.63	10	1.033
B2	3.30	11	1.264
B3	3.97	4	1.066
B4	3.67	4	0.884
B5	4.10	2	0.712
B6	3.83	6	0.874
B7	3.67	9	0.959
B8	3.87	5	0.860
B9	3.97	4	0.964
B10	4.20	1	0.805
B11	4.07	3	0.907
B12	3.83	6	0.913
B13	3.87	5	0.860
B14	3.83	6	0.834
B15	3.77	8	0.935
B16	3.80	7	0.805
B17	4.07	3	0.785
B18	4.10	2	0.662
B19	3.77	8	0.728
B20	4.07	3	0.691

4.2 *Ranking of BIM Obstacle Mean Scores*

The mean scores for individual BIM obstacles were computed using descriptive statistics. The objective was to assess whether any BIM obstacles had mean scores falling below 3. Notably, both the lowest observed mean (3.30) and the highest (4.20) are above 3, confirming the accuracy of the data's mean and the absence of anomalies. A detailed presentation of the mean score rankings for BIM obstacles among C&S engineering consultants, along with the mean, standard deviation (SD), and rankings, is given in Table 3.

4.3 *Validation Assessment*

Validation of the specific parameters under scrutiny for BIM obstacle decision-making was conducted using the Kaiser–Mayer–Olkin (KMO) and Bartlett tests. The KMO values, all exceeding 0.5, indicate compliance with the essential conditions for factor analysis, as recommended by Kaiser and Rice (1948). The validation

Table 4 Barlett and KMO tests on BIM obstacles for C&S engineering consultants

Independent variable	Bartlett	KMO
BIM obstacles	0.00	0.571

Table 5 Normality tests for BIM obstacles for C&S engineering consultant

BIM obstacle	Skewness		Kurtosis	
Variable	Statistic	Std. E	Statistic	Std. E
Process	– 0.473	0.427	– 0.646	0.833
Knowledge	0.350	0.427	– 0.721	0.833
Policy	0.371	0.427	– 0.639	0.833
Cost	– 0.544	0.427	– 0.544	0.833
Technology	– 0.471	0.427	– 0.174	0.833

* *Std. E* standard error

tests' outcomes, including Barlett test of sphericity and KMO values, are presented in Table 4. Notably, all KMO values exceed 0.5, and a significant Barlett value of 0.000 ($p < 0.001$) is observed for all BIM obstacles.

4.4 Assessment of Normality

To establish the normality of variables in this pilot study, an evaluation of data distributions was performed for skewness and kurtosis. The standard error values for skewness and kurtosis, depicted in Table 5, are 0.427 and 0.833, respectively. These values are indicative of good standard errors (< 1.0). This validates the assumption of normality for each variable, confirming the normal distribution of all BIM obstacle factors.

4.5 Factor Analysis

Factor analysis was applied to explore correlations among a set of variables, aiming to reduce data complexity using a smaller set of factors or components. This analysis, as described by Pallant (2013), also helps establish construct validity by identifying the most suitable items or factors for each dimension. With twenty (20) factors initially considered, the analysis revealed underlying relationships that required factor rotation. Each factor was classified within one of five (5) specific obstacle groups: process, knowledge, policy, cost, and technology. The group classification was based on the common influence obstacle in BIM by previous local researchers (Zahrizan et al. 2013; Samad et al. 2018; Al-Ashmori et al. 2020; Idris et al. 2021; Manzoor et al.

2021). The component matrix after rotation, with factor loadings exceeding 0.5 as recommended by Kaiser and Rice (1948), is presented in Table 6.

The obstacles were effectively categorized into these five groups. First, there are obstacles tied to the BIM implementation process itself, involving factors like the absence of BIM standards and challenges in coordinating BIM activities. Second, the knowledge and experience group encompasses factors such as insufficient BIM knowledge and lack of incentives for BIM adoption. The third group pertains to policy-related obstacles, including reluctance to change and software usability challenges. The fourth group highlights cost factors like financial constraints and high capital expenditure. Lastly, the technology group covers factors associated with technical aspects of BIM adoption, such as data interoperability issues and software compatibility.

This categorization sheds light on the diverse challenges C&S engineering consultants faced during BIM implementation. It provides a foundation for informed

Table 6 Rotated component factor analysis for BIM obstacle factors for C&S engineering consultant

Factor	Group				
	1	2	3	4	5
B9—lack of BIM standard	0.856				
B10—lack of BIM best practice	0.846				
B11—no BIM workflow	0.844				
B8—time barrier of BIM	0.793				
B12—BIM coordination difficult	0.789				
B18—lack of BIM experience		0.831			
B17—lack of BIM knowledge		0.775			
B20—lack of BIM Training		0.732			
B16—lack of BIM clear policy		0.698			
B15—lack of BIM incentive		0.610			
B19—reluctance to change			0.764		
B4—BIM software difficult			0.746		
B13—lack of BIM demand			0.735		
B14—no legal contractual of BIM			0.579	0.503	
B7—unwillingness to invest				0.743	
B6—financial constraint				0.724	
B5—high cost on CAPEX				0.720	
B1—interoperability data exchange					0.851
B3—high specification of hardware					0.799
B2—current software not compatible					0.756

1 process, *2* knowledge, *3* policy, *4* cost, *5* technology

Table 7 Ranking of BIM obstacles based on loaded factor

Ranking	BIM obstacle factors	Loaded factor
1	Lack of BIM standard	0.856
2	Interoperability data exchange	0.851
3	Lack of BIM best practice	0.846
4	No BIM workflow	0.844
5	Lack of BIM experience	0.831

strategies to overcome these hurdles and encourage successful BIM adoption in the Malaysian construction industry.

4.6 BIM Obstacle Ranking

The summary of findings regarding the BIM obstacle factors, derived from Table 6, is presented in Table 7. The factors are ranked based on the five highest factor loadings, offering insight into their significance.

4.7 Design Stage C&S BIM Process

Based on observation and preliminary interview session with Case Study A C&S engineering consultant in BIM process in design stage as shown in Fig. 1, a strategic leap was taken to foster seamless multi-disciplinary consultant coordination and optimize document storage through the establishment of a Common Data Environment (CDE) platform. Autodesk's BIM360, a paid database, assumed a pivotal role in this endeavor. At the helm, the architect consultant, often the lead consultant, led the charge in creating the project CDE. This platform democratizes access for all stakeholders, ranging from consultants and clients to JKR. Within this ecosystem, the architect consultant shoulders the responsibility of constructing model LOD 100–200.

The initial LOD100 conceptual model paves the way for the subsequent LOD200 preliminary design model. C&S engineering consultants then embark on a journey to extract the LOD200 Architectural Model from the CDE, which serves as a foundational reference for the creation of the LOD200 Civil Model and LOD200 Structure Model. The LOD200 Civil Model, serving as an infrastructure blueprint encompassing earthwork, road and drainage, water, and sewerage reticulation, is crafted using Civil 3D BIM software.

For Case Study A, the earthwork model was further augmented with a retaining wall component. Rigorous design inputs for each LOD 200 within the Civil Model are facilitated either through Excel spreadsheets or dedicated civil engineering software. This parallel design process ensures integrity. The final 3D model, rigorously

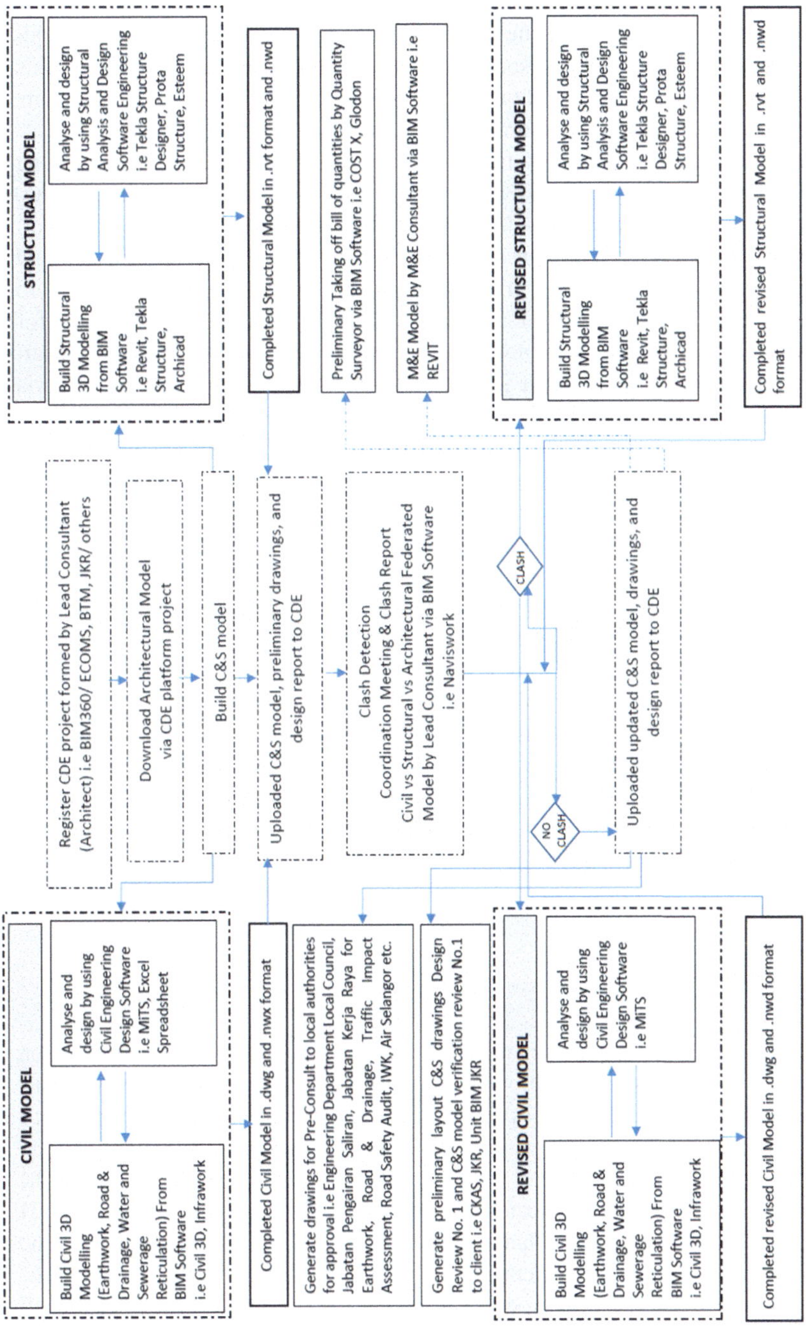

Fig. 1 BIM process in design stage

vetted against the JKR Integrity Model and Quality Checklist, finds a home in formats like .dwg (Civil 3D) and .nwc (Navisworks) as it graces the digital halls of BIM360. Simultaneously, the LOD 200 Structural Model undergoes similar refinement, capitalizing on engineering software such as Tekla Structure Designer and Autodesk's Revit. Architect-led Clash Detection Meetings, aided by Navisworks, lay bare any inconsistencies between architectural and structural models.

In instances necessitating modifications, technical consultations ensue, driving iterative refinement before models are re-uploaded to BIM360. With LOD200 models at hand, Clash Detection Meetings and subsequent reports shine a spotlight on disparities. Technical consultations pave the way for iterative design and modeling revisions, culminating in the triumphant reentry of refined models into BIM360. Quantity Surveyor (QS) consultants seize upon the LOD200 models for cost estimation through Cost X software, while Mechanical and Electrical engineering consultants embark on a journey to further embellish these models via Autodesk's Revit software. BIM360 emerges as a trove of working drawings, a valuable resource for pre-consultation with local authorities like the Engineering Department of Majlis Perbandaran Selayang and Jabatan Pengairan dan Saliran for Air Selangor submissions.

Subsequent to these phases, Cawangan Kejuruteraan Awam dan Struktur (CKAS), JKR, and Unit BIM, JKR perform comprehensive design reviews and meticulous model quality checks on the LOD 200 Civil and Structural Models. Design Stage I and II unfurl as a continuum, involving the development of the LOD 300 model. The methodological ethos remains akin to that of the preliminary stage, encompassing design, modeling, submission, quality checks, design reviews, and more. The orchestration of multi-disciplinary models and clash reports persists through Navisworks, as augmented versions make their home within BIM360. Should incongruities between disciplines endure, iterative design refinements rerun, serving as the crucible for conflict resolution. In the final reckoning, LOD300 models, poised for tendering, are entrusted to designated BIM360 folders.

5 Conclusion

BIM is often acclaimed as a powerful driver of innovation in the construction sector. However, a meticulous examination of the pilot study data underscores a robust piece of evidence: the BIM process itself is a prominent impediment faced by C&S engineering consultants when engaging with BIM. The responsibility rests with organizations, particularly those within the domain of C&S engineering, to decode the mechanics of effective BIM project implementation. With a significant pool of 321 targeted C&S engineering consultants registered under BEM, it is prudent to extend this investigation with a more extensive sample. Future research endeavors hold the potential to craft a tailored BIM process and framework for C&S engineering consultants, ensuring self-sufficiency within their existing operational frameworks, without relying on external modeling entities.

The establishment of BIMIF along with the devised BIM procedures tailored for C&S engineering consultants holds the potential to offer a substantial contribution. It can play a pivotal role in enhancing comprehension of the BIM implementation process within conventional consultant contract projects for government building construction on a broader scale. Furthermore, this chapter can serve as a valuable reference for private projects and academic researchers, thereby indirectly fostering a positive impact. As BIM becomes more integrated into the present C&S work processes, it can potentially catalyze an improvement in the overall BIM adoption rate in Malaysia moving forward.

Acknowledgements The authors would like to express their appreciation to the MTC Engineering Consultancy Sdn. Bhd. (MTCEC) as the C&S engineering consultant for Case Study A for their contribution in sharing the information.

References

Al-Ashmori YY, Othman I, Rahmawati Y, Amran YHM (2020) Civil engineering BIM benefits and its influence on the BIM implementation in Malaysia, vol 11, pp 1013–1019. https://doi.org/10.1016/j.asej.2020.02.002

Alreshidi E, Mourshed M, Rezgui Y (2017) Factors for effective BIM governance. J Build Eng 10:89–101. https://doi.org/10.1016/j.jobe.2017.02.006

Construction Industry Development Board Malaysia (2020) Construction 4.0 strategic plan (2021–2025): next revolution of the Malaysian construction industry. Construction Industry Development Board Malaysia (CIDB)

Idris O, Al-Ashmori YY, Rahmawati Y, Mugahed Amran YH, Al-Bared MAM (2021) The level of building information modelling (BIM) implementation in Malaysia. Ain Shams Eng J 12(0):455–463. https://doi.org/10.1016/j.asej.2020.04.007

Kaiser HF, Rice J (1948) Educational and psychological measurement. In: Educational and psychological measurement

Krejcie RV, Morgan DW (1970) Educational and psychological measurement, vol 38

Manzoor B, Othman I, Gardezi SSS, Altan H, Abdalla SB (2021) BIM-based research framework for sustainable building projects: a strategy for mitigating BIM implementation barriers. Appl Sci (Switz) 11(12). https://doi.org/10.3390/app11125397

Muñoz-La Rivera F, Vielma JC, Herrera RF, Carvallo J (2019) Methodology for building information modeling (BIM) implementation in structural engineering companies (SECs). Adv Civ Eng. https://doi.org/10.1155/2019/8452461

Pallant J (2013) SPSS survival manual: a step by step guide to data analysis using IBM SPSS. Aust N Z J Public Health 37(6):597–598. https://doi.org/10.1111/1753-6405.12166

Porwal A, Hewage KN (2013) Building information modeling (BIM) partnering framework for public construction projects. Autom Constr 31:204–214. https://doi.org/10.1016/j.autcon.2012.12.004

Roscoe JT (1975) Fundamental research statistics for the behavioral sciences, 2nd edn. Holt Rinehart and Winston, New York

Samad SA, Harun AN, Nawi MNM, Haron NA (2018) The potential use of BIM through an electronic submission: a preliminary study. Malays Constr Res J 3(Special Issue 1):82–96

Saunders MN (2007) Research methods for business students, 5/e. Pearson Education India

Unit Perancang Ekonomi JPM (2021) Rancangan Malaysia Kedua Belas. Unit Perancang Ekonomi, Jabatan Perdana Menteri

Waqar et al (2023) Barriers to building information modeling (BIM) deployment in small construction projects: Malaysian construction industry. Sustainability (Switz) 15(3). https://doi.org/10.3390/su15032477

Yasser Yahya A-A, Idris O, Rahmawati Y, Amran YHM, Sabah SHA, Rafindadi ADU, Mikić M (2020) BIM benefits and its influence on the BIM implementation in Malaysia. Ain Shams Eng J 11(4):1013–1019. https://doi.org/10.1016/j.asej.2020.02.002

Zahrizan Z, Ali NM, Haron AT, Marshall-Ponting A, Hamid ZA (2013) Exploring the adoption of building information modelling (BIM) in the Malaysian construction industry: a qualitative approach. IJRET Int J Res Eng Technol. EISSN: eISSN pISS, 2319-1163

Investigation on Factors Contributing to Public Project Delays in Kota Kinabalu Based on Human Errors Using the RII Method: A Pilot Study

Nazaruddin Abdul Taha, Hazwani Halid, Siti Jahara Matlan, Mohamad Shakri Mohmad Shariff, and Che Maznah Mat Isa

Abstract Delays are a common occurrence in the construction industry across the globe. The purpose of this study was to investigate the primary causes of construction delays and examine methods for minimizing delays in public projects in Kota Kinabalu using the RII method. Based on a review of secondary data, four categories were identified as contributing to the primary causes of construction delays, along with 50 causes of delay and 17 techniques for minimizing construction delays. Using IBM SPSS Statistics 26, the 140-survey data were analyzed and ranked by Relative Importance Index for each cause. The results indicate that poor communication and coordination, slowness in the decision-making process, unclear and insufficient detail in drawings, land and site problems, poor management and supervision of the site by the contractor, late payment, poor communication and coordination, and a delay in obtaining municipal permits were the primary causes of the delay. Proper planning prior to the start of the project is the most effective method for minimizing delay; regular coordination between the parties involved is required; and approved and confirmed design concepts, construction drawings, material selection, and logistics management must be in place well before construction.

Keywords Delay · Public project · Mitigation · Causes · Construction industry · Relative importance index

N. A. Taha · H. Halid · S. J. Matlan
Faculty of Engineering, Universiti Malaysia Sabah, 88400 Kota Kinabalu, Sabah, Malaysia

M. S. M. Shariff
INTI International University, 71800 Nilai, Negeri Sembilan, Malaysia

C. M. M. Isa (✉)
Civil Engineering Centre of Studies, Universiti Teknologi MARA Pulau Pinang Branch, Permatang Pauh Campus, 13500 Permatang Pauh, Pulau Pinang, Malaysia
e-mail: chema982@uitm.edu.my

E. Shaffie et al. (eds.), *Sustainable Green Infrastructure*,
https://doi.org/10.1007/978-981-96-1486-8_12

1 Introduction

Delay can be defined as the amount of time it takes to complete an undertaking relative to the planned or contractual timeline (Alajmi and Ahmed Memon 2022; Alaghbari et al. 2007). It occurs when the contract's progress falls behind schedule. This may be caused by any party to the contract and may be the direct result of one or more events. Thus, to accomplish economic growth, the authority increased construction spending (Alkhalifah et al. 2023). The enormous quantity of money used to build highways, railroads, hospitals, schools, residential and non-residential buildings, and airports. If these projects are delayed, the country's economic development will slow, leading to an increase in government expenditures. Delay has substantial effects on the completion cost and duration of a construction project. If the root causes of delays are identified, they can be reduced. Al-Tami (2015), Bekr (2018), Dehdasht et al. (2022) and Jadhav (2018) investigated the top ten causes of construction delays in Malaysia. Based on the data analysis, there are ten most significant causes were improper contractor planning, poor contractor project execution, insufficient contractor experience, insufficient client accountancy and expenses for work done, difficulties with subcontractors, material shortages, labor demand, equipment availability and failure, poor communication among stakeholders, and construction process errors (Kamaruddeen et al. 2020; Khan et al. 2014).

Based on a previous research's findings and discussions, it is recommended that financial and technical support be a very necessary and urgent step for construction investment, as the results of the analysis indicate that financial issues are the primary cause of construction project delays by Memon et al. (2012). However, other researchers suggested that contractors pay more attention to weather conditions, trust between the parties is essential for any construction project, and contractors should pay more attention to the five phases (Nassar 2016; Nor et al. 2023; Othman and Ismail 2014). In addition, other previous research suggested effective site management and supervision, an effective and strategic schedule, clear information and communication skills, collaborative construction work, proper planning and scheduling of projects, frequent coordination between the parties, a comprehensive and timely design, the use of appropriate construction methods, and a genuine emphasis on past experiences (Oyegoke and Al Kiyumi 2017; Rahman et al. 2020; Sambasivan and Soon 2007; Suleiman 2022). Consequently, the purpose of this study was to investigate the primary causes of construction delays and examine methods for minimizing delays in public projects in Kota Kinabalu using the RII method.

2 Methodology

To achieve the objectives of this research, secondary data from previous studies, journals, reports, and online articles were utilized. The primary data came from a survey questionnaire. The questionnaire devised a five-point ordinal Likert scale.

Table 1 Cronbach's alpha reliability test for all respondents

Section		Cronbach alpha	Number of questions
Section B: public project delay factor	Contractor's related	0.927	22
	Client's related	0.934	10
	Consultant's related	0.945	11
	External factor related	0.874	7
Section C	Method on minimizing delay	0.933	17

The scale ranges from 1 to 5 based on the extent of contribution. The data consists of 50 construction industry delay causes and 17 strategies for minimizing delay causes. There was a total of 140 respondents in the survey, including contractors, clients, and consultants. The Cronbach's Alpha analysis results in Table 1 show that all the items in the sections are correlated. Internal consistency exists, as do the collections of items that comprise each section. This is the case because the Cronbach's Alpha coefficient for each question set is exceedingly close to 1. Consequently, this chapter concluded that the test and questions were reliable. Using the rank system, the Relative Important Index equation was calculated in SPSS software to analyze the data in this study. The data comprises causes related to four groups of delay causes and methods for minimizing delay, considering contractors', clients', and consultants' perspectives.

2.1 Calculation of Relative Important Index

Many previous studies applied RII to determine the ranking of different factors from the point of view of contractors, clients, and consultants (Alkhalifah et al. 2023; Al-Tami 2015; Sambasivan and Soon 2007; Taofeeq et al. 2019; Yap et al. 2021). Similarly, with this study, the Relative Important Index (RII) is computed by SPSS software as:

$$\mathrm{RII} = \frac{\sum W}{A \times N} \tag{1}$$

where

W: Weighting given by the respondents to each factor (ranging 1–5, respectively),
A: The highest weight (5 in this case),
N: The total number of respondents.

The Relative Important Index (RII) ranged from 0 to 1 (with 0 being excluded); the higher the RII value, the more essential the cause or method for minimizing delay. Using RII, this study ranks the various causes and methods for minimizing delay in public initiatives in Kota Kinabalu. These rankings have enabled an evaluation and comparison of the significance of the factors considered by the three respondent categories (clients, consultants, and contractors). The RII of each individual cause considered by all respondents was used to evaluate the general and overall rankings to provide a holistic view of the causes of construction delays in Kota Kinabalu's public construction projects. Kota Kinabalu's public initiatives will minimize delays by employing the same procedure to determine the optimal solution.

3 Result and Discussion

Table 2 presents the demographics of the respondents. The primary data for this study were gathered from sections B and C of the questionnaire and analyzed by considering the contractor's, client's, and consultant's perceptions. This study was able to determine the most influential contributing factors to delays in the public construction sector in Kota Kinabalu based on the rankings assigned to each cause of delay and the best method to minimize time overruns in Kota Kinabalu.

Table 3 presents the result of the survey analysis of causes of contractor-related problems. The results were divided into four perceptions: contractor, client, consultant, and overall perceptions. The causes of delay were ranked based on the Relative Important Index value, as mentioned in the previous chapter. Referring to Table 3, poor communication and coordination have the highest value of RII (0.771) and rank in the first place among the causes of contractor-related problems.

The reason for the result was also highly contributed by the client's participants, as poor communication and coordination are the most important factors in the client's perception such in shown in Fig. 1. Based on the overall perception from contractors, clients, and consultants, poor management and supervision of the site by contractors have the second highest RII value (0.763). Both contractor and client respondents agreed that poor management and supervision of the site by the contractor was a strong factor that caused time overruns in public projects in Kota Kinabalu, while consultants believe that the factor is the main reason for contractor-related delays as it has the highest value of RII (0.787). Conflicts between the contractor and other parties are located at the third rank of RII (0.754), considering overall perceptions. Contractor participants agreed that shortage of labor is the most important reason for time overruns in public project delays in Kota Kinabalu, which has the highest value of RII (0.808). Shortage of construction materials in the market and conflicts between the contractor and other parties are the two main reasons for delay in contractor-related work with a RII value of 0.775. Based on the client's perceptions, other than conflicts between the contractor and other parties, inadequate planning and scheduling also ranked in second place, as the value of RII is the same (0.750). As for consultant perception, subcontractor delays were the third reason that led to a time overrun in a public project in Kota Kinabalu.

Referring to Table 4, the overall perceptions reveal that late payment is the major problem causing delay in public projects for client-related factors, as it has the highest RII (0.780) value among all causes. The result was mostly contributed by consultant

Table 2 Demographic of respondent

Demographic characteristic	Frequency	Percent
Respondent organization type		
Public	32	45.7
Private	38	54.3
Respondent position in organization		
Director	9	12.9
Designer	2	2.9
Project manager	2	2.9
Engineer	36	51.4
Other	21	30
Number of year respondents has experience in construction industry		
Less than 5 years	36	51.4
Between 5 and 10 years	15	21.4
Between 11 and 20 years	14	20.0
More than 20 years	5	7.1
Number of project delays which faced by respondents		
1–5 projects	52	74.3
6–10 projects	13	18.6
More than 10 projects	5	7.1
Types of projects with the most frequent delays		
School	10	14.3
Hospital	8	11.4
Other	14	20.0
Housing estate	15	21.4
Public facilities	23	32.9
% of delay from estimated project duration		
1–5%	12	17.1
6–10%	23	32.9
11–15%	9	12.9
16–20%	6	32.9
More than 20%	20	28.6

and contractor perception, as it is also the major problem among all causes. Based on the overall perception, late revision and approval of design documents by the owner contributed to the second main reason for delay in public project Kota Kinabalu, with a RII value of 0.746. This outcome was mostly contributed by consultant perceptions, as it also has the second highest value of RII among all causes. Next, as for overall perception, agree that change orders by the client during construction should be in the third rank of causes that lead to cost overruns in public projects. Both client and

Table 3 Result of factors of contractor-related delays

Factor	Contractor		Client		Consultant		Overall	
	RII	Rank	RII	Rank	RII	Rank	RII	Rank
Poor communication and coordination	0.742	6	0.813	1	0.773	2	0.771	1
Slow to give instructions	0.750	4	0.700	7	0.667	19	0.703	10
Conflicts between of the contractor and other parties	0.775	2	0.750	2	0.740	6	0.754	3
Poor management and supervision of the site by contractor	0.750	4	0.738	4	0.787	1	0.763	2
Subcontractor delays work	0.717	10	0.700	7	0.760	3	0.731	6
Inadequate planning and scheduling	0.742	6	0.750	2	0.720	9	0.734	5
Transportation delays	0.625	20	0.725	5	0.673	16	0.669	19
Contract modifications	0.617	22	0.688	10	0.627	21	0.637	22
Labor dispute and strike	0.708	13	0.650	19	0.627	21	0.660	20
Problem with subcontractors	0.742	6	0.688	10	0.700	11	0.711	8
Shortage labor	0.808	1	0.663	16	0.680	14	0.720	7
Discrepancies in design documents	0.692	16	0.713	6	0.687	13	0.694	15
Equipment availability and failure	0.717	10	0.663	16	0.680	14	0.689	16
Late issues of EOT	0.625	20	0.600	22	0.673	16	0.640	21
Lack of contractor experience	0.708	13	0.675	14	0.673	16	0.686	17
Rework for errors during construction	0.725	9	0.638	21	0.720	9	0.703	10
Inadequate contractor work	0.675	19	0.675	14	0.733	8	0.700	13
Poor qualification of the contractor's technical staff	0.683	17	0.650	19	0.747	5	0.703	10
Delays in site mobilization	0.700	15	0.700	7	0.700	11	0.700	13
Incompetent site manager and supervision	0.683	17	0.688	10	0.740	6	0.709	9

(continued)

Table 3 (continued)

Factor	Contractor		Client		Consultant		Overall	
	RII	Rank	RII	Rank	RII	Rank	RII	Rank
Shortage of construction materials in market	0.775	2	0.688	10	0.753	4	0.746	4
Changes in material types during construction	0.717	10	0.663	16	0.640	20	0.671	18

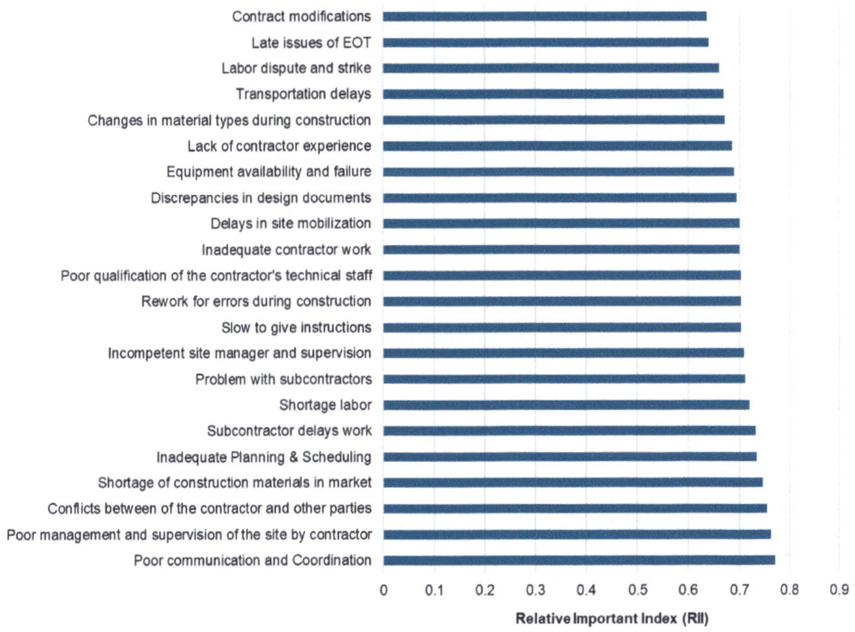

Fig. 1 Overall factors of contractor versus RII

consultant participants agree that change orders by the client during construction are the top three factors that contribute to delay, ranking second for the client and third for the consultant with RIIs of 0.650 and 0.767, respectively. Aside from late payment, contractor participants also agree that financial difficulties from clients contributed the most to construction delays, as it has the highest value of RII (0.825) together with late payment. The contractor also put slowness in the decision-making process in the top three most contributing factors to delay in the public project Kota Kinabalu with an RII value of 0.800. While client participants believe that slowness in the decision-making process is the major cause of time overruns as it is the highest value of RII (0.663). Next, client participants agreed that change orders by the client during construction and poor communications and coordination are the second-most important causes with a RII of 0.650. Figure 2 indicates the overall rank of RII for the factors of client.

Referring to Table 5, the overall perceptions reveal that poor communication and coordination, as well as delay in approving major changes in the scope of work, are the main reasons why delays occur in the public construction industry in Kota Kinabalu, with a value of RII of 0.709. Consultant participants with an RII of 0.700 agree that poor communication and coordination are the most important factors that contribute to delay. According to the different opinions of client participants, at rank 11, with a RII of 0.588, poor communication and coordination are the least likely factors to contribute to delays in public projects in Kota Kinabalu. Based on contractor perception, delay in approving major changes in the scope of work is the second-most important cause with a RII of 0.792, while client participants rank it third with a RII of 0.713. From overall perception, delays in the production of design documents are the third most important cause, with a RII of 0.706. Both contractor and client respondents agree that delays in the production of design documents are the second-most important factor. Mistakes and differences in design documents were

Table 4 Result of factors of clients related delays

Factor	Contractor		Client		Consultant		Overall	
	RII	Rank	RII	Rank	RII	Rank	RII	Rank
Change orders by the client during construction	0.758	7	0.650	2	0.767	3	0.737	3
Late revision and approval of design documents by owner	0.792	4	0.613	4	0.780	2	0.746	2
Financial difficulties from clients	0.825	1	0.613	4	0.733	4	0.737	3
Late payment	0.825	1	0.600	6	0.840	1	0.780	1
Delay in approval of shop drawing and sample materials	0.775	5	0.600	6	0.707	5	0.706	7
Poor communications and coordination	0.767	6	0.650	2	0.700	6	0.711	6
Conflicts between the joint ownership of the project	0.733	8	0.525	10	0.600	10	0.629	10
Delay to furnish and deliver the site	0.733	8	0.588	8	0.660	8	0.669	8
Slowness in decision-making process	0.800	3	0.663	1	0.680	7	0.717	5
Suspension of work by owner	0.733	8	0.575	9	0.653	9	0.663	9

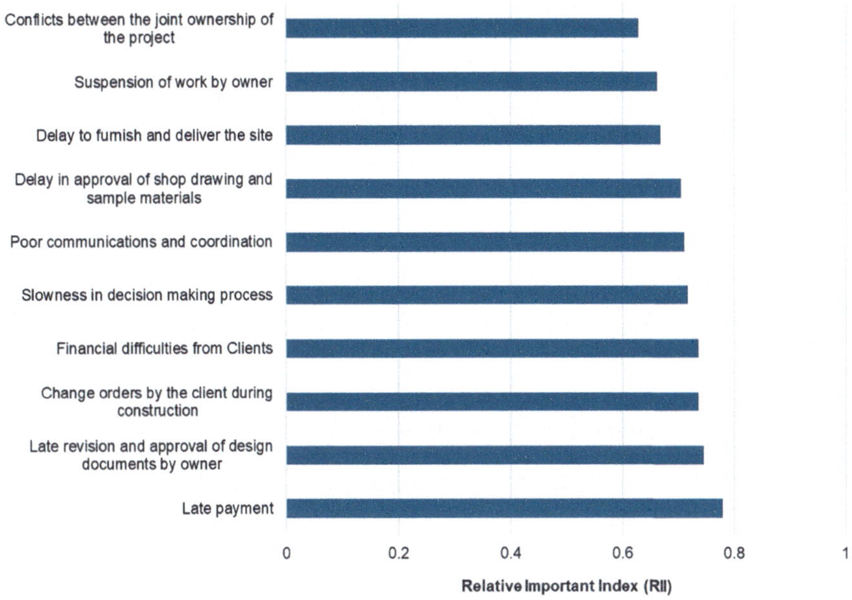

Fig. 2 Overall factors of client versus RII

ranked first by the contractor's perception with an RII value of 0.800, while clients agreed that the major problem that contributed to delay was unclear and insufficiently detailed drawings. Figure 3 shows the comparison for overall factors of consultant versus RII.

Referring to Table 6, the overall perceptions reveal that land and site problems are the main factor with the highest value of RII (0.794) for delay. Figure 4 presents the comparison for all overall factors of external. The outcome was contributed to the most by the contractor and client, as land and site problems ranked first in each of the group factors. According to a slightly different opinion by the consultant, land and site problems ranked third among all external factors (0.793). Delay in obtaining permits from municipalities ranked second based on the overall perception with a RII of 0.774 and first on the consultant's perception with a RII of 0.813. The changes in government regulations and laws ranked third by overall perception with a RII of 0.793 and second by consultant participants with a RII of 0.800. Based on the contractor's view, fluctuation in material prices is the second-most important cause of external factor delay with a RII of 0.817, followed by traffic control and restrictions at the job site in third place with a RII of 0.800.

According to Table 7, both contractor and client participants agree that proper planning before the start of the project is the best method to reduce delay in public projects in Kota Kinabalu, with the highest RII of 0.867 and 0.875, respectively.

The outcome influences the most overall perceptions, as proper planning before the start of the project also ranked first among all methods with a RII of 0.857 such

Table 5 Result of factors of consultant-related delays

Factor	Contractor		Client		Consultant		Overall	
	RII	Rank	RII	Rank	RII	Rank	RII	Rank
Poor communication and coordination	0.783	4	0.588	11	0.700	1	0.709	1
Inadequate consultant experience	0.742	9	0.613	10	0.573	8	0.640	10
Mistakes and differences in design documents	0.800	1	0.700	4	0.607	7	0.694	6
Delays in the production of design documents	0.792	2	0.725	2	0.627	6	0.706	3
Unclear and insufficient detail in drawings	0.758	7	0.738	1	0.647	3	0.703	4
Insufficient date of collection and survey prior to design	0.767	5	0.700	4	0.640	4	0.697	5
Lack of supervision	0.767	5	0.638	8	0.667	2	0.694	6
Delay in approving major changes in the scope of work	0.792	2	0.713	3	0.640	4	0.709	1
Unuse of advanced engineering design software	0.708	11	0.650	7	0.520	11	0.614	11
Poor estimation of quantity material	0.725	10	0.700	4	0.553	10	0.646	8
Conflicts between of the consultant and other parties	0.750	8	0.625	9	0.567	9	0.643	9

as shown in Fig. 5. Aside from that, coordination between the parties involved is required on a regular basis and is also the most effective way to reduce delays in public projects by overall perceptions. Consultant participants also agree that coordination between the parties involved is required on a regular basis and is the number one solution to reduce time overruns. Among contractors' group of factors, improvement in communications and the payment matter should be constantly followed up with the client, as it is the second-most effective way with a RII of 0.842. Overall perception results revealed that approved and confirmed design concepts, construction drawings, selection of materials, logistics management, etc., well before construction were the top three best ways to mitigate delay in public projects with a RII of 0.851. Both client and consultant participants also agreed that the method was the second-most effective, with RIIs of 0.863 and 0.860, respectively.

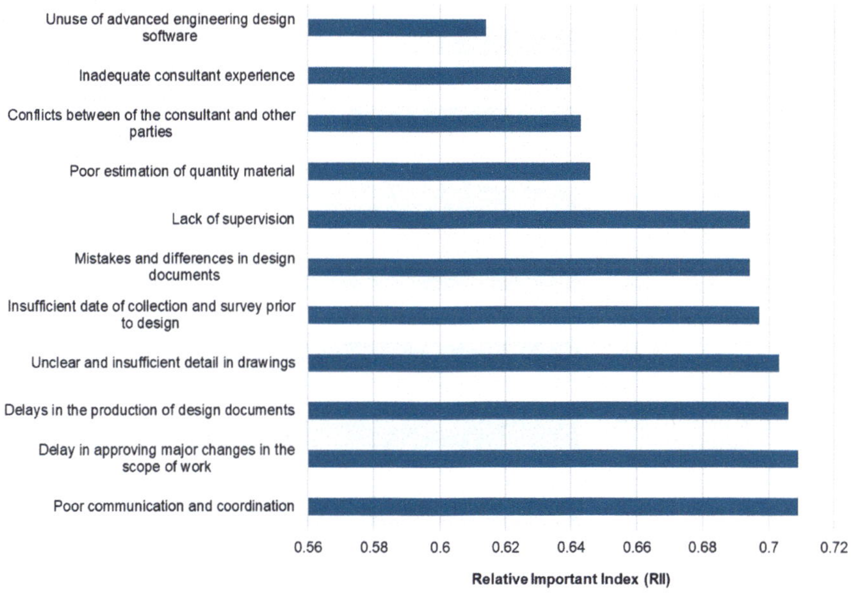

Fig. 3 Overall factors of consultant versus RII

Table 6 Result of external factor-related delays

Factor	Contractor		Client		Consultant		Overall	
	RII	Rank	RII	Rank	RII	Rank	RII	Rank
Land and site problems	0.825	1	0.750	1	0.793	3	0.794	1
Delay in obtaining permits from municipality	0.767	7	0.713	2	0.813	1	0.774	2
Weather conditions effect on construction activities	0.775	5	0.688	4	0.780	4	0.757	4
Changes in government regulations and laws	0.775	5	0.688	4	0.800	2	0.766	3
Traffic control and restriction at job site	0.800	3	0.613	7	0.687	7	0.709	7
Fluctuation in price material	0.817	2	0.663	6	0.720	5	0.740	5
Environmental uncertainty	0.783	4	0.713	2	0.693	6	0.729	6

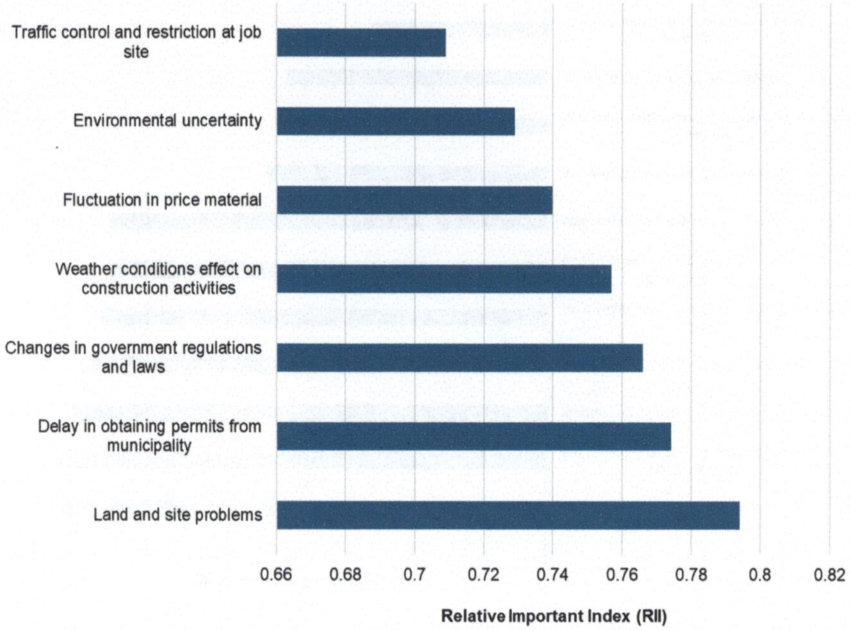

Fig. 4 Overall factors of external versus RII

4 Discussion

4.1 Summary of Discussion

Based on the 48 contractor's perceptions involved in the survey, for the contractor-related group of factors, shortage labor was the major cause of public project delay, followed by conflicts between the contractor and other parties. Next were slow instructions and poor management and supervision of the site by the contractor. Based on client-related factors, financial difficulties from clients and late payments were the most important causes, followed by slowness in the decision-making process. Based on consultant-related factor groups, mistakes and differences in design documents were the most contributing causes of delays in the construction industry, followed by delays in the production of design documents and delays in approving major changes in the scope of work. For the external factor-related problems, land and site problems were the primary causes, followed by fluctuations in material prices and traffic control and restriction at the job site.

Looking from the 32 clients' perceptions who involved in the survey, poor communication and coordination were the most important factors that caused delays in the construction industry, followed by conflicts between the contractor and other parties and inadequate planning and scheduling. For client-related delays, slowness in the decision-making process was the top cause, followed by poor communications and

Table 7 Result of method on minimizing delay

Factor	Contractor		Client		Consultant		Overall	
	RII	Rank	RII	Rank	RII	Rank	RII	Rank
A realistic proper work program must be prepared by the contractor	0.792	10	0.838	4	0.807	13	0.809	9
Proper technical training	0.833	4	0.850	3	0.853	5	0.846	4
A guidelines on mitigating delays should be prepared	0.783	12	0.825	7	0.820	9	0.809	9
The latest progress of the work invoicing submitted must be accompanied by adequate documents	0.767	17	0.775	13	0.813	12	0.789	15
The payment matter should be constantly followed up with the client	0.842	2	0.775	13	0.787	16	0.803	11
Hire a competent contractor to carry out a project	0.783	12	0.838	4	0.860	2	0.829	6
Improves productivity through modern tools and software	0.775	16	0.800	11	0.847	6	0.811	8
Proper planning before the start of the project	0.867	1	0.875	1	0.840	8	0.857	1
Approved and confirmed design concept, construction drawings, selection of materials, logistics management, etc., well before to construction	0.825	5	0.863	2	0.860	2	0.851	3
Improvement in communications	0.842	2	0.825	7	0.860	2	0.846	4
Efficient control and monitoring of on-site workers	0.808	7	0.813	10	0.847	6	0.826	7
New laws and regulations should be always updated to all parties	0.800	9	0.738	16	0.800	14	0.786	16

(continued)

Table 7 (continued)

Factor	Contractor		Client		Consultant		Overall	
	RII	Rank	RII	Rank	RII	Rank	RII	Rank
Client to stick to the original plan and to allocate an adequate budget for any contract modifications	0.792	10	0.825	7	0.800	14	0.800	13
Recommend a "Bonus" scheme to be completed early	0.817	6	0.700	17	0.820	9	0.791	14
Accurate initial cost estimates	0.783	12	0.763	15	0.720	17	0.751	17
Clear information and communication channels	0.783	12	0.800	11	0.820	9	0.803	11
Coordination between the parties involved is required on a regular basis	0.808	7	0.838	4	0.907	1	0.857	1

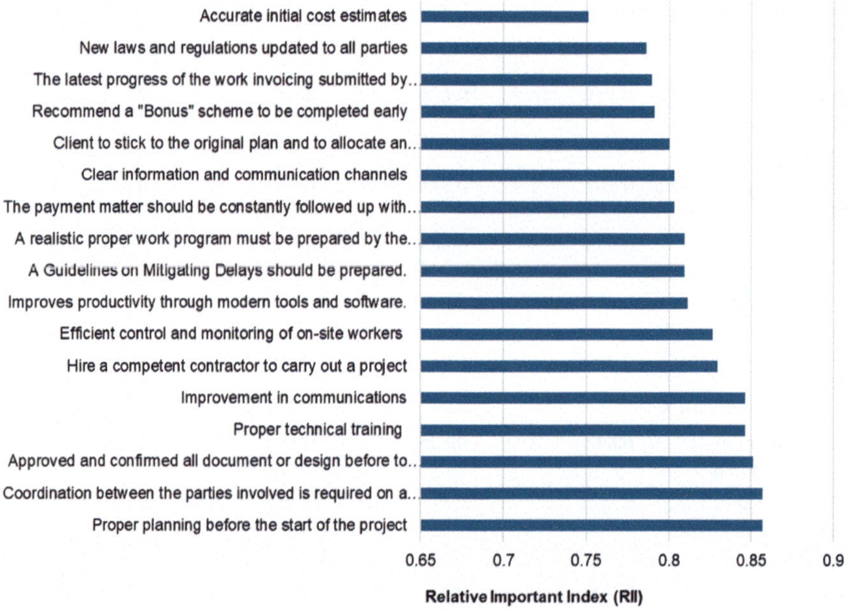

Fig. 5 Overall result recommendation for minimizing delay versus RII

coordination and change orders by the client during construction. For the consultant-related group of factors, unclear and insufficient detail in drawings was the most important factor, followed by delays in the production of design documents and delays in approving major changes in the scope of work. Finally, for external factor-related issues, similarly with views from contractors, land and site problems were the most crucial factors, followed by delays in obtaining permits from municipalities and environmental uncertainty. The next important view should be considered by the 60 consultants' respondents involved in the survey. Based on contractor-related factors, poor management and supervision of the site by the contractor were the number one reason for delays, followed by poor communication and coordination and subcontractor delays. Next will be a client-related factor, with late payment being the main reason for the delay in a public project, followed by late revision and approval of design documents by the owner and change orders by the client during construction. For consultant-related factors, poor communication and coordination were the main reasons for time overruns, followed by a lack of supervision and unclear and insufficient detail in drawings. Finally, based on external factors related, delay in obtaining permits from the municipality was the first rank of delay causes, followed by changes in government regulations and laws. Land and site problems are the top three important causes in the externally related factor group.

As a summary, the most important views should be considered by the three main stakeholders in the construction industry. By taking views from contractors, clients, and consultants, the average of the RII value will conclude the most important causes of each group of factors. It will not only show the causes that affect the construction industry in Kota Kinabalu, but it will also show the overall picture of the construction industry in Kota Kinabalu. Based on the contractor-related causes, poor communication and coordination were the crucial causes of delay, similarly from the client's point of view, followed by poor management and supervision of the site by the contractor and conflicts between the contractor and other parties. Next in the client-related group of factors, mostly contributed by contractors and consultants, late payment has become the top cause of time overruns in the construction industry, followed by late revision and approval of design documents by the owner. Both financial difficulties from clients and change orders by the client during construction were in the top three most important causes. Based on consultant-related factors, poor communication and coordination and a delay in approving major changes in the scope of work were the two most important causes, followed by delays in the production of design documents. Lastly, for external factor-related problems, land and site problems were the most crucial factors. Among the issue on land was the problem of land acquisition which would take longer time for application before construction started, followed by delays in obtaining permits from municipalities and changes in government regulations and laws.

4.2 Suggestion on Minimizing the Causes of Delay

Suggestions were also highlighted in this chapter to address the second objective for this study which is to analyze suggestions on minimizing the causes of public project delays in Kota Kinabalu. Four perceptions have been taken to indicate the best method to reduce or mitigate delay in the construction industry, especially public projects in Kota Kinabalu. Based on the results, the most effective ways to minimize delay are determined from the stakeholders' perspectives.

Based on the contractor's perception, three top suggestions are to carry out proper planning before the start of the project, and to ensure that the payment matter should be constantly followed up with the client and to improve communications. Next, based on the client's perceptions, the top three suggestions are to have a proper planning before the start of the project, to get approval and confirmation of the design concept, construction drawings, selection of materials, logistics management, etc., well before construction and finally to have a proper technical training is required, and the recruitment of skills and experience will help to reduce delays. Finally, based on the consultant's perceptions, the top three suggestions are to coordinate have regular coordination between the parties involved, to get approval and confirmation on the design concept, construction drawings, selection of materials, logistics management, etc., well before to construction and to improve communication aspects.

5 Conclusions, Limitations, and Recommendations for Future Research

This chapter accomplished the objectives of the study, which consisted of applying the RII method to examine suggestions for minimizing the causes of delay in public projects in Kota Kinabalu and investigating the causes of delay in public projects in Kota Kinabalu. A quantitative approach was used based on survey instrument administered to the construction stakeholders namely consultant, contractors, and clients. Then, the causes of project delays in the construction industry in Kota Kinabalu from the perspectives of contractors, clients, and consultants are examined together with the potential suggestions to mitigate these delays. The findings and suggestions offer valuable insights into the challenges faced by stakeholders in the construction industry and provide a foundation for improving project management and execution.

First, some of the contractor-related factors are shortage of labor, conflicts between contractors and other parties, slow instructions, and poor site management and supervision were identified as the primary causes of delays from the contractor's viewpoint. Therefore, effective communication and coordination emerged as a crucial factor to address these issues. Second, some of the client-related factors identified are clients' financial difficulties, late payments, and slow decision-making processes which were the leading contributors to project delays. Thus, proper planning, early design concept approval, and technical training were recommended to alleviate these issues. Third, the consultant-related factors found are mistakes and differences in design documents, along with delays in document production and approval of major changes in scope, which were the major factors affecting project timelines. Hence, enhanced coordination, early design confirmation, and improved communication were suggested to tackle these problems. Finally, other external factors identified are the land and site problems, fluctuations in material prices, and traffic control and restrictions were the top external causes of delays. Thus, effective permit acquisition, awareness of changing regulations, and proactive site management were proposed to minimize these challenges.

In summary, poor communication and coordination emerged as a common theme among all stakeholders as a critical factor contributing to project delays. This suggests that improving communication and fostering effective collaboration among contractors, clients, and consultants should be a priority for the construction industry in Kota Kinabalu. One of the suggestions can be made is enforcement of implementation of Building Integrated Modeling (BIM) for any public project in future. The suggestions provided by each stakeholder group offer practical ways to address these issues. By implementing these recommendations, stakeholders can work toward minimizing project delays, enhancing efficiency, and ultimately improving the construction industry's performance in Kota Kinabalu. It is imperative that all parties involved in construction projects actively consider these findings and work collaboratively to implement these measures for the betterment of the industry.

6 Limitations and Recommendations for Future Research

Some limitations of this study are related to the sample size which may not fully represent the diversity of perspectives within the construction industry in this region. Future research should aim to include a larger and more diverse sample of construction industry stakeholders from multiple regions in Malaysia or even internationally. This will enhance the robustness and generalizability of the findings. Another limitation is that the research focuses specifically on the construction industry in Kota Kinabalu, which may limit the generalizability of the findings to other regions with different socio-economic and regulatory environments. Hence, combining quantitative data with qualitative insights through interviews or focus group discussions can provide a more comprehensive understanding of the causes of project delays and potential mitigation strategies. Another limitation is related to response bias where the respondents may provide answers that they perceive as socially desirable or may not accurately reflect their actual experiences. Thus, conducting longitudinal studies that track the progress of construction projects over time can help identify trends and changes in delay factors. This can lead to more targeted interventions. In terms of the methodology, while the quantitative approach through the Relative Importance Index (RII) method was employed effectively, a complementary qualitative analysis could provide a deeper understanding of the underlying reasons for project delays. Hence, utilizing advanced data analytics and modeling techniques can help predict and prevent project delays more effectively, allowing for proactive management. Finally, the study primarily involves contractors, clients, and consultants. Other stakeholders such as regulatory authorities, local governments, and suppliers, whose actions or decisions can also impact project delays, were not included in the analysis, thus can be considered in future research. Overall, investigating and documenting successful case studies where project delays were minimized or avoided can provide valuable insights into best practices that can be adopted industry wide. In conclusion, while this study has provided valuable insights into the causes of project delays in the construction industry in Kota Kinabalu, there is room for further research that can broaden the scope, enhance the methodology, and provide a deeper understanding of this critical issue. Addressing these limitations and pursuing these recommendations will contribute to more effective project management and execution in the construction industry, not only in Kota Kinabalu but also globally.

References

Alaghbari, W, Kadir, MRA, Salim A, Ernawati (2007) The significant factors causing delay of building construction projects in Malaysia. Eng Constr Arch Manag 14(2):192–206

Alajmi AM, Ahmed Memon Z (2022) A review on significant factors causing delays in Saudi Arabia construction projects. Smart Cities 5(4):1465–1487

Alkhalifah SJ, Tuffaha FM, Al Hadidi LA, Ghaithan A (2023) Factors influencing change orders in oil and gas construction projects in Saudi Arabia. Built Environ Proj Asset Manag 13(3):430–452

Al-Tami SA (2015) Contemporary causes of construction delay in Saudi Arabia, according to contractors, consultants, and owners

Bekr GA (2018) Study and assessment of causes and effects of delay in large public construction projects in Jordan. Int J Appl Eng Res 13(8):6204–6210. www.ajer.org

Dehdasht G, Ferwati MS, Abidin NZ, Oyedeji MO (2022) Trends of construction industry in Malaysia and its emerging challenges. J Financ Manag Prop Constr 27(2):161–178

Jadhav DS (2018). Delay analysis in construction project. Int J Res Appl Sci Eng Technol 6(3):2244–2247. https://doi.org/10.22214/ijraset.2018.3355

Kamaruddeen AM, Sung CF, Wahi W (2020) A study on factors causing cost overrun of construction projects in Sarawak, Malaysia. Labour (Human) 2(7):13

Khan RA, Liew MS, Ghazali ZB (2014) Malaysian construction sector and Malaysia vision 2020: developed nation status. Proc Soc Behav Sci 109:507–513. https://doi.org/10.1016/j.sbspro.2013.12.498

Memon AH, Abdul Rahman I, Abdul Azis AA (2012) Time and cost performance in construction projects in southern and central regions of Peninsular Malaysia. Int J Adv Appl Sci 1(1):52–57. https://doi.org/10.11591/ijaas.v1i1.537

Nassar AH (2016) Causes and effects of delay in the construction industry in Egypt. Int J Eng Res Technol (IJERT) 5(04)

Nor SM, Subramaniam G, Sahudin Z (2023) The effects of foreign labor on labor productivity in construction industries. Inf Manag Bus Rev 15(2 (I) SI):116–127

Othman A, Ismail S (2014) Delay in government project delivery in Kedah, Malaysia. Rec Adv Civ Eng Mech 248–254. http://www.syuhaidaismail.com

Oyegoke AS, Al Kiyumi N (2017) The causes, impacts and mitigations of delay in megaprojects in the Sultanate of Oman. J Financ Manag Prop Constr 22(3):286–302. https://doi.org/10.1108/JFMPC-11-2016-0052

Rahman RA, Radzi AR, Saad MSH, Doh SI (2020) Factors affecting the success of highway construction projects: the case of Malaysia. IOP Conf Ser Mater Sci Eng 712(1):012030

Sambasivan M, Soon YW (2007) Causes and effects of delays in Malaysian construction industry. Int J Project Manage 25(5):517–526. https://doi.org/10.1016/j.ijproman.2006.11.007

Suleiman A (2022) Causes and effects of poor communication in the construction industry in the MENA region. J Civ Eng Manag 28(5):365–376

Taofeeq DM, Adeleke AQ, Hassan AK (2019) Factors affecting contractors risk attitude from Malaysia construction industry perspective. Soc Sci Human J 3(6):1281–1298

Yap JBH, Goay PL, Woon YB, Skitmore M (2021) Revisiting critical delay factors for construction: analysing projects in Malaysia. Alex Eng J 60(1):1717–1729

Enhancement in the Application of Value Management Through Project Management in the Malaysian Construction Industry

Mohd Hilmi Malek, Che Maznah Mat Isa, and Aini Jaapar

Abstract The application of VM has played a significant role in improving project outcomes in the Malaysian construction industry (MCI). Despite clear evidence of the benefits of VM, this discipline has significant setbacks in its application because of the theory–practice gap. Therefore, this study identifies the PM activities that can be integrated into the VM process. A qualitative method was employed, involving semi-structured interviews with eleven professionals and practitioners in the MCI, and provides evidence of the advantages of this integration, offering a framework for better implementation of VM in the construction industry. The results found similar objectives in the application of PM and VM to ensure the projects met the project performance by integrating PM activities into VM study and, at the same time, establishing the innovation of the concepts of VM. However, the findings are based on a relatively small sample size, which may limit the generalisability of the results to the broader Malaysian construction industry. Thus, future research should consider employing a mixed-methods approach, combining qualitative insights with quantitative data to validate and generalise the findings. Larger sample sizes and diverse participant groups from different regions and sectors of the construction industry would provide a more comprehensive understanding.

Keywords Project management activities · Value management · Integration

M. H. Malek · C. M. M. Isa (✉) · A. Jaapar
School of Civil Engineering, College of Engineering, Universiti Teknologi MARA, Shah Alam, Selangor, Malaysia
e-mail: chema982@uitm.edu.my

Center for Civil Engineering Studies, Universiti Teknologi MARA, Permatang Pauh, Pulau Pinang, Malaysia

Center of Studies for Quantity Surveying, College of Built Environment, Universiti Teknologi MARA, Shah Alam, Selangor, Malaysia

E. Shaffie et al. (eds.), *Sustainable Green Infrastructure*,
https://doi.org/10.1007/978-981-96-1486-8_13

1 Introduction

Value management, also known as value engineering, is a systematic and structured approach that aims to achieve the best value for money while meeting a project's required performance and quality standards (Kelly et al. 2015). The concept of VM offers a structured methodology to identify opportunities for value enhancement and cost-effective solutions. However, in practice, the implementation of VM faces various challenges and barriers that prevent its full potential from being realised. It involves collaborative processes that help project teams identify, evaluate, and optimise opportunities to enhance value, reduce costs, and improve project performance (Kelly et al. 2015; Ojo and Ogunsemi 2019; Surlan et al. 2016). The fundamental principle of VM is not merely cost-cutting but rather enhancing value by aligning project objectives with stakeholder expectations and project constraints (Jaapar et al. 2009, 2018; Khodeir and El Ghandour 2019; Maznan et al. 2012). Today's rapidly evolving construction industry (Ojo and Ogunsemi 2019; Oke and Aghimien 2018; Othman et al. 2020) emphasises that achieving project success goes beyond completing structures on time and within budget. As the industry evolves, it faces increasing pressure to deliver projects that meet high-quality standards and maximise stakeholder value (Kineber et al. 2021a, b, 2022). In this pursuit, VM has emerged as a powerful tool to optimise project outcomes, enhance efficiency, and achieve sustainable development.

The application of VM in the construction industry has grown steadily, driven by the need for efficient resource utilisation and enhanced project outcomes (Kineber et al. 2021a, b). Examining value management (VM) diffusion in the Egyptian residential building industry found that professionals in Egyptian building projects agree that VM activities can contribute to the success of building projects in terms of cost, quality, and time, and it shows that the implementation of VM contributes 55.5% to the success of the project. Further, (Mahinkanda et al. 2019) investigate how to bridge the theory–practice gap in value management (VM) in the Sri Lankan construction industry and suggest that VM application ensures value for money in construction projects, contributing to the economic, social, and environmental aspects of a country, and using resources efficiently and effectively (Yu et al. 2018). On the other hand, it found that VM provides opportunities for multidisciplinary professionals and stakeholders to focus on issues relating to society and the environment. However, integration presents threats, including the additional time and costs required for achieving sustainability targets. Similarly, In Malaysia, VM has been introduced since 1986 (Mohamad Ramly and Shen 2012), and the authors found that VM application can provide extraordinary outcomes towards project success when the project team has clear objectives, client participation and support, discipline and attitude, team mix, and decision-making authority are critical components for successful VM workshops.

While VM adoption has become more widespread, the theoretical underpinnings of VM champion a structured and collaborative methodology for enhancing value, and its real-world application encounters a series of challenges. Lin et al. (2022), examines the implementation of VM in small and medium construction projects in Malaysia and finds that building and large-scale projects are the most favoured subjects of VM studies to date. Further, (Jaapar et al. 2012) reviewed the gaps in the practice of VM in Malaysia and found an unclear description of current VM applications and a need for more exploration of the behavioural elements of VM workshops. The findings were criticised by (Mahinkanda et al. 2019), where many researchers have addressed the VM concept theoretically with the published body of knowledge, and there needs to be more VM implementation in the construction industry (Mahinkanda et al. 2019). Further, there needs to be more understanding of the VM concept and a lack of formal guidelines, and a prominent stumbling block is the resistance to change that permeates the construction industry, where established conventional project management practices often reign. Therefore, integrating project management (PM) activities into the VM methodology involves a synergistic approach that harmonises the structured principles of both disciplines.

The inherent similarities between the two disciplines facilitate the integration of PM activities into the VM methodology (Mahinkanda et al. 2019). Both approaches share a focus on achieving efficient and successful outcomes. PM's emphasis on defining objectives, setting timelines, and allocating resources aligns seamlessly with VM's goal of optimising value while controlling costs (PMI 2017). The structured planning and execution phases of PM correspond to the systematic analysis and evaluation stages of VM, enabling the identification of opportunities for enhancement (Oke and Aigbavboa 2017). Risk assessment in project management dovetails with VM's emphasis on identifying potential improvements and fostering a proactive approach to value creation. Moreover, the collaborative nature of VM resonates with PM's need for effective communication and stakeholder engagement. Project teams can collaboratively brainstorm innovative solutions, mirroring the cross-functional teamwork promoted by project management. Ultimately, project monitoring and review align with VM's continuous assessment, ensuring ongoing value enhancement (Kelly et al. 2015). By recognising these commonalities, integrating project management activities into the VM methodology becomes a natural extension, creating a cohesive framework that enhances project efficiency, quality, and value.

2 Literature Review

2.1 Value Management

Value management, often called value engineering, is a methodical and strategic approach to optimise the value of projects, products, or processes while minimising costs (Kelly et al. 2015). It revolves around the idea that value is the ratio of function to cost. It seeks to enhance worth by dissecting subjects into their essential functions and identifying avenues for overall improvement (Jaapar et al. 2012). The process involves understanding objectives, analysing functions, generating creative ideas, evaluating proposals, implementing chosen ideas, and constant review. This approach encourages collaboration, creativity, and efficient resource use. Industries such as construction and manufacturing benefit from value management's emphasis on value enhancement over mere cost reduction, fostering innovation and cross-functional synergy. Ultimately, value management is a dynamic process that systematically enhances value while maintaining cost efficiency, offering a holistic framework for achieving objectives in various sectors.

2.2 Impact of Value Management

Value management presents a plethora of compelling advantages spanning various industries. It starts by fostering a comprehensive outlook that transcends the mere reduction of costs, prioritising the optimisation of value across projects, products, or processes. Value management stimulates innovative problem-solving by dissecting subjects into their fundamental functions and assessing how these functions contribute to the overarching objectives. It fosters an environment of creativity and advancement. Moreover, this methodology actively encourages collaboration among individuals from diverse backgrounds, converging their insights to brainstorm and evaluate ideas. This collective effort often results in fresh solutions that might have yet to emerge within isolated departmental contexts, enriching the quality and viability of proposed enhancements (Kelly et al. 2015).

The systematic nature of the approach guarantees a meticulous examination of functions and their relationship to costs. This not only aids in identifying inefficiencies or redundancies but also enables informed decision-making regarding which suggested improvements are worth pursuing by evaluating ideas based on their potential impact, feasibility, and cost-effectiveness (Kineber et al. 2021a, b). Value management empowers organisations to make strategic choices aligned with their objectives and available resources. Furthermore, value management introduces adaptability to project development and process enhancement. Continuous monitoring and review mechanisms facilitate timely adjustments, ensuring anticipated value enhancements materialise and any deviations or challenges are promptly addressed. This flexibility enhances the nimbleness of projects and processes, enabling proactive responses to changing conditions (Kelly et al. 2015).

Another merit of value management lies in financial prudence. While its primary focus is enhancing value, the methodical scrutiny of functions often needs to include avenues for cost savings without compromising quality or performance. This dual benefit of optimising value and managing costs contributes to overall increased organisational efficiency and profitability. For instance, value management optimises structural functionality and durability in the construction industry while adhering to budgetary and timeline constraints. In manufacturing, it can lead to product improvements that align more closely with customer needs, thereby enhancing market competitiveness (Kelly et al. 2015). The methodology's prowess extends to project management by mitigating risks and maximising project success rates through targeted value enhancements. Value management is a potent approach that bridges innovation, collaboration, adaptability, and fiscal prudence, profoundly impacting diverse sectors.

2.3 Project Management

Project management is the disciplined application of knowledge, skills, tools, and techniques to efficiently plan, execute, monitor, control, and close a temporary endeavour with defined objectives, such as creating a product, providing a service, or achieving specific outcomes. It involves carefully orchestrating resources, including time, cost, scope, quality, human resources, communication, and risk, to achieve project goals within constraints. A project manager leads this process, ensuring clear objectives, stakeholder alignment, and effective communication while adapting to changes and challenges. By effectively managing the project lifecycle from initiation through planning, execution, monitoring, and closure, project management facilitates the delivery of results on time, within budget, and to the desired quality, thus enabling organisations to drive innovation, improve efficiency, and meet strategic objectives (PMI 2017).

2.4 Project Management Knowledge Areas

The project management body of knowledge (PMBOK) outlines a comprehensive framework for effective project management, delineated into ten distinct knowledge areas. These knowledge areas collectively encapsulate the critical facets of project management, offering project managers a structured approach to navigating the complexities of their projects. Each knowledge area represents a domain of expertise, encompassing specific processes, tools, and techniques that address essential aspects of project planning, execution, monitoring, and closure. These interconnected knowledge areas provide a roadmap for project managers to skillfully manage resources, risks, stakeholders, schedules, costs, quality, and more, ensuring that projects are delivered successfully while meeting objectives and adhering to constraints. Understanding and proficiently applying these knowledge areas empower project managers to orchestrate tasks with precision and agility, fostering a higher likelihood of project success and stakeholder satisfaction (PMI 2017).

2.4.1 Project Integration Management

Project integration management in construction projects encompasses harmonising and consolidating diverse processes, tasks, and technologies within the project management structure (Hussain Khahro et al. 2021). This entails aligning, formalising, and integrating project attributes and behaviours to achieve completion. Moreover, integration management is paramount in construction project performance because it facilitates efficient collaboration and coordination among stakeholders. Integration management ensures the seamless flow of information, resources, and activities through the fusion of various project processes, thereby enhancing overall project performance (Hussain Khahro et al. 2021).

The achievement of effective integration management brings about the coordinated execution of project tasks. It empowers project managers to define, comprehend, amalgamate, unite, and synchronise diverse processes, culminating in the triumphant conclusion of construction endeavours (Demirkesen and Ozorhon 2017; Hussain Khahro et al. 2021). Integration management guarantees the harmonious functioning of multiple project elements, resulting in enhanced performance dimensions such as time, cost, quality, safety, and client satisfaction. Furthermore, integration management contributes to the knowledge base by formulating a conceptual framework for managing integration and proposing tools and strategies to ensure its efficacy throughout the project lifecycle (Hussain Khahro et al. 2021). By comprehending and quantifying the factors influencing construction work, professionals can assess their projects and implement strategies to adeptly handle intricate endeavours (Demirkesen and Ozorhon 2017).

2.4.2 Project Scope Management

Project scope management refers to the practices and processes necessary to define, control, and manage the scope of a project to ensure its successful completion. It involves identifying and documenting what needs to be done to achieve the desired project outcomes and ensuring that the project is focused on delivering only the necessary work. This includes defining the project's objectives, deliverables, and boundaries and establishing a plan to manage any changes or modifications to the scope. Project scope management aims to provide clear communication and understanding among project stakeholders about the project's extent and functionality (Ogunberu et al. 2018). It is considered a critical function in project management as any changes or modifications in scope can lead to additional costs and affect the overall project's success (Al-Rubaiei et al. 2018).

2.4.3 Project Schedule Management

Project schedule management refers to the processes and techniques utilised to effectively plan, develop, control, and monitor the timeline of a project. It involves carefully coordinating and sequencing activities, resources, and milestones to ensure the project is completed within the allocated timeframe. Project schedule management aims to establish a realistic schedule, identify potential risks and delays, allocate resources efficiently, track progress, and make necessary adjustments to keep the project on track. It encompasses activities such as estimating activity durations, creating a project schedule, defining dependencies between tasks, assigning resources, and continually monitoring and updating the program as the project progresses. By effectively managing the project schedule, project managers can optimise time utilisation, meet project deadlines, and achieve project objectives (Cho et al. 2021; Sackey et al. 2020).

2.4.4 Project Cost Management

Project cost management refers to planning, estimating, budgeting, and controlling the costs associated with a construction project. It involves systematically managing project expenses to ensure that the project is completed within the allocated financial resources. This includes activities such as determining project objectives, sequencing project activities, producing detailed cost estimates and cost plans, monitoring project progress, identifying variations to time and cost plans, analysing project data to determine if the project is behaving as predicted, and calculating the size and impact of any variances. Project cost management is an essential part of overall project management and is crucial for ensuring the profitability and success of construction projects (Nor et al. 2022; Olawale and Sun 2015).

2.4.5 Project Quality Management

Project quality management involves extensive processes and activities to ensure a project aligns with designated quality standards and stakeholders' expectations. This structured approach extends across the entirety of the project's lifecycle, seamlessly integrating measures for quality control. These measures are designed to guarantee that project deliverables and outcomes conform to stipulated requirements and the needs of stakeholders. Project quality management encompasses diverse aspects, encompassing the establishment of quality objectives, formulation of quality plans, execution of quality assurance endeavours, and execution of assessments for quality control. Cumulatively, these undertakings meticulously assess and appraise the quality performance of the project. The primary objective of project quality management is the proactive identification and mitigation of potential issues, imperfections, and deviations that might jeopardise the project's quality. This endeavour strives to ensure the successful culmination of the project and the realisation of its intended value. Project quality management revolves around a comprehensive set of procedures and tasks that guarantee a project is in harmony with prescribed quality standards and stakeholders' expectations. This methodical approach spans the complete project lifecycle, seamlessly integrating measures for quality control. These measures ensure that project deliverables and results conform to stipulated requirements and the needs of stakeholders.

2.4.6 Project Resource Management

Project resource management refers to effectively and efficiently allocating, monitoring, and utilising resources in a project to ensure its successful completion. It involves identifying and acquiring the necessary resources, such as manpower, equipment, materials, and budget, and managing them throughout the project lifecycle. Project resource management aims to optimise resource utilisation, minimise cost and duration, and ensure that the right resources are available at the right time and in the right quantities to meet project objectives. It requires making informed decisions and prioritising resource allocation based on project requirements, constraints, and available resources (Khatun et al. 2021; Pinha and Ahluwalia 2019).

2.4.7 Project Procurement Management

Procurement management refers to planning, procuring, and managing the acquisition of goods, services, or works needed for a project or organisation. It involves identifying the needs, selecting the appropriate procurement method, conducting market research, evaluating suppliers, negotiating contracts, managing the procurement process, and ensuring that the goods or services meet the required quality standards, are delivered on time, and are within budget. It is an essential part of project management and aims to ensure that the right resources are acquired at the correct cost and time to meet project objectives and deliver value (Naoum and Egbu 2015).

2.4.8 Project Risk Management

Project risk management is a systematic process that focuses on identifying, assessing, and responding to the risks associated with a project. It involves identifying and evaluating risks and applying techniques and approaches to mitigate or reduce them tolerably. The main objective of project risk management is to maximise opportunities and minimise the adverse outcomes of risk occurrences in construction projects. It is an essential aspect of project management and critical to project success (PMI 2017).

Project risk management is proactive, aiming to anticipate and address potential risks before they occur (Hwang et al. 2014). It involves utilising various practices and techniques to control and manage risks throughout the life cycle of a project (PMI 2017). By identifying and monitoring threats, project managers receive early warnings if project objectives, such as time, cost, or quality, are at risk of not being achieved (Abazid and Harb 2018). Project risk management is critical in the construction industry, where complex projects involve significant budgets (PMI 2017). Project managers must reduce associated risks to increase the likelihood of project success. Risk management is not a guarantee of success but rather a tool to improve the chances of achieving project objectives (Hwang et al. 2014).

2.4.9 Project Communication Management

Project communication management refers to the processes and practices that facilitate the creation, distribution, receipt, acknowledgement, and understanding of information within a project. It encompasses various activities to ensure effective communication between project team members, stakeholders, and other relevant parties throughout the project lifecycle.

Effective project communication management involves developing a communication plan that outlines the strategies, methods, and channels for communication, as well as identifying the needs and requirements of stakeholders (Yap et al. 2017). It plays a critical role in project success by promoting collaboration, ensuring timely and accurate information exchange, and minimising the risk of miscommunication and misunderstandings.

The Project Management Institute (PMI) recognises communication management as one of the essential knowledge areas in project management. It includes processes such as planning, managing, and controlling communication. These processes involve formulating communication strategies, creating and disseminating project-related information, and monitoring and maintaining communication throughout the project (Taleb Al-Mayahi et al. 2018).

2.4.10 Project Stakeholder Management

Previous researchers described project stakeholder management refers to identifying, analysing, and effectively engaging with individuals or organisations interested in or can impact a project (Nguyen et al. 2020; Sanghera 2019; Srinivasan and Dhivya 2020; Yang and Shen 2015). Sanghera (2019) explained that it involves managing stakeholders' relationships, expectations, and needs throughout the project lifecycle to ensure their support and minimise any negative impact they may have on the project, which aims to create a positive and collaborative environment, enhance communication and decision-making processes, and ultimately contribute to the successful delivery of the project. Furthermore, Nguyen et al. (2020) explained that project stakeholder management involves understanding the stakeholders' interests, priorities, and concerns, incorporating their feedback and addressing their needs in project planning, execution, and evaluation. Also, effective stakeholder management is crucial for project success, as it helps build trust, manage conflicts, and ensure that the project meets stakeholders' diverse needs and expectations, where it requires continuous monitoring, adaptation, and proactive engagement to foster positive relationships and achieve stakeholder satisfaction.

2.5 Critical Project Management Activities

See Table 1.

Table 1 Critical project management activities

Knowledge areas	Critical project management activities	Authors
Integration (INTE)	Incorporates all the aspects of projects and provides a roadmap towards their success Coordinate, govern, schedule, budget, manage, and document for efficient projects Monitor, integrate, control, close, and reduce uncertainty for improved project outcomes	Hussain Khahro et al. (2021) Al-Kuhail et al. (2021), Demirkesen and Ozorhon (2017)
Scope (SCOP)	Develop a management plan defining, describing project scope, and outlining validation and control Gather, document, and manage stakeholder requirements to meet project objectives and specifications Clearly describe the project, deliverables, and limitations to establish achievable project boundaries Divide project components into hierarchical elements, curbing scope creep and linking deliverables, cost, time, and resources Verify project outputs match the defined scope, aiming for formal stakeholder approval of deliverables Monitor changes against baseline and assess new owner requests to ensure scope control	Al-Rubaiei et al. (2018), Lampa et al. (2017), Pheng (2018)

(continued)

Table 1 (continued)

Knowledge areas	Critical project management activities	Authors
Schedule (SCHE)	Assessing the interrelationship between schedule delay risk factors Evaluating the combined impact of project traits, organisational factors, and implementation on schedule delays Conducting focus groups of experienced managers to validate findings and delve deeper into schedule delay causes Set policies, procedures, and docs for project schedule from planning to execution Identify and document specific actions required to create project deliverables Identify and document relationships between project activities within this task Estimate work periods for activities based on resources and requirements within this task Analyse sequences, durations, resources, and constraints to build a project schedule model for execution Continuously monitor project status to update schedule and handle changes to the baseline Visual representation of project tasks, dependencies, critical path, and sequential order We are assessing duration using history, expert insights, and relevant factors for individual tasks	Cho et al. (2021), Sackey et al. (2020)
Cost (COST)	Set objectives, sequence tasks, and generate estimates and strategies in cost planning Monitor cost by tracking project progress and identifying variations in time and cost plans Reporting by sharing monitored data as agreed, forwarding to relevant teams for necessary steps Evaluate project performance, compare to predictions, and assess the size and impact of discrepancies	Olawale and Sun (2015), PMI (2017)

(continued)

Table 1 (continued)

Knowledge areas	Critical project management activities	Authors
Quality (QUAL)	Identify project quality standards and document how compliance will be shown for requirements Implement the plan, aligning the project with the organisation's quality policies and fulfilling requirements from planning Monitor and record quality activity results, assess performance and ensure outputs meet customer expectations by comparing requirements	PMI (2017)
Procurement (PROC)	Create a plan detailing procurement activities, timelines, budgets, and responsibilities for procurement management Identify the project's required goods or services, specifying quantity, quality, and specifications Assess and choose suppliers or contractors based on expertise, experience, capacity, and cost Draft binding agreements outlining terms, conditions, and expectations between the project owner and selected suppliers Negotiate with suppliers to establish favourable terms, encompassing pricing, delivery, and performance criteria Monitor execution, enforce terms, handle changes, and resolve disputes to ensure compliance Monitor performance, evaluate progress, and address issues to ensure project success and collaboration	Naoum and Egbu (2015), PMI (2017)
Communication (COMM)	Develop communication strategies, direct processes, and manage project information dissemination Enhance knowledge exchange by assessing project and stakeholder information requirements Update team on rules and operational norms Give performance feedback, share news, and offer support timely	Rostami et al. (2015), Subramaniam et al. (2021), Taleb Al-Mayahi et al. (2018), Yap et al. (2017)

(continued)

Table 1 (continued)

Knowledge areas	Critical project management activities	Authors
Risk (RISK)	Risk identification by determining the potential risks that may affect the project Risk evaluation–analysing the identified risks Risk response: developing and implementing strategies Monitoring and control: continuous monitoring of the identified risks and their impact on the project	Abazid and Harb (2018), Hwang et al. (2014)
Resource (RESO)	Evaluating resource availability and capacity for project requirement fulfilment Estimating the resource requirements for each project task or sub-task Allocating resources to specific tasks based on availability, skillsets, and project priorities Manage resource usage for efficiency and effectiveness Handle resource conflicts when tasks/projects vie for limited resources Optimise projects by adjusting resources to address constraints and enhance performance Review resource use to improve allocation and optimise strategies	Khatun et al. (2021), Perkins et al. (2020), Pinha and Ahluwalia (2019)
Stakeholders (STKH)	Identify stakeholders–gather information about the relevant parties involved Analyse and evaluate stakeholder involvement Secure higher authority support and set stakeholder management approach	Yang and Shen (2015)

3 Methodologies

This research aims to pinpoint the enhancement in the application of value management through project management in the Malaysian construction industry. It also seeks to propose a comprehensive framework for evaluating the adoption of project management activities into the value management methodology. A qualitative research methodology has been embraced in line with the exploratory nature of the research objectives (Cresswell 2013).

The research commenced with a systematic review of relevant literature from various sources, including journal articles, conference papers, research reports, and Internet information. This review aimed to comprehend the current landscape of PM knowledge and activities in the construction industry to ensure the project meets its objective and performance. Based on the PMBOK, 10 PM knowledge area entities were examined to extract key themes and topics related to implementing PM activities into the VM methodology.

Semi-structured interviews were subsequently conducted to extract insights and perspectives from the construction industry practitioners. This was done to identify the critical PM knowledge, processes, and activities necessary for effectively addressing critical PM activities that can be adopted in the VM methodology. The choice of interviews as the primary research method was due to their capability to elicit detailed and comprehensive information from participants with substantial knowledge, expertise, and experience in the subject matter (Cardano 2020). Semi-structured interviews were well-suited for this research as they allowed for probing and follow-up discussions, enabling a deeper understanding of contexts and underlying rationales (Denzin and Lincoln 2018).

A purposive sampling method was employed to select interview participants, ensuring individuals with rich and in-depth insights were chosen for thorough investigation (Rice and Ezzy 2005). The selection was based on criteria such as extensive experience managing the construction project, PM knowledge process and activities, and their roles in their respective organisations (detailed participant information is available in Table 2).

A total of eleven (11) interviews were conducted in this study. This sampling strategy was adopted to ensure the reliability and validity of the research by involving participants capable of providing meaningful insights. The number of interviews was considered adequate as it yielded substantial data for meaningful conclusions, satisfying the commonly accepted sampling criterion in qualitative research known as "reaching theoretical saturation" (Marshall et al. 2013). Both face-to-face and online interviews were conducted. Pilot interviews were conducted before the formal interviews, and participants were offered the opportunity to recap and clarify their points during the interviews to enhance research validity and reliability (Cresswell 2013; Creswell and Creswell 2018). Interview durations ranged from one to one and a half hours. The interviews were recorded, transcribed, and subjected to further analysis.

Table 2 Interviewees profiling

No.	Interviewee code	Position in the organisation	Type of the organisation	Years of professional experience
1	P1	Project Manager	Private construction firm	15–20
2	P2	Project Manager	Private construction firm	5–10
3	P3	Architect	Private consultant firm	10–15
4	P4	Architect	Private consultant firm	20–Over
5	P5	Design Engineer	Private consultant firm	15–20
6	P6	Project Engineer	Government agency	15–20
7	P7	Project Engineer	Government agency	15–20
8	P8	Project Engineer	Government agency	15–20
9	P9	Quantity Surveyor	Private construction firm	15–20
10	P10	Quantity Surveyor	Private consultant firm	15–20
11	P11	Academician	Local university	15–20

Content analysis was employed to analyse the qualitative data, as it provides a systematic and objective means of deriving inferences from data, describing and quantifying specific phenomena with minimal loss of information (Krippendorff 2018). This method suited the data collection protocol of the research, which involved documents and open-ended survey responses (Vaismoradi et al. 2013).

The critical PM activities and the PMBOK guideline provide an essential reference for analysing current practices. Insights and comments from the semi-structured interviews also contributed to the content analysis. Atlas.ti, a qualitative data analysis software, facilitated the development of emerging themes from the interviews and the reviewed PM activities. The software allowed for comparative analysis of text codes within the same category to ensure consistent coding and interpretations (Bazeley and Richards 2000).

4 Analysis and Discussion

An analysis of the PM activities, project performance attributes and VM objectives in the MCI, coupled with insights from semi-structured interviews, has unveiled critical PM activities that can nourish the VM application. These essentials can be classified into ten distinct PM knowledge: (1) Project Integration Management (INTE), (2) Project Scope Management (SCOP), (3) Project Schedule Management (SCHE), (4) Project Cost Management (COST), (5) Project Quality Management (QUAL), (6) Project Resource Management (RESO), (7) Project Procurement Management

(PROC), (8) Project Risk Management (RISK), (9) Project Communication Management (COMM), and (10) Project Stakeholder Management (STKH). The ensuing sections will elaborate on and discuss these ten project management.

4.1 Valued-Based Project Integration Management

The interview participant highlighted the significance of project integration management (INTE) that can be implemented in VM methodology. Interviewee P1 emphasised that project integration and value management are crucial aspects of successful project delivery. Integrating INTE into value management involves aligning the processes and activities of project integration with the overall goal of maximising value for the organisation. Interview P11 described that integrating INTE into the overall VM involves defining how value management activities fit into the project life cycle, from initiation to closeout. Besides, during the initiation stage, the project team should clearly outline the roles, responsibilities, and deliverables related to value management within the project management plan. Also, (Demirkesen and Ozorhon 2017; Hussain Khahro et al. 2021) suggested that integration management activities can be integrated into the value management methodology by incorporating the principles and processes of project integration within the existing value management framework. The researchers also suggested that project integration management ensures proper coordination and synchronisation of project activities, which is essential for project success. For example, the project team needs to develop a project integration plan based on the value analysis and outline how different project activities will be coordinated, unified, and synchronised. This plan should consider the dependencies, logical relationships, and sequencing of project activities.

4.2 Valued-Based Project Scope Management

Interviews P1, P2, P3, and P6 highlighted that integrating project scope management (SCOP) activities into the value management methodology involves aligning the systematic approach of VM with the structured scope definition and control practices of project management. Participant P1 explained that during the "Information" phase of VM, the project's scope should be thoroughly defined, including the boundaries, deliverables, and constraints. This ensures that the project's scope aligns with the intended objectives and sets the foundation for value optimisation. P3 and P6 further entailed integrating project scope management activities into the value management methodology which requires aligning the systematic approach of value management with the defined scope and control practices of project management. By ensuring that scope considerations are woven into each stage of the value management process, projects can achieve enhanced value while staying within the established scope boundaries.

Project scope management activities can be integrated into the value management methodology by following specific steps. Firstly, the project manager can utilise expert judgement or consult with building professionals with similar project experience to create the scope management plan. This step ensures that the necessary tasks and work are identified and defined (Pheng 2018). Secondly, the work breakdown structure (WBS) should be updated whenever there are changes to the project scope. The WBS helps identify the activities, allocate appropriate resources, and evaluate risks and cost implications associated with the actions (Pheng 2018). Controlling the scope is essential to monitor the project status, product scope, and scope changes (Corvello et al. 2017). Additionally, the project manager should perform activities such as collecting project requirements, defining the project scope, creating the WBS, verifying the scope, and controlling the scope. These activities are part of routine project scope management and contribute to ensuring the project is completed as planned (Corvello et al. 2017; Pheng 2018). Integrating project scope management into the value management methodology, project managers can ensure that the necessary work is accomplished while considering the value and benefits of the project and the integration helps optimise the project's scope to achieve the desired results efficiently (Al-Rubaiei et al. 2018).

4.3 Valued-Based Project Schedule Management

Interviewees P2, P3, P4, P7, and P9 explained that during the "Information" phase of VM, project time constraints and milestones should be clearly defined, aligning with the project's overall objectives. This ensures that the value-enhancing efforts are carried out within the project's time limitations. P4 and P7 further entailed that during the "Function Analysis" and "Creative Idea Generation", the breakdown of functions should consider their temporal dependencies and durations. Understanding how functions relate to each other in terms of time can aid in identifying opportunities for improving efficiency and optimising the project schedule. Ideas should not only focus on functional enhancements but also on strategies to streamline the project timeline. This could involve proposing innovative ways to expedite specific tasks or identifying opportunities to parallelise activities.

Sackey et al. (2020) highlighted that project schedule management activities can be integrated into the value management methodology by applying earned value management (EVM) techniques. EVM is a project management technique that measures progress by comparing the actual cost of work performed and the corresponding estimated earned value. For example, the authors suggested that EVM uses performance information, such as the actual cost of work performed and the estimated earned value, to measure a project's progress at a specific date. This includes tracking the completion status of project activities as per the planned schedule.

4.4 Valued-Based Project Cost Management

Interviewees P1, P10, and P11 explained that during the "Information" phase, project budget constraints should be clearly defined, aligning with the project's objectives. This ensures that the value-enhancing efforts remain within the project's financial limitations. Understanding the cost contribution of each function can aid in identifying opportunities for cost reduction or efficiency improvement while maintaining value. P2 recommended that ideas should not only focus on functional enhancements but also on strategies to reduce costs. This could involve proposing innovative approaches to resource allocation, materials sourcing, or process optimisation. Nor et al. (2022) have suggested several approaches that can be considered to integrate project cost management activities into the value management methodology. For example, the reseacher described that in the planning stage, the project begins with identifying the various functions or activities required to achieve the project's objectives. During this stage, cost estimation techniques can be utilised to estimate the financial implications of each function or activity.

4.5 Valued-Based Project Quality Management

Interviewees P5 and P6 similarly explained that during the "Information" phase of value management, the desired quality standards and benchmarks should be clearly defined, aligning with the project's overall objectives it ensure that the value-enhancing efforts uphold the required quality levels, while P2 and P3 explained that functions should be evaluated in terms of their value and the quality they contribute to the project. Understanding each function's quality implications can help simultaneously identify opportunities to enhance value and quality and the ideas should encompass functional improvements and strategies to enhance project quality. This could involve proposing innovative quality control, testing, or validation approaches. Integrating quality management activities into the value management methodology offers a holistic approach that enhances project outcomes. This synergy ensures both value realisation and the delivery of high-quality results that meet customer expectations. This process involves strategic steps combining quality management and value optimisation principles. To achieve successful integration, it is essential to establish quality objectives that align with broader project goals, guiding the project towards excellence while harmonising quality aspirations with value-driven ones. Additionally, identifying specific, measurable quality requirements for each project element is crucial, setting the stage for rigorous quality scrutiny throughout the project's lifecycle (PMI 2017).

4.6 Valued-Based Project Resource Management

Interviewees P1, P3, and P4 explained that project resource requirements should be clearly defined during the value study, aligning with the project's overall objectives. This ensures that the value-enhancing efforts are carried out with the necessary resources. Functions should be evaluated in terms of their resource implications. Understanding the resource demands of each function can aid in identifying opportunities for resource efficiency improvements while enhancing value, and the project team should consider the impact of proposed enhancements on project resources. Ideas that lead to improved value while efficiently using resources should be prioritised in alignment with the principles of value optimisation.

Previous researchers suggested several ways to integrate project resource management (RESO) activities into the value management methodology (Khatun et al. 2021; Perkins et al. 2020; Pinha and Ahluwalia 2019). For example, organisations can reduce project costs and duration by improving decision-making with flexible resource management, where this approach allows for better allocation and utilisation of resources, ensuring that they are effectively managed to achieve project goals and maximise value (Pinha and Ahluwalia 2019). Furthermore, integrating project resource management activities within the value management methodology can help identify and leverage these characteristics to enhance project performance, create value, and emphasise the importance of project management characteristics that constitute a source of competitive advantage (Perkins et al. 2020). Integrating project resource management activities into the value management methodology involves aligning resource allocation decisions with project objectives, enhancing resource utilisation, and leveraging project management characteristics as sources of competitive advantage. Khatun et al. (2021) suggested that dynamic resource allocation for optimising project performance can optimise project performance in a multi-project environment and organisations can strategically allocate resources to achieve optimal project outcomes and maximise value.

4.7 Valued-Based Project Procurement Management

Interviewees P1, P3, and P4 explained that project procurement management (PROC) needs should be identified during the value study, aligning with the project's objectives. This ensures that appropriate procurement strategies support the value-enhancing efforts. P5 and P11 further entail that functions should be evaluated regarding their procurement implications in the "Function Analysis" stage of value management. Understanding the procurement requirements of each function can aid in identifying opportunities for procurement efficiency improvements while enhancing value. Ideas should consider both functional enhancements and innovative procurement strategies. This might involve proposing novel approaches to supplier selection, contract negotiation, or sourcing that contribute to value enhancement. In contrast, the "Evaluation and Selection" step should assess the impact of proposed enhancements on project procurement. Ideas that enhance value while optimising procurement processes should be prioritised, aligning with the principles of value optimisation.

Naoum and Egbu (2015) suggested that integrating project procurement management activities into the value management methodology involves aligning the procurement process to deliver the required project functions at an optimal whole-life cost without compromising quality, performance, and reliability. The authors suggested early involvement by engaging key stakeholders, including the main contractor, during the design phase to allow for their expertise and input in reducing waste, improving buildability, and enhancing understanding of project requirements. Further, they highlighted collaboration and teamwork by fostering cooperation and effective communication by breaking away from the fragmented approach of traditional procurement methods. Implement integrated procurement methods, such as management contracting and partnering, that encourage shared ideas and improve the value of a project. Moreover, the researchers recommended that the project team incorporate value engineering workshops throughout the project to ensure active participation from all project stakeholders. This will encourage collaboration, innovation, improved constructability, enhanced communication throughout the supply chain, and the integration of sustainable practices. Also, the construction industry players must welcome change and facilitate innovative procurement methods to meet managerial, technical, and competative commercial challenges. This change should be driven by client requirements and supported by the project team.

4.8 Valued-Based Project Risk Management

Interviewees P1, P3, and P4 explained that potential risks should be considered and aligned with the project's objectives. This ensures that the value-enhancing efforts account for potential risks that might affect the project's success, and functions should be evaluated regarding their associated risks. Understanding the risks related to each function can aid in identifying opportunities to enhance value while mitigating potential negative impacts. Ideas should encompass both functional improvements and strategies to address project risks. This could involve proposing innovative approaches to risk mitigation, contingency planning, or alternative solutions that enhance value while reducing potential risks. It should include an assessment of how proposed enhancements impact project risks. Ideas that enhance value and contribute to risk reduction or mitigation should be prioritised, aligning with the principles of value optimisation. Integrating project risk management activities into value management requires ongoing risk monitoring and control throughout project execution. Proposed enhancements should be evaluated to determine their potential impact on project risks, and necessary adjustments should be made to ensure effective risk management.

Hwang et al. (2014) recommended that integrating project risk management (RISK) activities into the VM methodology can be achieved by incorporating risk management processes and techniques within the different phases. The researcher suggested that the integration ensures that risks are identified, assessed, and managed throughout the value management study, enhancing overall project outcomes. Also, the reseachers explained that one way to integrate project risk management activities into the value management methodology is by considering risk management as an essential component of the value management process, from the initial project briefing to the final implementation stage. Consequently, risks and uncertainties associated with different value alternatives can be assessed and managed effectively, leading to informed decision-making. Besides, integrating project risk management activities into the value management methodology involves identifying and evaluating critical success factors (CSFs) specific to risk management and value management. These CSFs guide the successful implementation of the integrated risk and value management process, ensuring that risk and value considerations are adequately addressed. Furthermore, the project team can establish a synergistic relationship between risk minimisation and value maximisation by aligning the objectives of risk management and value management to ensure that risk reduction efforts do not compromise the project's overall value. Nevertheless, the project teams can optimise risk and value trade-offs to achieve the best possible project outcomes by integrating risk management activities into the value management methodology.

4.9 Valued-Based Project Communication Management

Integrating project communication management (COMM) activities into the value management methodology involves aligning the structured communication practices of project management with the collaborative and cross-functional approach of value management. Interview P3, P5, P6, P7, P8, and P11 explained that clear communication plans should be established, ensuring stakeholders are informed about the value optimisation efforts and objectives during the value study. Further, P7 explained that functions should be assessed not only in terms of value but also in terms of their communication implications. Understanding how functions impact various stakeholders and communication channels can aid in identifying opportunities for effective communication strategies that enhance value. Ideas should encompass functional improvements and innovative ways to communicate value-enhancing efforts. This could involve proposing approaches to engage stakeholders, share progress, and gather insights.

Santalova et al. (2019) suggested several project communication management activities, which can be integrated into the value management methodology by following a systematic approach that ensures effective and efficient communication throughout the project. Based on the findings, the researchers suggested that the project team must identify the project's specific communication objectives and goals.

This can include disseminating project information, sharing updates, gathering feedback, and resolving issues. Besides, during the planning stage, it is also suggested that the project leader creates a comprehensive communication management plan that outlines the strategies, channels, and methods for communication. This plan should address the specific needs of the project and the stakeholders involved. It should also consider the frequency, format, and content of communication. Further, the reseachers recommended that the project team identifies the most appropriate communication channels and tools for effective communication and engage and involve all relevant stakeholders throughout the project. This can be achieved through regular communication, consultation, and active participation. Stakeholders should be informed, their feedback should be sought, and their concerns addressed.

4.9.1 Valued-Based Project Stakeholder Management

Interviewees P10 and P12 explained that during the "Information" phase of value management, stakeholder needs and expectations should be identified, ensuring that the value-enhancing efforts consider the perspectives and interests of relevant stakeholders. The project requirement and its functions should be evaluated not only in terms of value, but also in terms of their impact on different stakeholders. Understanding how functions affect various stakeholders can aid in identifying opportunities for enhancing value while addressing stakeholder concerns. Ideas should encompass functional improvements and strategies to engage and manage stakeholders effectively. This could involve proposing innovative stakeholder communication, involvement, and satisfaction approaches. P11 also entailed that the project team considers the impact of proposed enhancements on stakeholders. Ideas that enhance value while fostering positive stakeholder relationships and engagement should be prioritised, aligning with the principles of value optimisation.

The interdependent relationship between stakeholders and value management has gained significant attention in project management. Integrating these two essential aspects can pave the way for more efficient and successful project execution, benefiting all stakeholders and the strategic steps required to seamlessly merge project stakeholder management activities into the value management methodology, optimising project outcomes and ensuring value creation for all stakeholders.

The first cornerstone in this integration journey is stakeholder identification where this crucial step involves the thorough process of recognising and gathering pertinent information about all stakeholders. While applying the value management methodology, stakeholders need to be identified based on their relevance to the project, potential impact, and the values they bring to the table. This proactive identification sets the stage for a comprehensive understanding of the key players and what they can contribute to the project's objectives. The subsequent step, stakeholder assessment, further refines the integration process. The focus shifts to analysing the intricate dynamics of stakeholders' interests, influences, and behaviours. When incorporating value management, aligning stakeholders' interests and results with the broader project objectives is imperative. By acknowledging these factors, project teams can navigate project decisions in ways that resonate with stakeholders, thereby establishing a bridge between their expectations and the project's overarching goals.

The dynamic aspect of stakeholder engagement constitutes another linchpin in this integration narrative. Engaging stakeholders in decision-making processes and project activities is fundamental for fostering a collaborative and inclusive project environment. Within the value management framework, the level of stakeholder engagement—ranging from being informed to empowered—becomes a critical consideration. Within this sphere, stakeholders actively define project values, offer input, and contribute to decisions aligned with their interests. This active engagement amplifies the authenticity and relevance of the project's value proposition.

5 Conclusion and Recommendation

The theory–practice gap in applying value management in the Malaysian construction industry lies in effectively integrating value management principles with project management practices. Despite recognising both disciplines independently, it needs more research and implementation focus on combining them to maximise the value derived from applications and technology in construction projects. Bridging this gap could lead to significant advancements in project delivery, cost-effectiveness, and innovation within the industry. Integrating project management activities into the value management methodology involves a synergistic approach that harmonises the structured principles of both disciplines. Firstly, project management activities, such as defining objectives, creating schedules, and allocating resources, should be seamlessly aligned with the overarching goals of value management. This ensures that project milestones and deliverables are strategically positioned within the value optimisation process. Secondly, value management's systematic analysis of functions and their relationship to costs can be integrated into project planning by assessing how various project components contribute to value and identifying areas for enhancement. Additionally, project risk assessment can dovetail with value management's focus on identifying potential improvements, allowing risks to be proactively transformed into opportunities for value creation. Moreover, the collaborative nature of value management resonates with project management's need for effective communication and stakeholder engagement. Teams can collaboratively brainstorm innovative solutions and assess their feasibility, fostering cross-functional teamwork. Lastly, project monitoring and review align with value management's continuous assessment, enabling ongoing value enhancement throughout the project lifecycle. Integrating these activities ensures that project management becomes integral to value management, streamlining processes, enhancing outcomes, and facilitating a holistic approach to delivering maximum project value. Integrating enhancement in the application of value management presents a transformative opportunity for the Malaysian construction industry. By addressing integration challenges and merging technology with traditional practices, the industry can unlock new levels of efficiency, innovation, and continuous improvement. Adopting VM as a core practice in construction projects promises cost-effective solutions, a culture of innovation, and a prosperous future for the Malaysian construction industry. While the qualitative approach provided in-depth insights into integrating PM activities into VM, it has limitations. Reliance on qualitative data alone may not capture the full breadth of the phenomena, and the small sample size limits the generalisability of the findings to the broader Malaysian construction industry. Qualitative data is also subjective, relying on participants' perspectives and the researcher's interpretations. Future research should employ a mixed-methods approach, combining qualitative insights with quantitative data for validation and generalisation. Larger, more diverse sample sizes and longitudinal studies could provide a more comprehensive and long-term understanding of integrating PM activities into VM.

References

Abazid M, Harb H (2018) An overview of risk management in the construction projects. Acad Res Int 9(2). www.savap.org.pk

Al-Kuhail MY, Al-Dafiry HA, Barakat TA, Al-Ansi A (2021) Developing a diamond framework model based on "integration" of project success measures for construction project management in Yemen. Pertanika J Sci Technol 29(3):1787–1809. https://doi.org/10.47836/pjst.29.3.23

Al-Rubaiei QHS, Nifa FAA, Musa S (2018) Project scope management through multiple perspectives: a critical review of concepts. AIP Conf Proc 2016. https://doi.org/10.1063/1.5055427

Bazeley P, Richards L (2000) The Nvivo qualitative project book, 1st edn. SAGE Publications Ltd.

Cardano M (2020) Defending qualitative research: design, analysis and textualization. Routledge. www.routledge.com/Routledge

Cho K, Ahn S, Park K, Kim TW (2021) Schedule delay leading indicators in precast concrete construction projects: qualitative comparative analysis of Korean cases. J Manage Eng 37(4). https://doi.org/10.1061/(asce)me.1943-5479.0000915

Corvello V, Javernick-Will A, La Ratta A (2017) Routine project scope management in small construction enterprises. Int J Project Org Manage 9(1):18–30

Cresswell JW (2013) Qualitative inquiry and research design: choosing among five approaches. SAGE Publication

Creswell JW, Creswell JD (2018) Research design: qualitative, quantitative, and mixed methods approaches, 5th edn. SAGE Publication, Inc.

Demirkesen S, Ozorhon B (2017) Impact of integration management on construction project management performance. Int J Project Manage 35(8):1639–1654. https://doi.org/10.1016/j.ijproman.2017.09.008

Denzin NK, Lincoln YS (2018) The SAGE handbook of qualitative research, 5th edn. SAGE Publication

Hussain Khahro S, Ahmed Babar A, Hussain Siddiqui F, Raza Khoso A, Babar A, Hussain Ali T, Hussain S, Hussain F, Hussain T, Raza A (2021) Integration management: a bibliometric study of research trends

Hwang B-G, Asce M, Zhao, Xianbo S, Ong Y (2014) Value Management in Singaporean building projects: implementation status, critical success factors, and risk factors. https://doi.org/10.1061/(ASCE)ME

Jaapar A, Azmi N, Bari A, Takim R (2009) The impact of value management implementation in Malaysia. J Sustain Dev 2(2)

Jaapar A, Maznan NA, Zawawi M (2018) Current state of value management implementations in Malaysian public projects. Asian J Environ-Behav Stud 3(8):71–78. https://doi.org/10.21834/aje-bs.v3i8.280

Jaapar A, Zawawi M, Bari NAA, Ahmad N (2012) Value management in the Malaysian construction industry: addressing a theory and practice gap. Procedia Soc Behav Sci 35:757–763. https://doi.org/10.1016/j.sbspro.2012.02.146

Kelly J, Male S, Graham D (2015) Value management of construction projects, 2nd edn. Wiley. www.wiley.com/go/construction

Khatun MT, Hiekata K, Takahashi Y, Okada I (2021) Dynamic modeling of resource allocation for project management in multi-project environment. Adv Transdiscipl Eng 16:223–232. https://doi.org/10.3233/ATDE210101

Khodeir LM, El Ghandour A (2019) Examining the role of value management in controlling cost overrun [application on residential construction projects in Egypt]. Ain Shams Eng J 10(3):471–479. https://doi.org/10.1016/j.asej.2018.11.008

Kineber AF, Othman I, Oke AE, Chileshe N, Zayed T (2021a) Exploring the value management critical success factors for sustainable residential building—a structural equation modelling approach. J Clean Prod 293. https://doi.org/10.1016/j.jclepro.2021.126115

Kineber AF, Othman I, Ayodeji, Oke E, Chileshe N, Kamil Buniya M (2021b) Impact of value management on building projects success: structural equation modeling approach. https://doi.org/10.1061/(ASCE)CO.1943

Kineber AF, Othman IB, Oke AE, Chileshe N (2022) Modelling the relationship between value management's activities and critical success factors for sustainable buildings. J Eng des Technol 20(2):414–435. https://doi.org/10.1108/JEDT-03-2021-0159

Krippendorff K (2018) Content analysis: an introduction to its methodology, 4th edn. SAGE Publications Ltd.

Lampa IL, De GoContessoto A, Amorim AR, Zafalon GFD, Valêncio CR, De Souza RCG (2017) Project scope management: a strategy oriented to the requirements engineering. In: ICEIS 2017—proceedings of the 19th international conference on enterprise information systems, 2, 370–378. https://doi.org/10.5220/0006318603700378

Lin X, Mazlan AN, Ismail S, Durdyev S (2022) Status of value management implementation in small and medium construction projects in Malaysia. Buildings 12(5). https://doi.org/10.3390/buildings12050658

Mahinkanda MMMP, Sandanayake YG, Ekanayake BJ (2019) Bridging the theory-practice gap in value management in sri lankan construction industry. In: World construction symposium, 147–157. https://doi.org/10.31705/WCS.2019.15

Marshall B, Cardon P, Poddar A, Fontenot R (2013) Does sample size matter in qualitative research? A review of qualitative interviews in is research. J Comput Inf Syst 54(1):11–22. https://doi.org/10.1080/08874417.2013.11645667

Maznan NA, Jaapar A, Bari NAA, Zawawi M (2012) Value management: private sector's perception. Procedia Soc Behav Sci 50:383–391. https://doi.org/10.1016/j.sbspro.2012.08.043

Mohamad Ramly Z, Shen GQ (2012) Value management in Malaysia: past, present and future. In: Proceedings of the international conference on value engineering and management: innovation in the value methodology, ICVEM 2012, 105–110

Naoum S, Egbu C (2015) Critical review of procurement method research in construction journals. Procedia Econ Fin 21:6–13. https://doi.org/10.1016/s2212-5671(15)00144-6

Nguyen PV, Duc Bui T, Thi Sa Do H, Chi Minh City H, Chi Minh H (2020) The relationship between project management performance and stakeholder satisfaction in Vietnam: perspectives from the construction industry. Int J Innov Creat Change 1. www.ijicc.net

Nor MSM, Rahmat M, Azmi N, Rahim MHIA (2022) Challenges of cost management technique implementation during pre-tender in Klang valley construction project. E3S Web Conf 347. https://doi.org/10.1051/e3sconf/202234705009

Ogunberu AO, Akintelu SO, Olaposi TO (2018) Application of project scope management practices on project success among telecommunication organizations in Nigeria. Int J Dev Sustain 7(2). www.isdsnet.com/ijds

Ojo LD, Ogunsemi DR (2019) Critical drivers (CDs) of value management adoption in the Nigerian construction industry: a Delphi study. J Eng des Technol 17(1):250–264. https://doi.org/10.1108/JEDT-09-2018-0143

Oke AE, Aghimien DO (2018) Drivers of value management in the Nigerian construction industry. J Eng des Technol 16(2):270–284. https://doi.org/10.1108/JEDT-02-2017-0012

Oke AE, Aigbavboa CO (2017) The concept of value management. In: Sustainable value management for construction projects. Springer International Publishing, pp 13–29. https://doi.org/10.1007/978-3-319-54151-8_2

Olawale Y, Sun M (2015) Construction project control in the UK: current practice, existing problems and recommendations for future improvement. Int J Project Manage 33(3):623–637. https://doi.org/10.1016/j.ijproman.2014.10.003

Othman I, Kineber AF, Oke AE, Khalil N, Buniya MK (2020) Drivers of value management implementation in building projects in developing countries. J Phys Conf Ser 1529(4). https://doi.org/10.1088/1742-6596/1529/4/042083

Perkins D, Mathur G, Jugdev K (2020) Project management resources and outcomes: a confirmatory factor analysis. Int J Manag Proj Bus 13(3):600–615. https://doi.org/10.1108/IJMPB-07-2019-0170

Pheng LS (2018) Project scope management, pp 63–77. https://doi.org/10.1007/978-981-10-6992-5_5

Pinha DC, Ahluwalia RS (2019) Flexible resource management and its effect on project cost and duration. J Industr Eng Int 15(1):119–133. https://doi.org/10.1007/s40092-018-0277-3

Project Management Institute (2017) A guide to the project management body of knowledge: (PMBOK guide), 6th edn. Project Management Institute

Rice PL, Ezzy D (2005) Qualitative research methods: a health focus, 1st edn. Oxford University Press

Rostami A, Sommerville J, Wong LI, Lee C (2015) Engineering, construction and architectural management article information. Eng Constr Archit Manage 22(1):91–107. https://doi.org/10.1108/09699981111098711

Sackey S, Lee DE, Kim BS (2020) Duration estimate at completion: improving earned value management forecasting accuracy. KSCE J Civ Eng 24(3):693–702. https://doi.org/10.1007/s12205-020-0407-5

Sanghera P (2019) Managing communication and stakeholder engagement. In: PMP® in depth, 419–436. https://doi.org/10.1007/978-1-4842-3910-0_11

Santalova MS, Lesnikova EP, Nechaeva SN, Borshcheva AV, Charykova OG (2019) Information hindrances and communication barriers in project interactions. In: Lecture notes in networks and systems, vol 57, pp 273–281. Springer. https://doi.org/10.1007/978-3-030-00102-5_29

Srinivasan NP, Dhivya S (2020) An empirical study on stakeholder management in construction projects. Mater Today Proc 21:60–62. https://doi.org/10.1016/j.matpr.2019.05.361

Subramaniam C, Ismail S, Durdyev S, Rani WNMWM, Bakar NFSA, Banaitis A (2021) Overcoming the project communications management breakdown amongst foreign workers during the covid-19 pandemic in biophilia inveigled construction projects in malaysia. Energies 14(16). https://doi.org/10.3390/en14164790

Surlan N, Cekic Z, Torbica Z (2016) Use of value management workshops and critical success factors in introducing local experience on the international construction projects. J Civ Eng Manag 22(8):1021–1031. https://doi.org/10.3846/13923730.2014.945950

Taleb Al-Mayahi H, Ismail S, Hussaini Wahab M, Nurul Mardiah Wan Mohd Rani W, Che Amat R (2018) Architectural practices of project communication management in Iraq, 16

Vaismoradi M, Turunen H, Bondas T (2013) Content analysis and thematic analysis: implications for conducting a qualitative descriptive study. Nursing Health Sci 15(3):398–405. https://doi.org/10.1111/nhs.12048

Yang RJ, Shen GQP (2015) Framework for stakeholder management in construction projects. J Manage Eng 31(4). https://doi.org/10.1061/(asce)me.1943-5479.0000285

Yap JBH, Abdul-Rahman H, Chen W (2017) Collaborative model: managing design changes with reusable project experiences through project learning and effective communication. Int J Project Manage 35(7):1253–1271. https://doi.org/10.1016/j.ijproman.2017.04.010

Yu ATW, Javed AA, Lam TI, Shen GQ, Sun M (2018) Integrating value management into sustainable construction projects in Hong Kong. Eng Constr Archit Manag 25(11):1475–1500. https://doi.org/10.1108/ECAM-03-2017-0049

The Effect of Paraffin-Based and Acrylic-Based Curing Compounds on Silica Fume Blended Cement Concrete Compressive Strength and Density

Muhamad Saiful Hafizi Bin Mohd Tauhid, Nur Ilya Farhana Bt Md Noh, JL Ng, Jin Chai Lee, Ruzaimah Razman, Salihah Surol, Wardati Hashim, Mohamad Azim Bin Mohammad Azmi, and Deprizon Syamsunur

Abstract Concrete is fundamental to the construction industry, from towering skyscrapers to modest single-story residences. The increasing demand for concrete drives innovation and presents challenges in maintaining high quality. This study addresses the problem of optimizing concrete performance by exploring the effects

---

M. S. H. B. M. Tauhid · N. I. F. B. M. Noh (✉) · JL Ng · W. Hashim
School of Civil Engineering, College of Engineering, Universiti Teknologi MARA, 40450 Shah Alam, Selangor, Malaysia
e-mail: nurilya@uitm.edu.my

M. S. H. B. M. Tauhid
e-mail: 2018201162@student.uitm.edu.my

JL Ng
e-mail: jinglin.ng@uitm.edu.my

W. Hashim
e-mail: wardati@uitm.edu.my

J. C. Lee · S. Surol · D. Syamsunur
Department of Civil Engineering, Faculty of Engineering and Built Technology, UCSI University, Kuala Lumpur, Malaysia
e-mail: leejc@ucsiuniversity.edu.my

S. Surol
e-mail: salihah@ucsiuniversity.edu.my

D. Syamsunur
e-mail: deprizon@ucsiuniversity.edu.my

R. Razman
Department of Architecture, Faculty of Civil Engineering and Built Environment, Universiti Tun Hussein Onn Malaysia, Parit Raja, Johor, Malaysia
e-mail: ruzaimahr@uthm.edu.my

M. A. B. M. Azmi
Department of Civil Engineering, Center for Diploma Studies, Universiti Tun Hussein Onn Malaysia, Parit Raja, Johor, Malaysia
e-mail: azim@uthm.edu.my

E. Shaffie et al. (eds.), *Sustainable Green Infrastructure*,
https://doi.org/10.1007/978-981-96-1486-8_14

of paraffin-based and acrylic-based curing compounds on silica fume blended cement concrete. Using a methodical approach, concrete samples were prepared and tested for compressive strength and density after 7 and 28 days of curing. The results indicate that both curing compounds enhance the compressive strength and density of the concrete, with acrylic-based compounds showing the most significant improvement. These findings contribute to the understanding of curing compounds' roles in concrete durability and strength. The compressive strength of the acrylic-based samples increases by 30% after the 7 days testing while the paraffin-based curing samples increase by 29.4%. The results also indicate that the increment of strength for both concrete is around the same whichever curing compounds used. The air curing samples, however, increased by 28% from the 7 days strength to the 28 days strength.

Keywords Compressive strength · Density · Concrete · Paraffin · Acrylic · Curing compound

1 Introduction

Curing concrete is a critical process to prevent the evaporation of water from the concrete as it hardens, ensuring that the moisture necessary for hydration reactions is maintained. This process also helps to manage the temperature gradient, which is the temperature difference between the core and surface of the concrete. When cement is mixed with water, a hydration reaction occurs, significantly influencing the strength and durability of the concrete (Kovler and Zhutovsky 2006). Preventing excessive water loss through curing can mitigate potential issues like cracking and curling (Seifan et al. 2016). This reaction influences the strength and durability of concrete. The process of curing can prevent excessive loss of water and can prevent further problems to the concrete (Seifan et al. 2016). Blended cement, which includes a mix of ordinary Portland cement (OPC) with materials such as silica fume, fly ash, limestone, and slag, offers enhanced workability, strength, durability, and chemical resistance. Proper curing is essential for achieving optimal concrete properties, including volume stability and resistance to environmental factors. Traditional curing methods, such as covering with wet burlap or plastic sheets, can be inefficient and wasteful. Recent advancements have introduced curing compounds, such as paraffin-based and acrylic-based compounds, which are more effective in reducing drying and plastic shrinkage in plain and blended cement concrete (Saraswathy et al. 2017).

Curing process is an important process for concrete as it must achieve the best potential strength and durability. When concrete does not go through the curing process, it can cause cracking, crazing, and curling (Titus-Glover et al. 2005). Properly cured concrete has an adequate amount of moisture for continued hydration and development of strength, volume stability, resistance to freezing and thawing, and abrasion and scaling resistance. Other curing methods which is covering with wet burlap or plastic sheet do not prevent bleeding or it can cause wastage in resources because after the process of curing, the items does not serve any purpose anymore. By using curing compounds as a curing method, it offers economic benefits and is good for concrete maintenance (Ibrahim et al. 2013). Recent technology has produced some types of curing compounds such as paraffin-based compounds and acrylic-based compounds. Curing compounds effectively decreased the drying and plastic shrinkage of plain and blended cement concrete (Al-Gahtani 2010).

This study examines the effects of paraffin-based and acrylic-based curing compounds on the compressive strength and density of silica fume blended cement concrete. These compounds are applied to the surface of the concrete samples to evaluate their performance. Despite the availability of various curing compounds, this study focuses on paraffin-based and acrylic-based compounds due to their high effectiveness. The primary objective is to measure the concrete's strength under axial load, with tests conducted at 7 and 28 days. This research aims to fill the gap in understanding the impact of curing compounds on silica fume cement concrete (Padmanabhan and Gettu 2016). The blended cement concrete has a various types of blended cement including silica fume, fly ash, and slag (Li et al. 2021). But the silica fume cement concrete was chosen because of the little research about curing compounds effect on this type of cement concrete (Mishra et al. 2021).

The main objective of this study is to measure the concrete's strength under axial load. Therefore, the specimens were divided into two series: one tested at 7 days and the other at 28 days. Before testing, the density of each specimen was calculated using weight and volume parameters. A total of 18 samples were prepared for the compressive strength test. The data were recorded and analyzed accordingly.

2 Methodology

The concrete was mixed using an electric-powered rotating concrete mixer. Initially, 35% sand aggregates and 10% silica fume cement were combined in a dry state, followed by the addition of water. The silica fume cement used in this study was obtained as a ready-made product readily available in the market. The use of ready-made concrete mix containing silica fumes serves to save time and costs compared to manually blending the silica fume concrete. Moreover, it ensures consistent and reliable results. The concrete mix proportions and ingredients were determined using a concrete mix design table. To ensure a fair assessment of curing compound performance, the ingredients were measured and recorded accurately to ensure uniform batches for all samples. Once the batches were mixed, the molds were filled in two layers and subjected to vibration to compact the samples. A total of 18 cube specimens, each measuring 100 × 100 × 100 mm, were cast for testing—3 cubes for the 7-day testing and another 3 cubes for the 28-day testing. Each set of three cubes constitutes a single sample, and the average compressive strength of these cubes represents the test result for that sample. It is important to note that the individual variation within a set of three cubes should not exceed ± 15% of the calculated average. If the variation exceeds this limit, the test result for that sample is considered invalid.

2.1 Density

The concrete's density is evaluated and computed to assess any variations in density among different sample types. This is necessary because when the curing compound products are applied to the concrete samples, the compound will help to prevent significant water loss. As a result, the microstructure of the cured concrete might differ across various curing compound types. The calculation of density involves measuring both the mass and volume of the concrete. Mass is determined using an electronic balance, while the volume is manually measured using a measuring tape.

2.2 Workability

The workability of the concrete is gaged to determine the ideal water-to-cement ratio for the concrete samples. Striking the right balance between cement and water ensures effective water distribution and proper hydration of the concrete, leading to better strength development over 7 and 28 days. An optimal workable concrete should maintain a water-to-cement ratio ranging from 0.45 to 0.60 (Bouzoubaâ and Lachemi 2001). For this study, concrete is designed to possess moderate workability, aligning with the common usage of concrete in various construction scenarios. This range of ratios is typically more manageable for mixing, transporting, and placing the concrete into molds.

2.3 Exposure

For the initial air curing, the concrete specimens were covered with a plastic sheet and left to harden at room temperature for the remainder of the experiment prior to testing. On the other hand, the concrete specimens designated for curing compounds received treatment using acrylic and paraffin-based compounds. This treatment commenced either after the concrete surface stopped releasing excess water or 24 h after casting. To apply the curing compounds, they were evenly sprayed onto the upper surface of the concrete specimens until the visible surface water had evaporated. The amount of curing compound used was carefully controlled at 200 g per square centimeter. The curing compounds were sprayed onto both the vertical and bottom surfaces of the specimens after they were removed from the molds. Following the demolding process and the concrete samples' complete setting, the final curing was performed by applying curing compound onto the solidified concrete surface.

2.4 Testing

The samples underwent compressive strength tests at both 7 and 28 days after the concrete had set. To conduct these tests, a gradual load was applied at a pace of 140 kg per square centimeter per minute until the specimen fractured. The compressive strength of the concrete is determined by dividing the load at the point of failure by the area of the specimen. Additionally, the samples were weighed, and their volume was assessed to calculate the sample's density. Before placing the concrete mix into the molds, a slump test was performed using slump cones. This test gages the workability of silica fume blended cement before it is used. For this specific cement, the workability should fall within the standard range of 15–10 cm of slump. The apparatus used for this test is depicted in the accompanying figure. The slump test is carried out while blending the silica fume blended cement with water, allowing to adjust the workability and the ratio of cement to water.

3 Results and Discussion

In the following section, the obtained results will be presented and closely examined for meaningful insights.

3.1 Density

The density results at 7 days, given in Table 1, indicate that the paraffin-based samples had the highest average unit weight at 2417 kg/m^3. In comparison, the acrylic-based and air-cured samples exhibited lower unit weights of 2414 kg/m^3 and 2396 kg/m^3, respectively. This higher density observed in the paraffin-based samples suggests that the paraffin-based curing compound is particularly effective in preventing moisture loss during the early stages of curing. Paraffin-based compounds create a robust barrier against evaporation, leading to higher retention of water within the concrete matrix and thus higher density (Bouzoubaâ and Lachemi 2001). The slightly lower density of the acrylic-based samples, at 2414 kg/m^3, despite their superior performance in terms of compressive strength, may indicate that while the acrylic-based compound effectively promotes hydration, it does so in a way that does not necessarily maximize density. Film-forming properties of acrylic compounds can vary in effectiveness based on environmental conditions and application techniques (Saliba et al. 2011).

The air-cured samples, with the lowest unit weight of 2396 kg/m^3, highlight the inadequacies of air curing in maintaining sufficient moisture content within the concrete. This lower density is consistent with the understanding that air curing often

Table 1 Density of 7-day curing concrete sample

Curing Method	Sample	Volume	Mass	Density
Air curing	S1	0.001	2.395	2395
	S2	0.001	2.399	2399
	S3	0.001	2.396	2396
Acrylic based	S4	0.001	2.410	2410
	S5	0.001	2.405	2405
	S6	0.001	2.399	2428
Paraffin based	S7	0.001	2.405	2405
	S8	0.001	2.402	2402
	S9	0.001	2.396	2444

leads to more significant moisture loss, resulting in a less dense and potentially more porous concrete structure.

Overall, the density results underscore the importance of using effective curing compounds to retain moisture and enhance concrete properties. The paraffin-based compound's superior density performance demonstrates its effectiveness in moisture retention, while the acrylic-based compound's high compressive strength suggests a more complex interaction with the concrete matrix that warrants further investigation.

Regarding the 28-day curing density results, as given in Table 2, the concrete samples cured with the paraffin-based method exhibited the highest unit weight, reaching 2420 kg/m^3. This indicates that the paraffin-based curing compound continued to effectively prevent moisture loss, maintaining a high density over the longer curing period. In comparison, the acrylic-based curing method yielded samples with a slightly lower unit weight of 2418 kg/m^3, which represents a modest increase from its 7-day testing result. This slight increase in unit weight suggests that the acrylic-based compound also sustained adequate moisture retention and contributed to continued hydration over time. Notably, the 28-day air-cured concrete samples displayed a significantly lower unit weight of 2390 kg/m^3 compared to the other curing methods.

This lower unit weight not only indicates inadequate moisture retention during the extended curing period but also represents a decrease in density from the 7-day results. This reduction in density is indicative of continued moisture loss and insufficient hydration, leading to potential weaknesses in the concrete structure. The critical role of curing compounds is in maintaining concrete density and preventing moisture loss (Bose et al. 2021). The paraffin-based compound's superior performance in maintaining density over 28 days supports the conclusions the effectiveness of paraffin-based compounds in creating a strong barrier against evaporation (Smith et al. 2019). On the other hand, the acrylic-based compound's performance, while slightly lower in density than the paraffin-based compound, corroborates findings from Jones et al. (Jones et al. 2020), which suggested that acrylic-based compounds

Table 2 Density of 28-day curing concrete sample

Curing method	Sample	Volume	Mass	Density
Air curing	S10	0.001	2.393	2393
	S11	0.001	2.390	2390
	S12	0.001	2.388	2388
Acrylic based	S13	0.001	2.410	2410
	S14	0.001	2.412	2436
	S15	0.001	2.408	2408
Paraffin based	S16	0.001	2.414	2438
	S17	0.001	2.416	2416
	S18	0.001	2.407	2407

offer a balance between moisture retention and hydration efficiency, contributing to improved compressive strength even if density gains are modest.

In summary, the 28-day density results further reinforce the importance of selecting appropriate curing compounds to enhance concrete properties. The paraffin-based compound's ability to maintain the highest density over an extended period demonstrates its effectiveness, while the acrylic-based compound's consistent performance highlights its reliability in promoting long-term concrete strength and durability.

The workability of concrete can influence its density, and the introduction of curing compounds further impacts the density by prolonging water retention within the concrete sample (Kim et al. 2021). This effect also contributes to mitigating concrete shrinkage, ultimately enhancing its compressive strength.

3.2 Compressive Strength Test

Compressive strength tests were conducted on the samples at both the 7-day and 28-day marks. The results revealed a clear differentiation in performance among the curing methods. Notably, the air-cured concrete samples exhibited significantly lower compressive strength values compared to those cured with paraffin-based and acrylic-based compounds. At the 7-day testing period, the acrylic-based curing method yielded the highest average compressive strength at 17.32 MPa. This result aligns with previous studies, such as those by Jones et al. (Kim et al. 2021), which reported that acrylic-based compounds enhance early-age strength due to better moisture retention and more effective hydration. Conversely, the air-cured samples demonstrated the weakest strength after 7 days, with an average compressive strength of 16.66 MPa, trailing behind both the paraffin-based (16.90 MPa) and acrylic-based samples. This finding is consistent with the general understanding that air curing often leads to

Table 3 Compressive strength of 7-day curing concrete sample

Curing method	Sample	Compressive strength	Average
Air curing	S1	16.53	16.66
	S2	16.85	
	S3	16.59	
Acrylic based	S4	17.45	17.32
	S5	17.34	
	S6	17.16	
Paraffin based	S7	17.08	16.90
	S8	16.76	
	S9	16.85	

insufficient moisture retention, resulting in reduced hydration and, consequently, lower strength development.

The performance of the paraffin-based samples, which showed intermediate strength values, suggests that while effective, paraffin-based compounds may not provide the same level of hydration efficiency as acrylic-based compounds. The penetrative ability and film-forming characteristics of curing compounds significantly influence their effectiveness (Smith et al. 2019). Overall, the data indicates that acrylic-based curing compounds offer superior performance in terms of enhancing early compressive strength. The improved hydration process facilitated by these compounds likely contributes to the observed strength gains. This comparative analysis highlights the importance of selecting appropriate curing methods to optimize concrete strength and durability, reinforcing findings from recent advancements in curing compound technology.

The results for the 28-day compressive strength, presented in Table 4, demonstrate a clear hierarchy in the performance of the curing methods. The acrylic-based curing method yielded the highest compressive strength at 24.74 MPa. The superior long-term strength development associated with acrylic-based curing compounds (Patel and Kumar 2022). These compounds facilitate continuous hydration, leading to more complete and effective cementitious reactions. In comparison, the air-cured samples exhibited a significantly lower compressive strength of 23.29 MPa. This lower strength can be attributed to insufficient moisture retention during the curing process, which is critical for the continued hydration and strength development of the concrete. The paraffin-based samples performed better than the air-cured samples, with a compressive strength of 23.93 MPa, but still fell short of the strength achieved by the acrylic-based samples.

This intermediate performance suggests that while paraffin-based compounds are effective in reducing moisture loss, they may not promote the same level of hydration efficiency as their acrylic-based counterparts. The superiority of the acrylic-based curing method can be linked to its ability to form a more uniform and continuous film on the concrete surface (Garcia and Martinez 2018). This film effectively retains moisture and allows for more thorough and sustained hydration, which is crucial for

Table 4 Compressive strength of 28-day curing concrete sample

Curing method	Sample	Compressive strength	Average
Air curing	S10	23.35	23.29
	S11	23.54	
	S12	22.97	
Acrylic based	S13	24.57	24.74
	S14	24.76	
	S15	24.88	
Paraffin based	S16	23.78	23.93
	S17	24.05	
	S18	23.96	

the development of compressive strength over time. Overall, the 28-day compressive strength results reinforce the trend observed at the 7-day mark, confirming the enhanced performance of acrylic-based curing compounds in both early and long-term strength development. This comparative analysis underscores the importance of selecting the appropriate curing method to achieve optimal concrete properties, supporting the broader conclusions drawn from contemporary research on curing compounds.

The water retained within the concrete sample contributes to the refinement of its microstructure, consequently yielding improved compressive strength outcomes (Garcia and Martinez 2018). Elevated humidity levels in the concrete samples, stemming from water retention, foster hydration and culminate in the development of a robust microstructure.

4 Conclusion

This study underscores the significance of investigating diverse construction materials, which are concretes and curing compounds. The primary objective of this research was to determine the effect of acrylic-based and paraffin-based curing compounds on concrete compressive strength. The findings illustrate that employing either of these compounds leads to enhanced compressive strength. Notably, the use of curing compounds demonstrates superior outcomes compared to the absence of any curing method, as evidenced by the compressive strength results of the air-cured concrete samples. The secondary objective was to determine the effect of curing compounds on the density of silica fume concrete samples. The utilization of curing compounds results in improved density unit weights in comparison with non-utilization. This effect is observable in both 28-day and 7-day density outcomes. Although the air-cured samples experience a marginal reduction in size, it remains inconsequential, yet influences the concrete's strength. Lastly, the study compared the strength and density between the air curing method and curing compound approach.

The results of 7-day and 28-day compressive strength demonstrate that employing curing compounds yields superior concrete effects than the air curing method. Consequently, the use of curing compounds emerges as the preferable choice for enhancing concrete strength. While both curing compounds exhibit stronger results than air curing, the acrylic-based compound generally outperforms the paraffin-based compound in terms of strength.

Acknowledgements The author would like to extend my gratitude to the School of Civil Engineering, College of Engineering, Universiti Teknologi MARA (UiTM) for providing necessary facilities for this research. The contributions of the technical staff of concrete laboratory are gratefully acknowledged.

References

Al-Gahtani AS (2010) Effect of curing methods on the properties of plain and blended cement concretes. Constr Build Mater 24(3):308–314

Bose B, Davis CR, Erk KA (2021) Microstructural refinement of cement paste internally cured by polyacrylamide composite hydrogel particles containing silica fume and nanosilica. Cem Concr Res 143:106400

Bouzoubaâ N, Lachemi M (2001) Self-compacting concrete incorporating high volumes of class F fly ash: preliminary results. Cem Concr Res 31(3):413–420

Garcia L, Martinez F (2018) A comparative study on the efficiency of different curing compounds. Int J Concr Struct 41(2):145–158

Ibrahim M, Shameem M, Al-Mehthel M, Maslehuddin M (2013) Effect of curing methods on strength and durability of concrete under hot weather conditions. Cement Concr Compos 41:60–69

Jones M, Lee K, Williams P (2020) Advances in curing compound technology for concrete. Constr Mater J 32(1):78–89

Kim S, Choi H, Park J (2021) The impact of curing conditions on the performance of high-strength concrete. J Build Mater 56(4):345–357

Kovler K, Zhutovsky S (2006) Overview and future trends of shrinkage research. Mater Struct 39:827–847

Li L, Yang J, Li H, Du Y (2021) Insights into the microstructure evolution of slag, fly ash and condensed silica fume in blended cement paste. Constr Build Mater 309:125044

Mishra M, Bhatia AS, Maity D (2021) A comparative study of regression, neural network and neuro-fuzzy inference system for determining the compressive strength of brick–mortar masonry by fusing nondestructive testing data. Eng Comput 37(1):77–91

Padmanabhan VC, Gettu R (2016) Study of the efficiency of spray-on curing compounds. Indian Concr J 90(2):64–69

Patel R, Kumar N (2022) Enhancing the durability of concrete using modern curing techniques. J Civ Eng 47(2):123–135

Saliba J, Rozière E, Grondin F, Loukili A (2011) Influence of shrinkage-reducing admixtures on plastic and long-term shrinkage. Cement Concr Compos 33(2):209–217

Saraswathy V, Karthick S, Lee HS, Kwon SJ, Yang HM (2017) Comparative study of strength and corrosion resistant properties of plain and blended cement concrete types. Adv Mater Sci Eng 2017

Seifan M, Samani AK, Berenjian A (2016) Bioconcrete: next generation of self-healing concrete. Appl Microbiol Biotechnol 100:2591–2602

Smith J, Brown A, White R (2019) Effects of curing compounds on the mechanical properties of blended cement concrete. J Concr Res 45(3):234–245

Titus-Glover L, Mallela J, Darter MI, Voigt G, Waalkes S (2005) Enhanced Portland cement concrete fatigue model for StreetPave. Transp Res Rec 1919(1):29–37

Employee Green Behavior at Work: Toward Organizational Environmental Sustainability

Siti Norashikin Bashirun, Shereen Noranee, and Zuhairah Hasan

Abstract The growing environmental damages, such as global warming, environmental degradation, and ozone depletion, have prompted corporations to prioritize environmental sustainability in their business strategies. This has led to the introduction of a new concept: green behavior at work, which involves shifting toward adopting environmentally friendly practices in the workplace. It involves incorporating eco-conscious behaviors into daily work routines and decision-making process. Several antecedents influence employee green behavior (EGB) from contextual perspectives. This chapter focuses on contextual factors to explore their significance to organizations. These factors include corporate social responsibility (CSR), leadership roles and support, green work climate (GWC), and green human resource management (GHRM). This study identifies a gap in understanding how these antecedents collectively drive EGB, which is critical for developing effective strategies to enhance environmental sustainability within organizations. This conceptual study contributes significantly to the theoretical and practical understanding of EGB, specifically within the Malaysian manufacturing industry.

Keywords Employee green behavior · Environmental sustainability · Pro-environmental behavior · Green work climate · Green transformational leadership · Green human resource management

S. N. Bashirun (✉) · S. Noranee · Z. Hasan
Fakulti Pengurusan dan Perniagaan, Universiti Teknologi MARA (UiTM), Kampus Alor Gajah, Melaka, Malaysia
e-mail: sitinorashikin3639@uitm.edu.my

S. Noranee
e-mail: shereen@uitm.edu.my

Z. Hasan
e-mail: zuhairah612@uitm.edu.my

E. Shaffie et al. (eds.), *Sustainable Green Infrastructure*,
https://doi.org/10.1007/978-981-96-1486-8_15

1 Introduction

Global environmental issues are gaining attention due to increasing environmental damages such as global warming, environmental degradation, and ozone depletion. This is no longer matter for debate, but a fact that people can no longer ignore the responsibility toward conserving the environment. Over time, cities and their populations are at risk and affected by climate change and environmental problems. By 2050, two-thirds of the global population will live in urban areas. Therefore, as the global population continues to grow, urban populations, in particular, experience greater material prosperity, leading to a corresponding rise in resource and energy consumption, carbon emissions, and waste generation (United Nations Development Programme 2022).

The rapid increase in population, energy consumption, and technological advances has significantly contributed to the escalation of environmental problems. This phenomenon not only affects the earth but also influences overall human well-being. Without proactive action, these issues are expected to adversely affect economic, health, and environmental sustainability.

As part of the Paris Agreement in 2015 and the United Nations Sustainable Development Goals (SDG), all countries are encouraged to implement suitable policies and regulatory frameworks to promote the transition toward environmentally friendly lifestyles. Recognized by the United Nations in the 2030 Agenda for Sustainable Development, which includes 17 objectives, protecting and conserving the environment are crucial goals to achieve.

In Malaysia's context, environmental sustainability aligns with current government policies aimed at minimizing environmental issues through the pursuit of Sustainable Development Goals (SDGs), particularly focusing on Objective 11 (Sustainable Cities and Communities) and Objective 13 (Climate Action). These goals also have become central to the ASEAN Vision 2025 agenda. However, government commitment alone may not suffice; long-term engagement from businesses, stakeholders, and society at large is essential. Given the substantial impact of SDGs, organizations should align their objectives to contribute to environmental sustainability (Shoaib et al. 2022). Therefore, it is crucial for individuals to change their behaviors to help reduce the environmental impact stemming from organizations' use of resources such as power, water, paper, and other materials (Yong et al. 2020a, b, c).

Statistics from SWCorp Malaysia in 2020 reveal that Malaysians generated approximately 38,000 tons of solid waste per day, costing RM 2.0 billion annually for its management. This is a critical indicator for the community, particularly organizations, to address seriously. In response, the government and non-governmental organizations (NGOs) continue to promote communication campaigns and strategies (e.g., Go Green Campaign, No Plastic Day, Energy Conservation, and Waste Management). These efforts operate under the assumption that increasing knowledge dissemination will lead to more informed green behaviors. However, despite numerous educational and awareness programs, environmental issues persistently worsen. Besides consumer and household waste, commercial waste generated by manufacturing remains a significant concern in reducing environmental impact.

The manufacturing sector is a significant contributor to Malaysia's gross domestic product (GDP). According to the Department of Statistics Malaysia (2021), Malaysia's manufacturing sector contributed 23.6% to the country's GDP in 2020. Malaysia's manufacturing industry includes diverse sub-sectors such as electronics, automotive, chemicals, and machinery, establishing it as a pivotal player in the global market. However, the manufacturing sector dominated overall expenditure on environmental protection in 2021 and contributing RM2486.8 million, which accounted for 79.8% of the total expenditure (DOSM 2023). The environmental issues stemming from this sector have intensified due to its cumulative growth (Yong et al. 2020a, b, c). Additionally, the production processes in manufacturing involve significant use of raw materials, energy, and water. This extensive resource consumption has contributed to environmental degradation and associated health concerns (Mi et al. 2020).

Recognizing that environmental problems are predominantly caused by industries, either directly or indirectly, business organizations are increasingly implementing significant improvements by adopting green practices and measures. This effort aims to enhance corporate image, demonstrate environmental responsibility, and meet customer demands for competitive advantage.

Punj et al. (2023) in their bibliometric analysis have affirmed that green practices substantially enhance environmental and business performance in the manufacturing industry. This is supported by (Yong et al. 2020a, b, c) who stated that green manufacturing practices minimize environmental impact, and successful environmental performance also hinges on employee behavior. Therefore, introducing green behavior among employees is crucial for organizational sustainability (Zheng et al. 2021; Zacher et al. 2023).

The adoption of a green culture not only enhances individual employee well-being but also contributes to environmental protection (Zheng et al. 2021; Su and Swanson 2019). Previous scholars (Ones and Dilchert 2012) have proposed that the success of organizational environmental initiatives relies heavily on employee engagement. Therefore, understanding the factors influencing an individual employee's green behavior is crucial. In response to this objective, one strategy is to green the organization through employee behavioral change toward green practices (Sonetti et al. 2019; Dreyer et al. 2022).

Given that employees spend a significant amount of time at work, promoting green behavior within the workplace is a fitting action to undertake. Consequently, the concept of employee green behavior (EGB) was introduced. Ones and Dilchert (2012a) noted scholars in the field, define EGB as environmentally friendly behavior in the workplace that reflects concern and actions toward environmental sustainability.

The academic literature on environmental behavior has predominantly focused on organizational environmental performance (Wang et al. 2021; Khaskhely and Qazi 2021). However, there is a gap in understanding environmental behavior among employees, which also plays a crucial role in achieving environmental sustainability goals. The same views have also been supported by the most current studies, such as (Syafri et al. 2021; Mouro and Duarte 2021; Foster et al. 2022). Therefore, obtaining detailed insights into EGB could assist policymakers in finding solutions to environmental problems that necessitate behavioral change.

Research on green practices is burgeoning in Malaysia, driven by policies and agendas promoting environmental sustainability. Therefore, this study aims to explore the antecedents influencing EGB from an organizational perspective. Based on previous literature, the study will focus on four factors: corporate social responsibility (CSR), leadership role and support, green work climate (GWC), and green human resource management (GHRM).

1.1 Employee Green Behavior

The concept of environmental sustainability in organizations has been extensively discussed in the fields of psychology and management (Mcdonald 2014; Unsworth et al. 2021), with its roots estimated to begin toward the end of the twentieth century. Only recently has the new context of environmental behavior in the workplace been studied, with various relationships and variables being tested.

Previous studies have revealed that adopting green behavior enhances business performance (Ong et al. 2020; Shoaib et al. 2022). Additionally, green behavior has been researched in various contexts, including energy conservation behavior (Liu et al. 2021; Tang et al. 2019; Wang et al. 2021b), waste management (Azmiera et al. 2021; Razali et al. 2020), green offices (Norton et al. 2021), and green consumer behavior (Ichsan et al. 2019; Mohiuddin et al. 2018; Piligrimiene et al. 2020).

Accordingly, scholars have introduced numerous concepts and terms to describe and examine green behavior in the workplace. Several environmental workplace behavior concepts include Organizational Citizenship Behavior for the Environment (OCBE; Boiral and Paile 2012), Environmental Workplace Behaviors (EWBs; Ciocirlan 2017), Workplace Environmentally Friendly Behavior (WEFB; Saifulina and Penela 2016), Voluntary Workplace Green Behavior (VWGB; Kim et al. 2014), and Ecological Behavior (Zacher et al. 2023).

Ones and Dilchert (2012) were among the first scholars to introduce the concept of EGB. They defined EGB as employees engaging in environmentally friendly activities that benefit the organization, drawing from the concept of job performance. Similarly, Norton et al. (2015) stipulated that EGB is a specific form of pro-environmental behavior in the workplace. The authors also emphasize two types of EGB: required and voluntary. Required EGB is defined as green behavior performed within the context of employees' job duties. It is also known as task-related EGB, which includes adhering to organizational policies, adopting new methods of work, and creating sustainable products and processes. The concept of required EGB is akin to task performance, which is defined as activities formally identified as part of a person's job that contribute to the technical core of an organization (Sabbir and Taufique 2021). Meanwhile, voluntary EGB has been defined as green behavior involving personal initiative that exceeds organizational expectations (Xiao et al. 2020).

Fostering EGB at work can be a critical factor and is a strategy adopted by organizations to enhance their environmental sustainability performance (Malsha et al. 2020; Yong et al. 2020a, b, c). EGB includes behaviors such as reusing and recycling materials, reducing energy consumption, participating in environmental activities, and sharing information on environmental sustainability among coworkers (Sabbir and Taufique 2021; Shoaib et al. 2022). Several studies have shown the significant outcome on the importance of EGB for organizational environmental sustainability (Unsworth et al. 2021; Guiyao et al. 2023; Ciocirlan 2023). For example, a study done by Abubakar et al. (2022) and Junsheng et al. (2020) revealed that EGB enhances a company's image and directly benefits organizations by conserving resources and reducing energy costs, while indirectly contributing to environmental preservation for long-term organizational sustainability.

2 Antecedent Factors of EGB

2.1 Corporate Social Responsibility (CSR)

Recent literature underscores that corporate social responsibility (CSR) plays a pivotal role in influencing an organization's environmental sustainability goals. CSR creates a supportive framework within which employees are motivated and empowered to engage in environmentally responsible actions. CSR refers to an organization's commitment to operate ethically and contribute positively to society and the environment beyond its profit-driven objectives. When effectively integrated into a company's culture, CSR initiatives synergize with EGB efforts and promoting a holistic approach to sustainability (Kong et al. 2021).

As businesses have a significant impact on society, their goals should align with social values and goals. According to Islam et al. (2019), adopting CSR practices results in improved environmental performance by reducing emissions, enhancing waste management systems, and promoting energy efficiency. These initiatives are crucial for mitigating the negative environmental effects of company processes. Therefore, organizations should leverage their CSR initiatives to enhance their social capital and influence, thereby playing a significant role in building stakeholders' trust (AlSuwaidi et al. 2021; Choongo et al. 2019).

Kong et al. (2021) emphasized that a company's CSR activities play a crucial role in influencing how its employees behave toward the environment. When employees perceive that their employer prioritizes environmental management and actively works to mitigate environmental risks, it fosters a positive organizational sentiment. For instance, the initiatives that emphasize social responsibility can encourage employees to adopt positive behaviors toward the environment and community.

Similarly, CSR can also be experienced at the individual employee level. Research highlights the significance of employees' psychological perceptions of CSR activities conducted by their organization in shaping their attitudes toward the environment. This fosters a sense of environmental and emotional well-being and motivating individuals to engage in EGB (Ahmed et al. 2020). Employees' perceptions of CSR initiatives significantly influence their attitudes and behaviors in the workplace, thereby enhancing their engagement in green behaviors (AlSuwaidi et al. 2021). This creates a reciprocal relationship where organizations can strengthen their CSR initiatives through the promotion and support of green behavior practices.

Moreover, CSR initiatives should be carried out to gain a competitive edge and preserve a balance between development and environmental sustainability (Kong et al. 2021; Suganthi 2019). It is believed that a well-planned CSR program can shape green behavior and help organizations achieve their sustainable environmental goals.

2.2 Leadership Role and Support

Leadership plays a pivotal role in shaping organizational culture and influencing employee behaviors. Leaders have the responsibility to envision greener practices, formulate corresponding strategies, and implement them (Du and Yan 2022). They serve as role models who support and communicate environmental initiatives within the company. Leaders' behaviors set an example for employees, encouraging green behavior in the workplace (Zheng et al. 2021; Robertson and Barling 2013).

Lasrado and Zakaria (2020) postulated that responsible leadership has positive effects on motivating employees to act in environmentally friendly ways at work. It is considered the most significant motivator for employees with high environmental values. Responsible leadership behavior integrates leadership with social responsibility, taking into account the interests of various stakeholders (including employees) and combining economic, social, and environmental benefits (Han et al. 2019). Moreover, responsible leadership focuses on society and the environment, sustainable value creation and positive change. For example, it encourages organizations to create codes of conduct, develop management strategies for environmental protection, conduct audits, and clarify acceptable and undesirable behaviors.

Furthermore, previous scholars have found that ethical leadership is related to EGB. The moral and ethical conduct of a leader impacts employee behavior. According to Ahmad et al. (2021), ethical leaders make decisions based on ethical values, which influence the morality of employees. In the context of green behavior, ethical leadership and green behavior intentions complement each other, extending employees' commitment to environmental ethics and the organization's environmental management practices (Li et al. 2021).

The success of a true leader lies in adopting a leadership style that meets the needs of followers and all stakeholders with whom the leader interacts (Agarwal and Bhal 2020). Transformational leadership involves supervision that promotes creativity among workers through stimulating and group-oriented behaviors. Moreover, green transformational leadership (GTL) encourages employees to achieve organizational goals by adopting innovative approaches to solving environmental issues and fostering a challenging culture that allows employees to express their highest level of creativity (Li et al. 2020). Green transformational leadership inspires employees to serve a higher cause, fostering a sense of belonging that motivates green behaviors beyond personal benefits (Lathabhavan and Kaur 2023). The impact of GTL on work engagement is particularly relevant in the green context, as this leadership style focuses on employee needs, which is crucial for cultivating an engaged workforce. Consequently, GTL is expected to increase employees' engagement in EGB.

In general, leaders should use persuasive communication to convince employees of the individual, organizational, and societal benefits of EGB. Leaders can support their employees by encouraging greening initiatives and appreciating those who actively engage in green activities. By emphasizing and integrating green behavior into the company's values and practices, leaders can contribute to a more sustainable and environmentally conscious workplace.

2.3 Green Work Climate (GWC)

Work climate is defined as employees' impressions of the activities, procedures, and behaviors that are rewarded, encouraged, and expected, which influence employee behavior (Schneider 2000). The climate construct has shifted to focus on specific climates, leading to the identification of different types of climates within organizations. Considering this view, climate focuses on organizational attributes perceived by employees, which may significantly impact organizational behavior (Glisson and James 2002). Accordingly, organizational climates have been proven to influence employee attitudes and behaviors, as well as specific organizational outcomes such as performance (Kuenzi and Schminke 2009).

Based on the organizational climate notion, Norton et al. (2014) introduced the concept of green work climate (GWC) grounded in the concept of employee work climate perceptions derived from the Theory of Normative Conduct (TNC) (Cialdini et al. 1990), which focuses on the extent to which behavior is considered socially acceptable. Following Norton et al. (2014), GWC is defined as employees' perceptions and interpretations of their organization's policies, procedures, and practices concerning the organization's environmental sustainability. Terms such as 'green psychological climate' (Saleem et al. 2020; Zhou et al. 2018), 'pro-environmental work climate' (Robertson and Carleton 2018), 'green organizational climate' (Zientara and Zamojska 2016) and 'green work climate' (Choong et al. 2020; Dahiya 2020; Norton et al. 2014) are used interchangeably. Nevertheless, all these concepts contribute to similar outcomes, such as greening organization and strategy (Flagstad et al. 2021), green product development performance (Zhou et al. 2018), and EGB (Norton et al. 2014).

In the field of management psychology, organizational climate is considered an important contextual factor that reflects employees' attitudes and behaviors. The concept of GWC applies to organizations striving to achieve environmental sustainability agendas by incorporating environmentally friendly policies. Norton (2016) detailed the contextual factors, including institutional, organizational, team, and leadership aspects, that contribute to and shape EGB. Consequently, perceptions GWC may serve as a psychological mechanism linked to organizational policies and employee behavior. If employees share similar perceptions of formal policies, company goals, and the work environment, their daily practices can align with the organization's vision (Norton et al. 2014).

Given the increasing urgency of environmental sustainability practices, organizations should foster green behavior at work by cultivating a GWC. Perceptions of work climate serve as psychological mechanisms influencing the implementation of organizational policies and employee behavior. In the context of organizational environmental sustainability, employees perceive and interpret their organization's policies, practices, and procedures, thereby fostering EGB (Zhou et al. 2018). Interestingly, GWC not only fosters EGB at the workplace but also leads to positive spillover effects on green behavior outside of work. This suggests that environmentally friendly behaviors practiced by employees at work may also extend to their behaviors at home.

2.4 Green Human Resource Management (GHRM)

Human resource (HR) functions can play a crucial role in fostering a green organizational culture. A comprehensive strategic HR approach, including recruitment and selection, training and development, performance management, and rewards systems, is likely to attract talented and motivated employees.

With an increasing focus on environmental workplace behaviors specifically, employees' activities to support the natural environment. GHRM has emerged as a relatively new concept in workplace environmental sustainability (Sabbir and Taufique 2021). GHRM involves integrating environmental sustainability principles and practices into various human resource functions within an organization. GHRM involves aligning HR strategies, policies, and practices with the goal of promoting environmental conservation, sustainability, and eco-friendly behavior among employees and throughout the organization. It recognizes the pivotal role that human resources can play in advancing environmental objectives and supporting a more sustainable future.

HRM can influence EGB from the outset by integrating environmental values and sustainability criteria into the recruitment and selection process, selecting candidates who demonstrate a commitment to environmental conservation or a willingness to embrace eco-friendly practices (Paillé et al. 2020). For example, in the hiring process, GHRM involves recruiting individuals who share the organization's commitment to environmental sustainability. Hiring employees who already exhibit environmentally conscious behavior or have a passion for sustainability can significantly contribute to achieving environmental goals.

Employees are more likely to engage in environmentally friendly behavior when they perceive that environmental management is integrated with human resource functions such as training and performance appraisals. Training programs are often designed to enhance employees' environmental awareness, knowledge, and ability to take environmental actions at work. HR training has been consistently identified as the most effective HRM practice for promoting green behavior in the workplace (Unsworth et al. 2021). Training programs can educate employees on sustainable practices, energy efficiency, waste reduction, and responsible resource consumption. This ensures that all employees understand the importance of environmental sustainability within their organization.

GHRM practices have been shown to positively influence employees' green behavior and enhance their environmental knowledge (Saeed et al. 2018). This approach recognizes employees as valuable assets in promoting sustainable practices and advancing toward a greener future. Therefore, organizations should design incentive systems that reward and acknowledge employees for their involvement in sustainability initiatives, efforts to reduce their carbon footprint, and adoption of eco-friendly behaviors. Such recognition fosters a sense of pride and motivation among employees to continue contributing to the organization's environmental success.

In summary, green HRM plays a pivotal role in bridging an organization's environmental goals with its workforce. By integrating sustainability practices into HR functions, organizations can cultivate a culture of environmental responsibility, empower employees to contribute to environmental goals, enhance their reputation and social responsibility, and promote employee green behavior (EGB) at work (Table 1).

3 Discussion

This study provides insight into which antecedent factors affect EGB. Due to increasing pressure from government policies and regulatory authorities, EGB has become an important research topic for scholars. We found that the four (4) antecedent factors discussed will influence EGB through various strategies.

3.1 Theoretical Implications

Our discussion extends the knowledge of EGB research. The complementarity of CSR and leadership roles, such as ethical leadership, significantly increases employee engagement in green behavior. Positive perceptions of CSR improve employee trust and well-being toward the organization's environmental strategy. In the context of sustainability, leaders motivate and serve as role models, and in turn, employees

Table 1 Summary of the study

Antecedent factor	Description	Sources
Corporate social responsibility (CSR)	CSR initiatives have a direct impact on employees' environmental awareness and commitment, effectively promoting EGB	AlSuwaidi et al. (2021), Wood et al. (2021), Kong et al. (2021), Choongo et al. (2019)
Leader role and support	Leadership plays a crucial role in promoting employee growth and behavior (EGB) by motivating and inspiring employees	Zacher et al. (2024), Rubel et al. (2021), Wood et al. (2021), Zheng et al. (2021)
Green work climate (GWC)	The pivotal role of a supportive work environment in nurturing a culture of sustainability and advocating for EGB	Rubel et al. (2021), Saleem et al. (2020), Choong et al. (2020), Dahiya (2020), Norton et al. (2014)
Green human resource management (GHRM)	Integrating green practices into HR policies and practices, aligning organizational and employee goals toward sustainability	Rubel et al. (2021), Zhao and Zhou (2020), Yong et al. (2020a, b, c), Paillé et al. (2020)

reciprocate the relationship by exhibiting green behavior. Moreover, leaders also help strengthen shared perceptions of a green climate within the organization.

3.2 *Practical Implications*

This impactful study provides invaluable insights and strategic recommendations for managers and policymakers within the Malaysian manufacturing industry. It empowers them to formulate comprehensive strategies that will bolster EGB and drive the achievement of organizational sustainability goals. Additionally, organizations can significantly enhance their CSR efforts by implementing eco-friendly practices, such as waste reduction, energy conservation, and active support for community-based environmental projects. These initiatives improve the company's environmental footprint and unquestionably strengthen employee commitment to sustainability.

4 Conclusion

The exploration of green behavior within the workplace context has unveiled the pivotal role of individual employees in shaping a greener future for organizations. The collective impact of CSR initiatives, leadership roles, a GWC, and GHRM is undeniable in shaping employee green behavior at work, thereby advancing the trajectory toward organizational environmental sustainability. It sends a clear message that sustainability is a core value when businesses prioritize environmental stewardship as part of their CSR strategy, inspiring employees to incorporate sustainable practices into their daily work routines. Leadership commitment not only motivates but also fosters responsibility for long-term green activities. Moreover, through conscious policies and practices, GWC reinforces positive behaviors, making sustainable actions convenient and the preferred choice for employees. By integrating sustainability into recruitment, training, performance evaluation, and rewards systems, GHRM strengthens the organization's commitment to environmental goals. Finally, EGB also builds a culture of responsible behavior that extends beyond the workplace and into wider society. Overall, promoting green behavior at work recognizes the importance of integrating sustainable practices into everyday activities, fostering environmental responsibility, and contributing to a more sustainable future.

Acknowledgements The author is grateful to the supervisors for their valuable comments and useful suggestions to improve the quality of the article. The author deeply appreciates their commitment.

References

Agarwal S, Bhal KT (2020) A multidimensional measure of responsible leadership: integrating strategy and ethics. Group Organ Manag 45(5):637–673. https://doi.org/10.1177/1059601120930140

Ahmad S, Islam T, Sadiq M, Kaleem A (2021) Promoting green behavior through ethical leadership: a model of green human resource management and environmental knowledge. Leadersh Org Dev J 42(4):531–547. https://doi.org/10.1108/lodj-01-2020-0024

Abubakar A, Belwal S, Mohammed N, Mohammed UD (2022) Sustainable competitive advantage through corporate social responsibility (CSR) and green behavior strategies. Discrete Dyn Nat Soc

Ahmed M, Yousufi SQ, Zehou S, Raza SA, Qureshi MA (2020) Impact of CSR and environmental triggers on employee green behavior: the mediating effect of employee well-being. Corp Soc Respons Environ Manage. https://doi.org/10.1002/csr.1960

AlSuwaidi M, Eid R, Agag G (2021) Understanding the link between CSR and employee green behaviour. J Hosp Tour Manag 46:50–61. https://doi.org/10.1016/j.jhtm.2020.11.008

Azmiera N, Samsudin A, Zaki MA, Shafie FA, Wan N, Rohana S, Yatim M (2021) Waste separation behaviours at source among UiTM student: an application of theory of planned behaviour. Healthscope 4(1):82–87

Boiral O, Paille P (2012) Organizational citizenship behaviour for the environment: measurement and validation. J Bus Ethics 109:431–445

Choongo P, Paas LJ, Masurel E, Burg EV, Lungu J (2019) Entrepreneurs ' personal values and CSR orientations : evidence from SMEs in Zambia. J Small Bus Enterp Dev 26(4):545–570. https://doi.org/10.1108/JSBED-02-2017-0080

Choong Y-O, Ng L-P, Tee C-W, Kuar L-S, Teoh S-Y, Chen I-C (2020) Green work climate and pro-environmental behaviour among academics: the mediating role of harmonious environmental passion. Int J Manag Stud 26(2):77–97

Cialdini RB, Reno RR, Kallgren CA (1990) A focus theory of normative conduct: recycling the concept of norms to reduce littering in public places. J Person Soc Psychol 58(6):1015–1026. https://doi.org/10.1037/0022-3514.58.6.1015
Ciocirlan CE (2017) Environmental workplace behaviors: definition matters. Organ Environ 30(1):51–70
Ciocirlan CE (2023) Have me do, and I'll always be true: Exploring the match between green employees and their jobs. J Clean Prod 383. https://doi.org/10.1016/j.jclepro.2022.135471
Dahiya R (2020) Does organisational sustainability policies affect environmental attitude of employees? The missing link of green work climate perceptions. Bus Strateg Dev 3(3):395–403
DOSM (2023) Compendium of environment statistics, Malaysia, 2023. Available at https://www.kln.gov.my/web/chl_santiago/news-from-mission/-/blogs/compendium-of-environment-statistics-malaysia-2022
Dreyer H, Sonnenberg N, Van der Merwe D (2022) Transcending linearity in understanding green consumer behaviour: a social–cognitive framework for behaviour changes in an emerging economy context. Sustainability (Switzerland) 14(22). https://doi.org/10.3390/su142214855
Du Y, Yan M (2022) Green transformational leadership and employees' taking charge behavior: the mediating role of personal initiative and the moderating role of green organizational identity. Int J Environ Res Public Health 19(7):4172
Flagstad I, Johnsen S, Rydstedt L (2021) The process of establishing a green climate: face-to-face interaction between leaders and employees in the microsystem. J Values Based Leadersh 14(1)
Foster B, Muhammad Z, Yusliza MY, Faezah JN, Johansyah MD, Yong JY, Saputra J, Ramayah T, Fawehinmi O (2022) Determinants of pro-environmental behaviour in the workplace. Sustainability 12:15–19
Glisson C, James LR (2002) The cross-level effects of culture and climate in human service teams. J Organ Beh 23(6):767–794. https://doi.org/10.1002/job.162
Guiyao T, Ren S, Li Y, Wang M, Zhang S (2023) Employee green behaviour: a review and recommendations for future research. Int J Manage Rev. https://doi.org/10.1111/ijmr
Han Z, Wang Q, Yan X (2019) How responsible leadership motivates employees to engage in organizational citizenship behavior for the environment: a double-mediation model. Sustainability (Switzerland) 11(605):1–13. https://doi.org/10.3390/su11030605
Ichsan IZ, Sigit DV, Miarsyah M, Azrai EP, Heryanti E (2019) Students' pro-environmental behavior and environmental learning outcomes based on green consumerism. Jurnal Pendidikan Biologi Indonesia 5(1):1–9
Islam T, Ali G, Asad H (2019) Environmental CSR and pro-environmental behaviors to reduce environmental dilapidation: the moderating role of empathy. Manag Res Rev 42(3):332–351. https://doi.org/10.1108/MRR-12-2017-0408
Junsheng H, Masud MM, Akhtar R, Rana S (2020) The mediating role of employees ' green motivation between exploratory factors and green behaviour in the Malaysian food industry. Sustainability 12(509)
Khaskhely MK, Qazi SW (2021) Contribution of employee green values in improving corporate sustainable environmental performance via organizational green culture and employee green behaviors—a conceptual nexus. J Tianjin Univ Sci Technol 11:88–115
Kim A, Kim YK, Han K, Susan EJ, Ployhart RE (2014) Multilevel influences on voluntary workplace green behavior: individual differences, leader behavior, and coworker advocacy. J Manag 43(5)
Kong L, Sial MS, Ahmad N, Sehleanu M, Li Z, Zia-Ud-Din M, Badulescu D (2021) CSR as a potential motivator to shape employees' view towards nature for a sustainable workplace environment. Sustainability (Switzerland) 13(3):1–14. https://doi.org/10.3390/su13031499
Kuenzi M, Schminke M (2009) Assembling fragments into a lens: a review, critique, and proposed research agenda for the organizational work climate literature. J Manag 35(3):634–717. https://doi.org/10.1177/0149206308330559
Lasrado F, Zakaria N (2020) Go green! Exploring the organizational factors that influence self-initiated green behavior in the United Arab Emirates. Asia Pacific J Manage 37(3):823–850. https://doi.org/10.1007/s10490-019-09665-1

Lathabhavan R, Kaur S (2023) Promoting green employee behaviour from the lens of green transformational leadership. Leadersh Org Dev J 44(8):994–1015. https://doi.org/10.1108/LODJ-05-2023-0233
Li M, Gong Z, Gilal FG, Van Swol LM, Xu J, Li F (2021) The moderating role of ethical leadership on nurses' green behavior intentions and real green behavior. BioMed Res Int. https://doi.org/10.1155/2021/6628016
Li D, Zhao L, Ma S, Shao S, Zhang L (2020) What influences an individual's pro-environmental behavior? A literature review. Resour Conserv Recycl 146:28–34
Liu X, Wang Q, Yi I, Chi H, Yang D, Chan EH (2021) Are you an energy saver at home? The personality insights of household energy conservation behaviors based on theory of planned behavior. Resour Conserv Recycl 174
Malsha KPHGN, Anton Arulrajah A, Senthilnathan S (2020) Mediating role of employee green behaviour towards sustainability performance of banks. J Gov Regul 9(2):92–102
Mcdonald FV (2014) Developing an integrated conceptual framework of pro-environmental behavior in the workplace through synthesis of the current literature. Admin Sci 4:276–303
Mi L, Sun Y, Gan X, Yang H, Lv T, Shang K, Qiao Y, Jiang Z (2020) Promoting employee green behavior through the person-organization fit: the moderating effect of psychological distance. Front Psychol 11:1–16. https://doi.org/10.3389/fpsyg.2020.568385
Mohiuddin M, Al Mamun A, Syed FA, Masud MM, Su Z (2018) Environmental knowledge, awareness, and business school students' intentions to purchase green vehicles in emerging countries. Sustainability (Switzerland) 10(5)
Mouro C, Duarte AP (2021) Organisational climate and pro-environmental behaviours at work: the mediating role of personal norms. Front Psychol 12:1–9. https://doi.org/10.3389/fpsyg.2021.635739
Norton T (2016) A multilevel perspective on employee green behaviour
Norton TA, Ayoko OB, Ashkanasy NM (2021) A socio-technical perspective on the application of green ergonomics to open-plan offices: a review of the literature and recommendations for future research. Sustainability (Switzerland) 13(15)
Norton TA, Parker SL, Zacher H, Ashkanasy NM (2015) Employee green behavior: a theoretical framework, multilevel review, and future research agenda. Organ Environ 28(1):103–125
Norton TA, Zacher H, Ashkanasy NM (2014) Organisational sustainability policies and employee green behaviour: the mediating role of work climate perceptions. J Environ Psychol 38:49–54. https://doi.org/10.1016/j.jenvp.2013.12.008
Ones DS, Dilchert S (2012) Environmental sustainability at work: a call to action. Ind Organ Psychol 5(4):444–466
Ong J, Noordin R, Jaidi J (2020) Environmental management accounting in the Malaysian manufacturing sector. Asian J Account Persp 13(1):85–103
Paillé P, Valéau P, Renwick DW, Valéau P, Renwick DW (2020) Leveraging green human resource practices to achieve environmental sustainability. J Clean Prod. https://doi.org/10.1016/j.jclepro.2020.121137
Piligrimiene Ž, Žukauskaite A, Korzilius H, Banyte J, Dovaliene A (2020) Internal and external determinants of consumer engagement in sustainable consumption. Sustainability (Switzerland) 12(4)
Punj N, Ahmi A, Tanwar A, Abdul Rahim S (2023) Mapping the field of green manufacturing: a bibliometric review of the literature and research frontiers. J Clean Prod 423. https://doi.org/10.1016/j.jclepro.2023.138729
Razali F, Daud D, Weng-Wai C, Anthony Jiram WR (2020) Waste separation at source behaviour among Malaysian households: the theory of planned behaviour with moral norm. J Clean Prod 271
Robertson JL, Barling J (2013) Greening organizations through leaders' influence on employees' pro-environmental behaviors. J Organ Behav
Robertson JL, Carleton E (2018) Uncovering how and when environmental leadership affects employees' voluntary pro-environmental behavior. J Leadersh Organ Stud 25(2):197–210

Rubel MRB, Kee DMH, Rimi NN (2021) Green human resource management and supervisor pro-environmental behavior: the role of green work climate perceptions. J Clean Prod 313:127669. https://doi.org/10.1016/j.jclepro.2021.127669

Sabbir MM, Taufique KMR (2021) Sustainable employee green behavior in the workplace: integrating cognitive and non-cognitive factors in corporate environmental policy. Bus Strat Environ. https://doi.org/10.1002/bse.2877

Saeed BB, Afsar B, Hafeez S, Khan I, Tahir M, Afridi MA (2019) Promoting employee's pro-environmental behavior through green human resource management practices. Corp Soc Responsib Environ Manag 26(2):424–438

Saifulina N, Carballo-Penela A (2017) Promoting sustainable development at an organizational level: an analysis of the drivers of workplace environmentally friendly behaviour of employees. Sustain Dev 25(4):299–310

Saleem M, Qadeer F, Mahmood F, Ariza-Montes A, Han H (2020) Ethical leadership and employee green behavior: a multilevel moderated mediation analysis. Sustainability 12(8):3314

Schneider B (2000) The psychological life of organizations. In: Ashkanasy N, Wilderom CPM, Peterson MF (eds) Handbook of organizational culture & climate. Sage, Thousand Oaks, pp 17–21

Shoaib M, Nawal A, Roman Z, Korsakien R (2022) Go green! Measuring the factors that influence sustainable performance. Clean Prod 366. https://doi.org/10.1016/j.jclepro.2022.132959

Sonetti G, Brown M, Naboni E (2019) About the triggering of UN sustainable development goals and regenerative sustainability in higher education. Sustainability 11(1):254

Su L, Swanson SR (2019) Perceived corporate social responsibility's impact on the well-being and supportive green behaviors of hotel employees: the mediating role of the employee-corporate relationship. Tour Manag 72:437–450

Suganthi L (2019) Examining the relationship between corporate social responsibility, performance, employees' pro-environmental behavior at work with green practices as mediator. J Clean Prod. https://doi.org/10.1016/j.jclepro.2019.05.295

Syafri W, Prabowo H, Ashari S, Muafi M (2021) The impact of workplace green behavior and green innovation on green performance of SMEs: a case study in Indonesia. J Asian Finance 8(5):365–0374

Tang Z, Warkentin M, Wu L (2019) Understanding employees' energy saving behavior from the perspective of stimulus-organism-responses. Resour Conserv Recycl 140:216–223

United Nation Development Programme (2022) Available at https://www.undp.org/eurasia/our-focus/environment

Unsworth KL, Davis MC, Russell SV, Bretter C (2021) Employee green behaviour: how organizations can help the environment. Curr Opin Psychol 42, 1–6. https://doi.org/10.1016/j.copsyc.2020.12.006

Wang H, Khan MAS, Anwar F, Shahzad F, Adu D, Murad M (2021a) Green innovation practices and its impacts on environmental and organizational performance. Front Psychol 11:1–15

Wang Q, Chang R, Xu Q, Liu X, Yi I, Ma Y, Wang Y (2021b) The impact of personality traits on household energy conservation behavioral intentions—an empirical study based on theory of planned behavior in Xi' an. Sustain Energy Technol Assessments 43(November 2020)

Wood BP, Eid R, Agag G (2021) A multilevel investigation of the link between ethical leadership behaviour and employees green behaviour in the hospitality industry. Int J Hosp Manag 97:102993. https://doi.org/10.1016/j.ijhm.2021.102993

Xiao J, Mao JY, Huang S, Qing T (2020) Employee-organization fit and voluntary green behavior: a cross-level model examining the role of perceived insider status and green organizational climate. Int J Environ Res Public Health 17(7)

Yong JY, Yusliza MY, Fawehinmi OO (2020a) Green human resource management: a systematic literature review from 2007 to 2019. Benchmarking 27(7):2005–2027. https://doi.org/10.1108/BIJ-12-2018-0438

Yong JY, Yusliza MY, Jabbour CJC, Ahmad NH (2020b) Exploratory cases on the interplay between green human resource management and advanced green manufacturing in light of the ability-motivation-opportunity theory. J Manage Dev 39(1):31–49. https://doi.org/10.1108/JMD-12-2018-0355

Yong JY, Yusliza MY, Ramayah T, Chiappetta Jabbour CJ, Sehnem S, Mani V (2020c) Pathways towards sustainability in manufacturing organizations: empirical evidence on the role of green human resource management. Bus Strateg Environ 29(1):212–228. https://doi.org/10.1002/bse.2359

Zacher H, Rudolph CW, Katz IM (2023) Employee green behavior as the core of environmentally sustainable organizations. Annu Rev Organ Psych Organ Behav 10:465–494. https://doi.org/10.1146/annurev-orgpsych-120920-050421

Zacher H, Kühner C, Katz IM, Rudolph CW (2024) Leadership and environmental sustainability: an integrative conceptual model of multilevel antecedents and consequences of leader green behavior

Zhao H, Zhou Q (2020) Socially responsible human resource management and hotel employee organizational citizenship behavior for the environment: a social cognitive perspective. Int J Hosp Manag. https://doi.org/10.1016/j.ijhm.2020.102749

Zheng S, Jiang L, Cai W, Xu B, Gao X (2021) How can hotel employees produce workplace environmentally friendly behavior? The role of leader corporate and coworkers. Front Psychol 12:1–13. https://doi.org/10.3389/fpsyg.2021.725170

Zhou S, Zhang D, Lyu C, Zhang H (2018) Does seeing "mind acts upon mind" affect green psychological climate and green product development performance? The role of matching between green transformational leadership and individual green values. MDPI J Sustain 10. https://doi.org/10.3390/su10093206

Zientara P, Zamojska A (2016) Green organizational climates and employee pro-environmental behaviour in the hotel industry. J Sustain Tourism 26(7):1142–1159

Transport and Air Pollution in ASEAN: Lessons Learnt from Countries Practicing a Sustainable Approach

Roszi Naszariah Nasni Naseri and Nur Asyikeen Kamarudin

Abstract Over the last few decades, the Association of Southeast Asian Nations (ASEAN) region has rapidly developed, seeing tens of millions of people become city dwellers. The region's urban population is estimated to be almost 405 million out of a total population of almost 726 million, despite the rapid urbanization faced by ASEAN. It is reported that most ASEAN cities are polluted by air. Some ASEAN countries also reported an alarming number of deaths from lower respiratory infections; trachea, bronchus, and lung cancers; ischemic heart disease; stroke; and chronic obstructive pulmonary disease due to the highest levels of PM2.5, most exceeding WHO's annual mean. Ambient air pollution sources come from transport, waste management, industry supply, agricultural practices, and household energy. However, air pollution in most countries in ASEAN comes from transportation. A high share of powered two- and three-wheelers adds to cities' congestion and air pollution problems. A recent review of sustainable development describes progress toward making cities inclusive, safe, resilient, and sustainable in Southeast Asia as very slow and stagnant, contributing to the urgent need to accelerate efforts in this area. Thus, the aim of this study is to review the reality of the current situation in ASEAN and describe the countries' practices of a sustainable approach. This aligns with SDG11 toward making cities safe and sustainable by focusing on achieving zero carbon emissions.

Keywords Transport · Air pollution · ASEAN cities · Sustainable · Approach

R. N. N. Naseri (✉)
Faculty of Business and Management, Universiti Teknologi MARA (UiTM) Cawangan Melaka, Malacca City, Malaysia
e-mail: roszinaseri@uitm.edu.my

N. A. Kamarudin
Academy of Language Studies, Universiti Teknologi MARA (UiTM) Cawangan Melaka, Malacca City, Malaysia

E. Shaffie et al. (eds.), *Sustainable Green Infrastructure*,
https://doi.org/10.1007/978-981-96-1486-8_16

1 Introduction

The cities of ASEAN countries have witnessed rapid urbanization in the last few decades. ASEAN refers to the Association of Southeast Asian Nations. It is also known as a fast-growing trade bloc of 10 member states: Brunei, Cambodia, Indonesia, Lao PDR, Malaysia, Myanmar, the Philippines, Singapore, Thailand, and Vietnam, with a combined population of over 660 million (see Fig. 1). ASEAN ranks as the third largest regional economy in Asia with a GDP of almost US$3.2 trillion and the fifth largest economy in the world, after the USA, China, Japan, and Germany (ASEAN 2023). According to United Nations (2018a, b), ASEAN cities indicated where city growth rates are the highest, with Singapore (100%) and Brunei Darussalam (78.2%) among the most urbanized as of 2020 and some of the less urbanized countries, such as the Lao People's Democratic Republic (Lao PDR) (36.3% urban in 2020, Cambodia (22.4% urban in 2020), but set to rise by 2030 (see Table 1).

Despite the rapid urbanization faced by ASEAN, a study by Haseeb et al. (2019) revealed that the increasing rate of urbanization in the ASEAN region has led these countries to focus on their health, which is largely influenced due to environmental pollution. In the same vein, Living ASEAN (2018) reported that most ASEAN cities are polluted by air (see Table 2). Air pollution is broadly understood as any material of any source within the atmosphere, which exists in either solid, liquid, or gas phase that has destructive effects or the ability to modify the natural characteristics of the

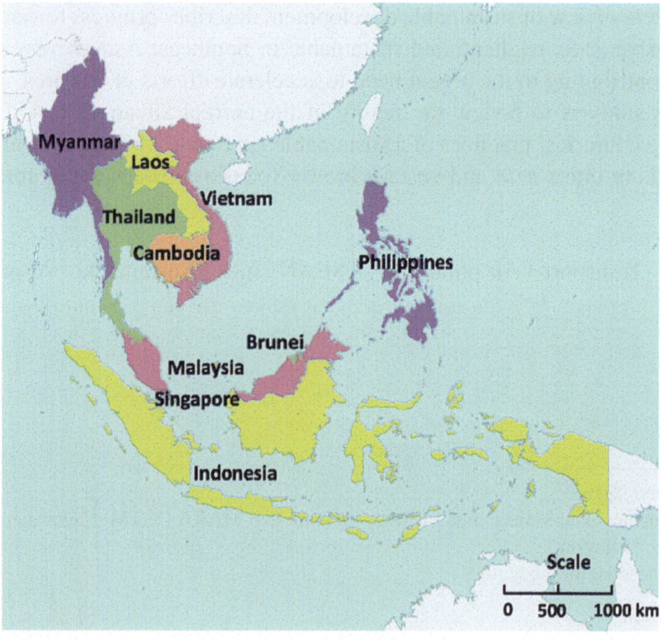

Fig. 1 Map of ASEAN countries. *Source* Lau et al. (2022)

Table 1 ASEAN annual percentage (%) of urban population, 2015–2030

	2015	2020	2025	2030	**% Total urban population growth**
ASEAN	**47.2**	**50.1**	**52.9**	**55.7**	+ *8.5*
Brunei	76.7	78.2	79.7	81.1	**+ 4.4**
Cambodia	22.2	22.4	26.5	29.0	**+ 6.8**
Indonesia	53.3	56.6	59.8	62.8	**+ *9.5***
Lao PDR	33.1	36.3	39.6	42.9	**+ *9.8***
Malaysia	74.2	77.2	79.7	81.8	**+ 7.6**
Myanmar	29.9	31.1	32.8	35.0	**+ 5.1**
Philippines	46.3	47.4	49.0	50.9	**+ 4.6**
Singapore	00.0	100.0	100.0	100.0	**+ 0.0**
Thailand	47.7	51.4	55.0	58.4	**+ *10.7***
Vietnam	33.8	37.3	40.9	44.5	**+ *10.7***

Source United Nations (2018a, b)
ASEAN is bolded in the first row to indicate it represents the aggregated or regional average of urban population growth for all ASEAN countries. This makes it visually distinct from the individual country data below, highlighting its summary nature.
Bold italics for % Total urban population growth column: The bold italics emphasize the growth rates, which provide a critical summary measure of urban population changes over time. The visual distinction draws attention to this data as the key conclusion of the table, summarizing long-term trends.

atmosphere and poses a health risk to living things, or which causes the deterioration of the environment and ecosystem (Ghorani-Azam et al. 2016). Typically, pollutants come from both natural sources (i.e., plant pollens, wind-blown dust, volcanic emissions, and lightning-induced forest fires) and anthropogenic sources (i.e., transportation emissions, industrial emissions, and emissions from other processing activities) (Pozzer et al. 2023). However, air pollution in most urban areas in ASEAN countries comes from transportation (Richmond-Bryant et al. 2017; Sicard et al. 2023).

Keyvanfar et al. (2018) and Güzel et al. (2024) reported that road transport is a major contributor to air pollution and climate change in general and contributes

Table 2 Air quality around the ASEAN by city

Country	Levels of PM 2.5	Indicator
Hanoi	256	Very unhealthy
Bangkok	186	Unhealthy for everyone
Chiang Mai	169	Unhealthy for everyone
Ho Chi Minh	169	Unhealthy for everyone
Jakarta	165	Unhealthy for everyone
Manila	91	Moderate
Singapore	91	Moderate

Source Living ASEAN (2018)

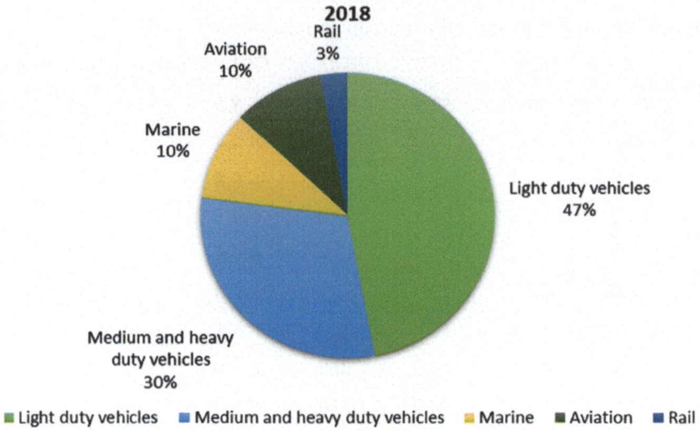

Fig. 2 Transport sector CO_2 emissions by mode. *Source* Dale and Lutsey (2019)

significantly to the emissions of greenhouse gases, particulate matter, and volatile organic compounds. Further, the International Energy Agency (2016) revealed that cars were the single largest category of emitter with a share of more than 34% of all transport emissions. It can be seen that rapid urbanization and economic growth, have accelerated the growth of motorization in cities in developing countries. For example, in the ASEAN region, the number of motor vehicles per one thousand people has more than tripled in the past 30 years (ASEAN 2019). In the same vein, ASEAN (2021) revealed that there is a high share of powered two and three-wheelers, which adds to the congestion and air pollution problem in cities. Figure 2 shows the development of CO_2 emissions for different transport modes, with light-duty passenger road vehicle transport being by far the largest contributor (Dale and Lutsey 2019; United Nations 2018a, b).

ASEAN was reported as having the highest total number of transportation-attributable and ozone deaths in 2015 (Anenberg et al. 2019). Some ASEAN countries also reported an alarming number of deaths from lower respiratory infections; trachea, bronchus, and lung cancers; ischemic heart disease; stroke; and chronic obstructive pulmonary disease due to the highest levels of PM2.5, most exceeding WHO's annual mean (ASEAN Sustainable Urbanization Report 2022) (see Table 3). In addition, the World Health Organization (2016) reported that air pollution is responsible for one of every nine deaths, or approximately three million deaths, annually. The ASEAN countries have made various efforts and adopted a variety of policies to address these problems over the years with varying degrees of effectiveness. Still, many of these environmental problems remain serious or are worsening. Further research is needed on the level of ambition and extent of implementation of these environmental policies in order to understand the reasons for insufficient progress (Elder and Ellis 2022). The ASEAN cities need to seek transport as a matter of urgency (Sibuea et al. 2021).

Table 3 Air pollution exposure and attributable death

Country	Annual mean concentration of fine particular matter in urban areas of Southeast Asia, 2016 (in μg/m^3)	Total deaths attributable to ambient air pollution, 2016
Brunei	7.44	36
Cambodia	27.21	5113
Indonesia	20.66	95,156
Lao PDR	21.78	2635
Malaysia	17.15	9495
Myanmar	33.73	25,483
Philippines	23.72	46,816
Singapore	17.24	2211
Thailand	31.88	33,546
Vietnam	23.75	34,232

Source ASEAN Sustainable Urbanization Report (2022)

2 ASEAN Countries' Environmental Policies for the Sustainable Development Goals (SDGs)

In terms of global development, sustainable development, or SD, has gained traction. According to Jaramillo-Nieves and Río (2010), it is conventionally characterized as development that attempts to satisfy the demands of the current generation without sacrificing the availability and capacity of needs for future generations. Many increasingly acknowledge that "healthy people" depend on a "healthy planet" (UNEP 2019). In order to promote human well-being and enhance the ecological and natural underpinnings of human civilization, the Sustainable Development Goals (SDGs) advocate an integrated approach (United Nations 2015; Elder and Olsen 2019). The idea of sustainable development is starting to gain traction on a worldwide scale. Long-term economic and environmental stability is the main objective of sustainable development, which can be attained by incorporating social, environmental, and economic factors into decision-making (Gold 2015).

Protecting the environment from degradation is one of the Sustainable Development Goals (SDGs). This can be achieved by managing natural resources sustainably, consuming and producing goods in a sustainable manner, and addressing climate change in a way that will ensure that future generations can continue to live on the planet and meet their needs (Navarette et al. 2020). In addition, a sustainable city keeps itself safe from long-term environmental threats that could jeopardize its progress toward development (United Nations 2001). The essential elements for a city to become sustainable are outlined in Table 4.

In 1992, the ASEAN Environmental Improvement Program (ASEAN-EIP) was established and funded by the United States Agency for International Development (USAID). This program aims to mitigate emissions and create an awareness of the

Table 4 Sustainable city

A city should strive to provide the following
A clean, safe physical environment of high quality
An ecosystem that is stable now and sustainable in the long term
A strong, mutually supportive, and non-exploitive community
A high degree of participation by the public over the decisions affecting their lives, health and well-being
High levels of positive health and low levels of disease

Source World Health Organization (1997)

environment (Stevenson 2007). Since then, ASEAN has been committed to pursuing a clean and green region with fully established mechanisms for sustainable development to ensure the protection of the regional environment, resources, and high quality of life. Among the initiatives undertaken is the implementation of emission standards for new or in-use motor vehicles in ASEAN countries. These standards serve as a benchmark, with compliance enforced by regulatory agencies within the respective governments (United Nations 2010). However, emissions from uncontrolled vehicles in use have detrimental effects on air quality, and thus, the citizens have to cover the costs related to health problems (Silitonga et al. 2012; Mustapa and Bekhet 2016).

Currently, most of the ASEAN countries face challenges in providing timely sustainable transport solutions to keep up with the rapid increase in transport demand and motorization rate. In response, ASEAN Transport Strategic Plan 2016–2025 has been implemented (ASEAN 2016). ASEAN member countries agreed to actively pursue sustainable transport and develop ‘Avoid’, ‘Shift’ and ‘Improve’ strategies at the regional and member states level (Attard et al. 2015). Despite the program and efforts, controlling vehicle emissions and their impacts on air quality is still a major challenge (Lyu et al. 2020; Sun et al. 2020). Motorized motorcycles are projected to see the highest growth rates in the transport sector. Existing regulations on vehicle emissions standards in this category are relatively weak. With the exception of Singapore, Thailand, and Vietnam, none of the other ASEAN countries has actively tackled emissions from motorcycles. Of particular note is that Vietnam has one of the strictest vehicle emissions standards for motorcycles in ASEAN, which may stem from the fact that Vietnam has a transportation fleet dominated by this mode of transport (Owen and Tao 2016) (refer Table 5).

EU-ASEAN Business Council (2021) reported, as can be inferred in Table 6 that most ASEAN countries are lagging behind other regions regarding vehicle standards. This is, for instance, the case compared with the European Union, in which Euro 6 standards are applied for four-wheelers, while for two-wheelers, Euro 5 applies. Despite the penetration of Euro I-compliant vehicles, the fleets in developing ASEAN countries are still dominated by decades-old vehicles with poorer fuel efficiency and emissions performance. However, inferior performance is not limited to old vehicles but is also often seen in new vehicles that violate existing emissions standards because of corruption or weakness in the mechanisms to enforce the standards (Kassim et al.

Table 5 Low- and zero-emission vehicles (LZEVs) policy ambition and characteristics in the ASEAN region

Country	Ambition and presence of policies promoting LZEVs	Type of policies in place
Brunei Darussalam	Comparatively low	Reluctant proponent of policies that mostly deal with local pollution and public transportation Economy of the country centered on fossil fuels
Cambodia	None	No policy in place
Indonesia	Comparatively high	Industrial policy to strengthen, maintain, and improve its standing in the production of automobiles and developing supply chains. It incorporates mechanisms intending to utilize local nickel reserves for the domestic electric car sector and sets lofty goals, such as having 20% of domestic vehicle sales be electric by 2025. Certain legislative tools, like the luxury sales tax exemption, are intended to increase demand
Lao PDR	None	No policy in place
Malaysia	Comparatively high	Industrial policy to uphold, protect, and grow the country's supply networks and auto manufacturing sector. Enables manufacturers to negotiate incentives. Policies aimed at various technological platforms. Lack of a distinct stance on electric cars. Little thought is given to developing the facilities needed for electricity
Myanmar	None	No policy in place
Philippines	Medium	Industrial policy aims to reinforce and broaden its presence in the production of automobiles and developing supplier networks. Tariff exemptions are intended to promote local assembly for electric car components. Give consideration for lighter cars (two- and three-wheelers). Very little has been done to solve the infrastructure required for access to energy and to address demand issues (which have just lately been addressed with tax incentives for investors)
Singapore	Comparatively low	Broader legislative initiatives that prioritize public transportation over private. Electric vehicle early adopter incentives have been put in place. Differentiated feebate systems do not target electric vehicles directly; instead, they are based on environmental performance

(continued)

Table 5 (continued)

Country	Ambition and presence of policies promoting LZEVs	Type of policies in place
Thailand	Comparatively high	Industrial policy to strengthen, maintain, and improve its standing in the production of automobiles and developing supply chains. It contains explicit guidelines for providing incentives to manufacturers. Incentives for customers are also included to boost regional demand. The only nation in ASEAN to have created policies for grid integration of electric vehicles. The strategy is laid out in three stages, with a concentration on research and demonstration through 2021 and then scaling up thereafter
Vietnam	Medium	Private sector projects with an emphasis on two-wheelers that come prior to public policy. A three-year period beginning on March 1, 2022, within which the registration charge for battery-electric vehicles will be reduced to 0%

Source Global Fuel Economy Initiative (2022)

2020). Serious institutional strengthening is necessary in ASEAN cities to improve inspection and maintenance regimes for vehicles. Despite that, ASEAN countries also lag behind the rest of the world in terms of fuel efficiency (ASEAN 2019). With respect to transport policy choices in Asia–Pacific countries, it is concluded that a crucial issue explaining differences in motorization and the success of public transport is the decision of whether or not to restrain private vehicle ownership and use (Bakker et al. 2017).

In addition, (ADB 2019) reported that there has been insufficient progress on the SDGs, especially in the environmental dimensions, related to SDGs 11 in the Asia–Pacific has been limited. A recent review of sustainable development describes progress toward making cities inclusive, safe, resilient, and sustainable in Southeast Asia as very slow and stagnant, contributing to the urgent need to accelerate efforts in this area (ESCAP 2021). Thus, it is important to analyze program practices by other regions in order to understand the sustainable practices approach.

Table 6 Standards throughout ASEAN for motorcycles, gasoline, and diesel

	Vietnam	Malaysia	Singapore	Indonesia	Thailand	Philippines	EU
Euro 6			Four-wheeler				Four-wheeler
Euro 5							Two-wheeler
Euro 4			Two-wheeler	Four-wheeler	Four-wheeler Two-wheeler		
Euro 3	Four-wheeler Two-wheeler	Four-wheeler Two-wheeler					
Euro 2				Two-wheeler			

Source EU-ASEAN Business Council (2021)
*The EU is included to the right for comparative purposes

3 A Review of Countries' Practices of a Sustainable Approach

Owing to heightened consciousness about the significance of addressing pollution and reducing greenhouse gas emissions and their influence on global warming, different cities across the globe have opted to formulate distinct approaches concerning the transportation domain. Sustainable urban transportation is one of them. Table 7 lists the sustainable approach practices of the various countries.

4 Conclusion

In densely populated ASEAN cities, air pollution is a major health hazard. Because of the rising motorization of cities, urban transportation is increasingly contributing to air pollution. Since the rate of motorization in ASEAN cities is already high at the current income levels, there is significant concern about the possible future effects that cities with rapidly expanding economies may have on local and global surroundings. If the current trend of unsustainable city development persists, it will be extremely difficult to create environmentally sustainable cities in ASEAN. In many ASEAN cities, there is an urgent need to improve environmental performance. The creation of new smart cities has the potential to solve numerous environmental concerns and challenges while also igniting new growth engines, lowering particulate matter emissions in order to create safe cities and communities. Notwithstanding this, tightening the laws governing vehicle emissions may offer opportunities for reduction while simultaneously enhancing the surrounding environment. Member nations of ASEAN, who are aware of global best practices for emission restrictions, may have a major influence in the creation and development of these standards. Nevertheless, ASEAN members need to make a commitment to equitable, safe, accessible, and sustainable development as their cities continue to expand. It is imperative that cities create well-defined action plans with short- and long-term objectives, assigned roles, and public support and involvement.

Table 7 Sustainable approach practices

Country	Practices of a sustainable approach
Berlin, Germany	Berlin put the low emission zones (LEZ) into effect. Because of the inner city's high population density and pollution exposure, a local economic zone (LERZ) is in place ("S-Bahn Ring") Only automobiles that adhere to strict exhaust pollution regulations are permitted to drive in the LEZ. Various stickers were created based on the class that the cars were in. There are four distinct emission classifications: class 1, which is sticker-free, and classes 2, 3, and 4. To be permitted to join the LEZ, a sticker known as a "Feinstaubplakette" needs to be placed on the windscreen, indicating that the car is Euro standard Foreign automobiles must purchase a sticker and are also impacted. According to Attard and Shiftan (2015), tourists who are driving into Berlin from outside the city must buy a sticker that is valid for the entire time they are there. The age of the car is the main factor affecting the emissions requirement for foreign cars The car registration office, technical certification bodies, and approved repair shops are the entities that issue the stickers, which are based on a national rule and cost between 5 and 15 euros
Santiago, Chile	A method for renting electric bicycles was implemented Except for deliveries that were scheduled, the city center was restricted to vehicles New trees were planted and sidewalks were extended The nation's busiest pedestrian zone was created by redesigning the old town center's streets Training in sustainable transportation has become mandated, beginning with preschool
Cairo, Egypt	Through the aid of a vehicle scrapping and recycling program, taxi owners may voluntarily trade in their old, highly polluting cars for new, more ecologically friendly ones through managed scrapping and recycling The pre-registered car vendors offered to the taxi owner at a discounted price along with financing options. In the end of 2018, almost 45,000 cabs had been exchanged, recycled, and demolished in Cairo alone, saving about 350,000 tons of emissions
Yichang, China	Thirty kilometers of the city's projected bike lane The overall condition of the walkways has been enhanced There were 700 trees planted The system of sharing bicycles was put into place Internal combustion engine-powered two-wheelers are not anymore permitted
Kayseri, Turkey	Designed a productive, high-occupancy, multimodal public transportation system that limits threats to public health and the environment

(continued)

Table 7 (continued)

Country	Practices of a sustainable approach
Bogota, Colombia	The very first cities globally to establish emergency bike lanes. With the 84 km constructed, 46 km are continuing in service today and 28 km were made permanent. The addition of bike lanes to several major roadways resulted in a triple in bike use. Solidified by government regulations like Acuerdo 804, which designated bicycling as the city's primary mode of transportation Built up a fleet of 1485 electric buses for its public transport system, positioning the city as one of the top three e-bus fleets in the world A program that exempts HOVs—vehicles carrying more than two passengers—from the odd–even license restriction scheme implemented by the city. Owing to this scheme, half of the cars in the city are not driven during peak hours The city has prioritized access and safety for women, children, people with disabilities, and senior citizens by developing more pedestrian zones, adding more permeable green areas, and growing its network of bike lanes
Paris, France	Changing congested arterial roads to pedestrian streets with restricted access for motorized vehicles; extending public cycling facilities generally; and converting temporary bike lanes created in response to public requirements during the pandemic into more permanent ones Reclaiming car parks and urban areas to create public areas for leisure activities, tree planting, and pedestrian traffic in the city
Lviv, Ukraine,	Establishing even more "walkable cities." Implemented an integrated street design that significantly improved cycle lanes, to the point where people using public transportation used to drive to bus stops
Medellin, Columbia	EnCicla, a public bicycle program: an integrated public bicycle system that links universities, mass transit, and other important city locations BRT—"Metroplús": The creation and implementation of a BRT project that is completely integrated (both fare-wise and physically) with the current mass transit system (cable car and metro). Control of vehicle exhaust emissions and enhancement of sulfur content
Tartu, Estonia	Intelligent street lighting, biogas buses, electric car and bike rentals, charging stations, and numerous ICT solutions all contribute to the town's increased liveability

(continued)

Table 7 (continued)

Country	Practices of a sustainable approach
Vitoria-Gasteiz, Spain	A low-emission zone has been implemented, along with the "Superblock Program," stricter parking laws, more infrastructure for safe cycling, electric vehicle charging available in all public parking lots, and better access for bicyclists and pedestrians in school zones The majority of sidewalks in the city are pedestrian-friendly, with over 2.5 m of width, a slope of less than 5%, and mechanical ramps that provide access to the steeper areas. A significant portion of the city is pedestrianized. Cross-street traffic calming has been introduced by the city to enhance the cohabitation of various mobility modes This included eliminating on-street parking, enlarging walkways, placing bollards, and painting markers. It also entailed limiting lanes and entrances. Additionally, all local roads in the city have a speed limit of 30 km/h (18 mph), whereas avenues and arterial highways have a 50 km/h speed limit. Speed limits on smaller local roads, such as those in the old downtown district, are 20 km/h
Sonderborg, Denmark	The approach includes four areas of intervention for a low-carbon transportation strategy: developing new biogas buses for public transportation; encouraging walking and bicycling; using electrical vehicles and carpooling; and sustainable transportation Biomethane buses and electric vehicle charging infrastructure are examples of green mobility options. Incorporate citizens in the goal of Smart Zero Carbon Sonderborg, which is to provide new ICT-based energy information and implement integrated solutions with thorough learning
California, USA	The state of California has set lofty targets to have 1.5 million zero-emission vehicles (ZEVs) on its roads by 2025 in an effort to reduce the emissions from transportation. California now holds the largest market share for electric vehicles (EVs) in the US, accounting for 49% of all EV registrations. The state has put in place a variety of laws, consumer incentive programs, and charging infrastructure measures, as detailed below, in order to maintain this growing trend and accomplish its goals California's policies regarding electric vehicles • To increase the goal to 5 million electric vehicles by 2030 and include infrastructure goals for hydrogen fueling stations and plug-in electric vehicle chargers to accommodate this increase in electric vehicle usage • The clean energy and pollution reduction act of 2015

Source World Bank (2018), Institute for Transportation and Development Policy (2020, 2023)

Acknowledgements The authors are indebted to Institute for Infrastructure Engineering and Sustainable Management, Universiti Teknologi MARA (UiTM), Shah Alam, Selangor, Malaysia, for providing the opportunity to publish this article. The authors also wish to thank the Faculty of Business and Management, Universiti Teknologi MARA (UiTM) Melaka for the opportunity given to run this study.

References

ADB (2019) Strengthening the environmental dimensions of the sustainable development goals in Asia and the Pacific. Asian Development Bank. https://www.adb.org. Accessed 4 July 2023

Anenberg S, Miller J, Henze D, Minjares R (2019) A global snapshot of the air pollution-related health impacts of transportation sector emissions in 2010 and 2015. https://theicct.org/sites/default/files/publications/Global_health_impacts_transport_emissions_2010-2015_20190226.pdf. Accessed 1 July 2023

ASEAN (2016) Kuala Lumpur transport strategic plan (ASEAN transport strategic plan) 2016–2025. The ASEAN Secretariat, Jakarta

ASEAN (2019) ASEAN fuel economy roadmap for the transport sector 2018–2025: with focus on light-duty vehicles. The ASEAN Secretariat, Jakarta

ASEAN (2021) VASEANstat. https://www.aseanstats.org/. Accessed 4 July 2023

ASEAN (2023) Investing in ASEAN 2023. https://asean.org/book/investing-in-asean-2023/. Accessed 2 July 2023

ASEAN Sustainable Urbanisation Report (2022) Sustainable urbanisation. https://asean.org/wp-content/uploads/2022/12/ASEAN-Sustainable-Urbanisation-Report_Final_Dec-2022.EPUB.pdf. Accessed 8 July 2023

Attard M, Shiftan Y (eds) (2015) Sustainable urban transport. Emerald Group Publishing, England

Bakker S, Dematera Contreras K, Kappiantari M, Tuan NA, Guillen MD, Gunthawong G, Van Maarseveen M (2017) Low-carbon transport policy in four ASEAN countries: developments in Indonesia, the Philippines, Thailand and Vietnam. Sustainability 9(7):1217–1225

Dale H, Lutsey N (2019) Estimating the infrastructure needs and costs for the launch of zero-emission trucks. International Council on Clean Transportation, Washington

Elder M, Ellis G (2022) ASEAN countries' environmental policies for the sustainable development goals (SDGs). Environ Dev Sustain. https://doi.org/10.1007/s10668-022-02514-0

Elder M, Olsen SH (2019) The design of environmental priorities in the SDGs. Global Pol 10(S1):70–82

ESCAP (2021) Asia and the Pacific SDG progress report 2021. https://www.unescap.org/kp/2021. 4 July 2023

EU-ASEAN Business Council (2021) Advancing ASEAN's fuel economy, emissions standards, post pandemic. https://www.eu-asean.eu/wp-content/uploads/2022/02/Advancing-ASEANs-Fuel-Economy-Emissions-Standards-Post-Pandemic-2021.pdf. Accessed 2 July 2023

Ghorani-Azam A, Riahi-Zanjani B, Balali-Mood M (2016) Effects of air pollution on human health and practical measures for prevention in Iran. J Res Med Sci 21(2016):1–12

Global Fuel Economy Initiative (2022) New analysis identifies priorities to help ASEAN countries improve vehicle efficiency. https://www.globalfueleconomy.org/blog/2022/march/new-analysis-identifies-priorities-to-help-asean-countries-improve-vehicle-efficiency. Accessed 1 July 2023

Gold R (2015) Brief for GSDR 2015 the concept of sustainable development: definition and defining principles. Florida International University, United States of America

Güzel TD, Alp K (2024) The effects of technological developments in transportation vehicles on air pollution mitigation of metropolitan cities: a case study of Istanbul. Sci Total Environ 912:168996

Haseeb M, Kot S, Hussain HI, Jermsittiparsert K (2019) Impact of economic growth, environmental pollution, and energy consumption on health expenditure and R&D expenditure of ASEAN countries. Energies 12(19):3598–3605

International Energy Agency (2016) Energy technology perspectives 2016—towards sustainable urban energy systems. International Energy Agency, Paris

Institute for Transportation and Development Policy (2020) STA receives record number of applicant cities for 2021 award. https://www.itdp.org/2020/10/02/stareceives-record-number-of-applicant-cities/. Accessed 1 July 2023

Institute for Transportation and Development Policy (2023) Sustainable transport award. http://staward.org/. Accessed 1 July 2023

Jaramillo-nieves L, Río P (2010) Contribution of renewable energy sources to the sustainable development of Islands: an overview of the literature and a research agenda. Sustainability. https://doi.org/10.3390/su2030783

Kassim KA, Husain NA, Ahmad Y, Jawi ZM (2020) End-of-life vehicles (ELVs) in Malaysia: time for action to guarantee vehicle safety. J Soc Autom Eng Malaysia 4(3)

Keyvanfar A, Shafaghat A, Muhammad N, Ferwati M (2018) Driving behaviour and sustainable mobility—policies and approaches revisited. Sustainability 10(1):1152–1165

Lau H, Zhang K, Bokka, Harsha K, Ramakrishna S (2022) A review of the status of fossil and renewable energies in Southeast Asia and its implications on the decarbonization of ASEAN. Energies. 15.2152.10.3390/en15062152

Living ASEAN (2018) Air quality around the ASEAN. https://livingasean.com/special-scoop/air-quality-asean-air-pollution-bangkok-hanoi-jakarta/. Accessed 3 July 2023

Lyu M, Bao X, Zhu R, Matthews R (2020) State-of-the-art outlook for light-duty vehicle emission control standards and technologies in China. Clean Technol Environ Policy 22(4):757–771

Mustapa SI, Bekhet HA (2016) Analysis of CO_2 emissions reduction in the Malaysian transportation sector: an optimisation approach. Energy Policy 89(1):171–183

Navarrete S, Borini FM, Avrichir I (2020) Environmental upgrading and the United Nations sustainable development goals. J Clean Prod 264(2020):121–135

Owen AD, Tao JY (2016) International best practice for emissions and fuel standards: implementation possibilities for ASEAN. Asia Pacific J Energy Environ 3(1):7–16

Pozzer A, Anenberg SC, Dey S, Haines A, Lelieveld J, Chowdhury S (2023) Mortality attributable to ambient air pollution: a review of global estimates. GeoHealth 7(1):e2022GH000711

Richmond-Bryant J, Owen RC, Graham S, Snyder M, McDow S, Oakes M, Kimbrough S (2017) Estimation of on-road NO_2 concentrations, NO_2/NO_X ratios, and related roadway gradients from near-road monitoring data. Air Qual Atmos Health 10(1):611–625

Sibuea MB, Sibuea SR, Pratama I (2021) The impact of renewable energy and economic development on environmental quality of ASEAN countries. AgBioforum 23(1):12–21

Sicard P, Agathokleous E, Anenberg SC, De Marco A, Paoletti E, Calatayud V (2023) Trends in urban air pollution over the last two decades: a global perspective. Sci Total Environ 858:160064

Silitonga AS, Atabani AE, Mahlia TMI (2012) Review on fuel economy standard and label for vehicle in selected ASEAN countries. Renew Sustain Energy Rev 16(3):1683–1695

Stevenson RS (2007) An assessment of the design and effectiveness of the ASEAN environmental improvement program. J Clean Prod 12(3):227–236

Sun S, Jin J, Xia M, Liu Y, Gao M, Zou C, Wang T, Lin Y, Wu L, Mao H, Wang P (2020) Vehicle emissions in a middle-sized city of China: current status and future trends. Environ Int. https://doi.org/10.1016/j.envint.2020.105514

UNEP (2019) Global environment outlook—GEO-6: healthy planet, healthy people. United Nations Environment Programme (UNEP). https://doi.org/10.1017/9781108627146

United Nation (2010) Mandatory disclosure of automotive emissions. http://www.un.org/jsummit/html/sustainabledev/p2managingresources/1908mandatorydisclosure.pdf. Accessed 1 July 2023

United Nations (2001) Sustainable cities programme 1990–2000—a decade of united nations support for broad-based participatory management of urban development. Nairobi, UN-Habitat

United Nations (2015) Transforming our world: the 2030 agenda for sustainable development. https://Sustainabledevelopment.un.org. Accessed 3 July 2023

United Nation (2018a) Sustainable transport, sustainable development interagency report. Paper presented at the transport conference, New York, United States of America

United Nations (2018b) World urbanization prospects: the 2018 revision. https://asean.org/wp-content/uploads/2022/12/ASEAN-Sustainable-Urbanisation-Report_Final_Dec-2022-EPUB.pdf. Accessed 2 July 2023

World Bank (2018) Egypt: scrapping and recycling old vehicles to lower pollution and improve livelihoods. https://www.worldbank.org/en/news/feature/2018/10/25/egypt-scrapping-and-recycling-oldvehicles-to-lower-pollution-and-improvelivelihoods. Accessed 2 July 2023
World Health Organization, Regional Office for Europe (1997) Twenty steps for developing a healthy cities project. WHO Regional Office for Europe, Copenhagen
World Health Organization (2016) Ambient air pollution: a global assessment of exposure and burden of disease. Fundamentals of air pollution engineering. Courier Corporation, United States of America

Properties of Cement Stabilized Cold In-Place Recycling-Base Course Incorporated with Ground Coal Bottom Ash and Calcium Chloride

Mohd Izzat Joohari, Ekarizan Shaffie, Nortasha Zulaika Mohd Khalil, and Hanna Natasya Hisham

Abstract Cold in-place recycling (CIPR) is a technique whereby the deteriorated pavement surface is recycled back to produce a new pavement base course by using a CIPR machine with the addition of a stabilizer. A chemical stabilizer is one of the stabilizers that can be used. In Malaysia, ordinary Portland cement (OPC) is the most widely used chemical stabilizer for CIPR. However, higher production of cement leads to higher pollution in the atmosphere. Researchers are constantly looking for cement-alternative materials, especially sustainable green materials. Therefore, this study was conducted using a waste coal bottom ash, ground to a very fine particle and mixed with OPC as a stabilizer with the presence of calcium chloride ($CaCl_2$). The primary goals of this research are to analyze how incorporating ground coal bottom ash (GCBA) as a partial substitute for OPC affects the characteristic and compaction properties of the proposed design mix. Sieve analysis, particle shape, specific gravity, and pH index were among the tests conducted. A modified proctor test was also done to determine the optimum moisture content (OMC), maximum dry density (MDD), and bulk density of the mix. The results show that the OMC was 5.60% with MDD at 1.93 Mg/m^3 while the mix with 4% of OPC and 1% of $CaCl_2$ recorded the highest bulk density at 2.05 Mg/m^3. This indicates that the presence of GCBA and $CaCl_2$ exerts a discernible influence on the properties of the base course.

M. I. Joohari (✉)
Civil Engineering Studies, College of Engineering, Universiti Teknologi MARA Cawangan Pulau Pinang, 13500, Permatang Pauh, Pulau Pinang, Malaysia
e-mail: izzatjoohari@gmail.com

E. Shaffie · N. Z. M. Khalil
School of Civil Engineering, College of Engineering, Universiti Teknologi MARA, 40450 Shah Alam, Selangor, Malaysia
e-mail: eka@uitm.edu.my

E. Shaffie
Institute for Infrastructure Engineering and Sustainability Management, Universiti Teknologi MARA, 40450 Shah Alam, Selangor, Malaysia

H. N. Hisham
College of Engineering, Universiti Teknologi MARA, 40450 Shah Alam, Selangor, Malaysia

E. Shaffie et al. (eds.), *Sustainable Green Infrastructure*,
https://doi.org/10.1007/978-981-96-1486-8_17

Keywords Cold in-place recycling · Pavement base course · Chemical stabilizer · Characteristic properties · Compaction properties

1 Introduction

Most roads nowadays are constructed with asphalt surfaces which makes them flexible and may bend under axial load. The excessive axial loads and the increased traffic frequency, however, could create stress on the asphalt pavements. These road pavements eventually approach the end of their functional structural lives. This is when road defects such as cracks, potholes, and uneven surfaces can be seen. When these defects are left without any action, they become worse and may endanger the traffic user.

Extending the lifespan of pavements is a crucial concern due to its significant impact on economic and various other factors (Susanto et al. 2018). To extend the service life of existing pavement surfaces and increase performance and pavement structural durability, restoration, resurfacing, and rehabilitation are the primary treatments that can be used. These treatments will have an immediate impact on the pavement's condition and will influence the rate at which it ages in the future. Pavement recycling is one of the road rehabilitation techniques in which the old, deteriorated pavement is recycled and transformed into a new base course layer (Gonzalo-Orden et al. 2019). There are two types of pavements recycling available which are hot in-place recycling (HIPR) and cold in-place recycling (CIPR) depending upon the recycling processes and mixing temperature. HIPR can be an excellent alternative to repairing pavement that has sustained damage to the upper layers and is in good bearing capability. Full-depth reclamation is recommended if the pavement has structural flaws and extensive damage, and CIPR can be helpful if the situation falls between these two categories (Ghanizadeh et al. 2018).

In Malaysia, CIPR was first put into practice in the late 1980s (Sufian et al. 2008). Since then, using recycled road pavements instead of conventional restoration techniques has gained popularity and approval from the Department of Work (JKR) Malaysia. One benefit of the CIPR is cost reductions of up to 40% compared to conventional methods (Sufian et al. 2008). Apart from cost saving, this method can eliminate rutting, cracking, uneven surfaces, and base issues (Fig. 1).

In the common practice of CIPR, stabilizing agents are added to the recycled section and mixed together to create a stabilized base course. There are two types of stabilization methods available which are mechanical stabilization and chemical stabilization. Mechanical stabilization includes compaction, modifying particle size distribution and adding reinforcement to the base structure while chemical stabilization is a method to stabilize the base layer through chemical reactions that could either be cement, lime, fly ash, or bituminous materials. Most of the CIPR pavement rehabilitation in Malaysia uses cement as stabilizing agent due to its availability and strength provided (Tan and Chan 2021). Higher demand for cement in the construction industries has led to higher production of cement. As the supply of cement continues to

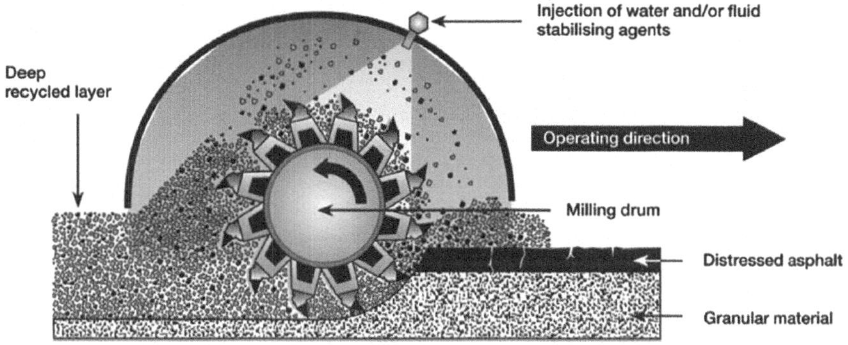

Fig. 1 Process for typical cold in-place recycling (Wirtgen GmbH 2004)

grow to meet demand, issues related to environmental pollution are unavoidable. During the production of cement, CO_2 will be released along with water vapor at high temperatures over the formation of CaO. CO_2 is one of the main greenhouse gas emissions that contributes significantly to global warming. Additionally, CO_2 accounts for 65% of greenhouse gases (Mohamad et al. 2021).

Researchers are continuously looking for sustainable green alternatives to cement. Sufian et al. (2009) in their study used cement and lime as stabilizing agents. Baghini et al. (2016) on the other hand used styrene–butadiene copolymer latex and Portland cement, while Li et al. (2013) applied cement-fly ash as a stabilizing agent in the evaluation of CIPR performance. As the search for new sustainable green alternatives continues, coal bottom ash (CBA) is one of the waste byproducts from coal power plants that can be looked into. Previous research has shown that CBA is mostly used as aggregate materials (Suárez-Macías et al. 2021; Jun et al. 2022; Dungca and Jao 2017; Nguyen et al. 2019). Nevertheless, its use as supplementary cementitious material (SCM) is still limited compared to fly ash. The pozzolanic properties of bottom ash make it appropriate to be used as SCM in the same way as fly ash. However, it must be first grind to a very fine size as unmodified bottom ash is not a good pozzolan material. Some researchers have determined that the pozzolanic characteristics of it can be enhanced through grinding (Arun et al. 2020).

This study aimed to analyze the influence of ground coal bottom ash (GCBA) to partially replace ordinary Portland cement (OPC) as a stabilizing agent with the addition of calcium chloride ($CaCl_2$) toward the characteristic properties of the proposed CIPR design mix. Besides, the compaction properties that included optimum moisture content (OMC), maximum dry density (MDD), and bulk density are also studied to understand the effects of the proposed stabilizing agent on the highest density that the base layer material can achieve when compacted to its optimum moisture content.

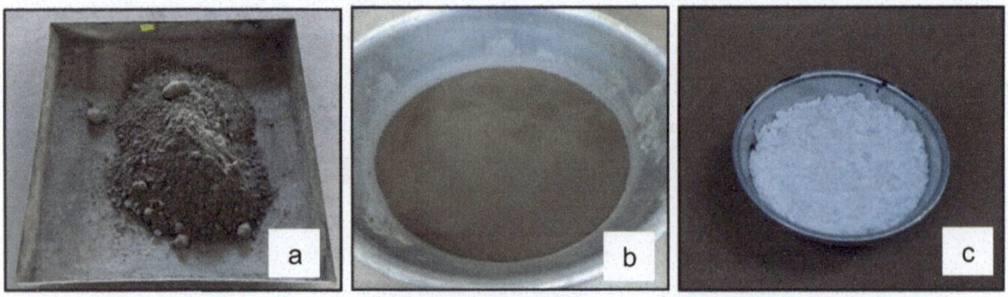

Fig. 2 Stabilizing agents used in this study **a** OPC, **b** GCBA and accelerator, **c** $CaCl_2$

2 Materials

Generally, the mix of CIPR consists of crushed aggregate (CA), recycled asphalt pavement (RAP), and stabilizing agent. A granite CA used in this study was obtained from Kajang Rock Quarry (Malaysia) and RAP was obtained from the pavement rehabilitation site in Putrajaya.

Two types of stabilizing agents were used, which are the Type I ordinary Portland cement (OPC) that complies with Malaysian Standard MS EN 197-1:2007 for CEM Type I and CBA that was obtained from TNB Jana Manjung, Perak. The GCBA was produced by grinding CBA into a smaller size passing 40-μm sieve, illustrated by the OPC size range. In addition, $CaCl_2$ industrial type was also used as an accelerator for the curing process (Fig. 2).

3 Design Mix

This study was carried out based on a 50:50 ratio of CA-RAP and different percentages of stabilizing agents between OPC and GCBA. The percentage of stabilizing agents was set at 4% as it might be considered as the optimum percentage for the OPC to be used at the site. 1% of $CaCl_2$ was used as an accelerator for the curing process. The control sample was prepared with the inclusion of 4% OPC only. Table 1 summarizes the various composition of stabilizing agents used for this study.

Table 1 Composition of stabilizing agent and accelerator

Percentage of stabilizing agent	Stabilizing agents		Accelerator $CaCl_2$(%)
	OPC (%)	GCBA (%)	
4%	1	3	0, 1
	2	2	
	3	1	
	4	0	

Table 2 Gradation limit for recycled layer material (JKR 2008)

BS sieve size (mm)	Cumulative % passing
50.0	100
37.5	85–100
20.0	60–100
10.0	40–90
5.0	30–75
2.36	20–60
0.425	10–45
0.075	2–20

4 Methods

Two stages were conducted in this study as the first stage focused on the characteristic properties such as sieve analysis, particle shape, specific gravity, and pH index. The second stage investigated the compaction properties that include OMC, MDD, and bulk density of the design mix. The properties of the road base are crucial because they directly impact the road performance, durability, and safety.

4.1 Sieve Analysis (Particle Size Distribution)

The particle size distribution of the base layer materials directly influences its stability and load-bearing capacity of the road. A balanced distribution of particle sizes provides better interlocking and compaction, leading to improved strength and stability under traffic loads. Therefore, sieve analysis was necessary to be carried out in the early stage of base layer design mix. An analysis on the site specimen was also done that was used as a reference for the material gradation. The material was subjected to a dry sieve analysis in accordance with the BS812-113-1 standard. Table 2 shows the gradation limit for the CIPR mix that must be fulfilled as specified by JKR Standard Specification for Roadworks Section 4: flexible pavement.

4.2 PH Index Value

The pH value is an essential factor to consider in the stabilization treatment process as it significantly influences the chemical properties of the mix and the effectiveness of the proposed stabilizing agent on the design mix. The proposed stabilizing agent in this study should be compatible with the pH of the CA and RAP to achieve proper bonding and stabilization. The test was carried out in accordance with ASTM D4972-19. In general, cement is unlikely to react normally with material that has an organic content greater than 2% or a pH lower than 5.3. If the pH value is greater than 12.0, then the organics present in the mix will not prevent the hardening process (Baghini et al. 2016). In this test, 20 g of air-dried specimens consisting of a design mix that passed through a 2.00-mm sieve were added to a glass container and thoroughly mixed with distilled water. The specimens were left for 1 h before readings were taken.

4.3 Specific Gravity

Specific gravity is another important factor in defining a material's quality or strength, which indirectly reflects its density. OPC and GCBA may contain a high moisture content if exposed to different conditions and humidity. Therefore, the specific gravity was calculated to determine their density. The Pycnometer method was used in this study where the specimen and distilled water were filled in the bottle. This method calculates the difference between the specific gravity of a material by dividing the volume of water in an empty and filled bottle by the same amount.

4.4 Particle Shape

In particle shape analysis, elongation and flakiness index of CA and RAP were the test conducted. This test is crucial for compaction, deformation resistance, and workability of the mix. The tests were carried out in accordance with BS 812-105.2: 1990 and MS 30.

Elongated particles are those whose length is significantly greater than their average width. These elongated particles can have a detrimental effect on the workability and strength of the road structure because they have a smaller surface area compared to their volume. Additionally, the presence of elongated particles may increase the amount of voids in the mix, lowering its overall density and possibly compromising its performance. In the elongation index test, any particle with a length that exceeds 1.8 times the average sieve size of the size fraction to which it belongs is considered elongated. This test was conducted using an elongation gauge with the size of aggregate used between 63 and 6.3 mm.

The flakiness index, on the other hand, was determined by comparing the particle dimensions of the aggregate with their thickness. This test helps to evaluate the proportion of flat and elongated aggregate of the mix which could also affect the workability and strength of the road structure. Similar to the elongation index test, the aggregate sizes used range from 63 to 6.3 mm and were measured using a flakiness thickness gauge.

4.5 OMC, MDD, and Bulk Density

The road base is the most crucial part of the pavement structure because it provides stability and helps in distributing the loads imposed by traffic. Achieving the OMC is essential because it affects the compactability and strength of the road-base structure. If the moisture content is too low, the mix will be too dry and difficult to compact, leading to inadequate density and reduced strength. On the other hand, if the moisture content is too high, the mix will become too saturated and compaction cannot be done properly, also resulting in lower strength and potential long-term issues such as settlement and rutting.

The MDD is related to OMC since it is the highest density that the road-base material can achieve when compacted to its OMC. The MDD value is important because it helps the compaction effort required during the road construction. Technically, the road-base material achieves its optimum strength at the proper MDD, providing a solid and long-lasting foundation for the pavement layers.

In this study, a modified proctor test was used to determine the OMC and MDD. The OMC and MDD were first obtained without any stabilizing agent added to it. The bulk density of the mix consisting of stabilizing agents and accelerator was then established using the obtained OMC. The test was conducted in compliance with BS EN 13286-2:2004 where 4.5 kg rammer was used to compact the specimen in 105 mm diameter × 115.4 mm height 1 L mold. After the compaction, three small portions (top, middle, and bottom part) were then extracted from the specimen of each design mix and placed in the oven for moisture content determination.

5 Results and Discussion

The characteristic and compaction properties of the proposed design mix are discussed as follows.

5.1 Verification of Particle Size Distribution

Particle sieve analysis was the first test conducted on the proposed design mix. Figure 3 illustrates the gradation ratio of 50% CA, and 50% RAP was within the limit of allowable particle size of the recycled base course as required by JKR specification. The particle size distribution of the design mix covered the range size between 20 and 0.075 mm. The result for the site specimen was also within the allowable limit and had nearly the same size distribution as the proposed gradation.

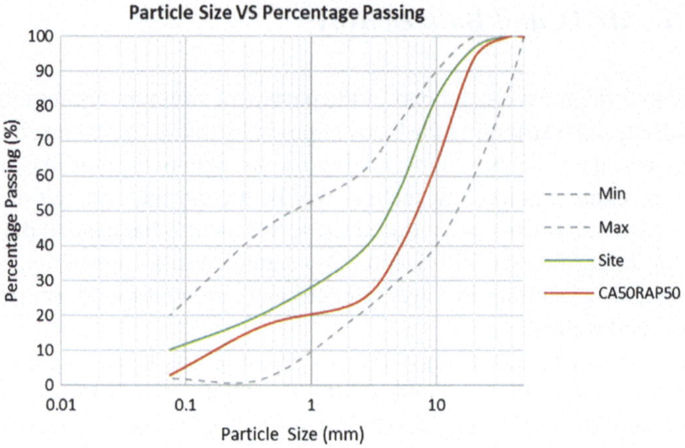

Fig. 3 Particle size distribution

5.2 PH Index

Results of the pH index value in Table 3 show that the design mix containing stabilizing agents of GCBA and OPC recorded pH values in the range of 14.0 which was high in alkali. pH value slightly decreased as the stabilized design mix was added with 1% of $CaCl_2$. This is normally due to the hydrolysis process when the water molecules react with the ions of $CaCl_2$ which cause the calcium ions (Ca^{2+}) and chloride ions (Cl^-) are released into the solution. This process results in the formation of hydrochloric acid (HCl), which is a strong acid that eventually lowers the pH value. As for the hardening process of the cementitious material from the stabilizing agents, Baghini et al. (2016) reported that the presence of organics in the mix will not interfere with the hardening process if the pH index value is higher than 12.0. The results show that the design mix composition with or without $CaCl_2$ recorded a pH value higher than 12.0. This indicated that the proposed design mix might experience a hardening process without any organic interference.

Table 3 pH index value

CA:RAP ratio	Stabilizing agent		pH index value	
			$CaCl_2$content	
	GCBA (%)	OPC (%)	0%	1%
50:50	3	1	14.68	14.47
	2	2	14.84	14.79
	1	3	14.78	14.83
	0	4	14.83	14.70

Table 4 Specific gravity

Material	Specific gravity, G_s
OPC	2.93
GCBA	2.56
RAP	2.00
Aggregate	2.43

A lower pH value (more acidic conditions) can potentially affect the hardening process of the mix. Although the pH value itself may not be a direct factor in the chemical reactions of the hydration and hardening process, extreme pH conditions can indirectly impact the process and affect the mix properties.

5.3 *Specific Gravity*

Table 4 exhibits the specific gravity of every material used in the mix of the CIPR-base course. It was found that the OPC had the highest G_s among all followed by GCBA with a value of 2.93 and 2.56, respectively. Generally, the specific gravity of OPC is around 3.15 and GCBA ranges from 2.00 to 2.60 as studied by previous researchers around the world (Marto and Soon Tan 2016). The results indicated that both of these materials which were used as stabilizing agents were denser and might provide better stability and structural strength to the road base. RAP on the other hand recorded the lowest Gs of 2.00. Since RAP is the layer that sits on the very top of the pavement structure, this might be due to the aging and oxidation process over time, me which made it brittle and less dense.

5.4 *Particle Shape*

According to JKR specification, the flakiness and elongation index of the aggregate base material when tested should not be more than 25%. In Table 5, it can be seen that both the aggregate and RAP recorded 16.31% and 18.62% respectively in total. As for the proposed design mix of CA50RAP50, the combined value recorded was 20.04%. These values were still within the requirement limit from JKR. From the analysis done on both flakiness and elongation index, it can be stated that the RAP has a slightly higher flakiness and elongation index due to the milling of old asphalt from existing pavements which leads to the generation of more elongated and flattened particles. High flakiness and elongation index beyond the limit may have an impact on the properties of the mix. It can reduce the cohesion of the stabilizing agent and the mix, potentially leading to issues like segregation, poor bond between particles, and reduced overall performance of the mix.

Table 5 Particle shape analysis

Material	Flakiness index (%)	Elongation index (%)
Aggregate	9.93	6.38
Combined value	16.31	
RAP	11.54	7.08
Combined value	18.62	
CA50RAP50	11.25	8.79
Combined value	20.04	

5.5 OMC, MDD, and Bulk Density

The relationship between the moisture content and the dry density is illustrated in Fig. 4. As mentioned earlier, the OMC and MDD of the mix are first conducted without the addition of stabilizing agents and accelerator. The results show that the OMC was 5.60% with MDD at 1.93 Mg/m^3. A study conducted by Sufian et al. (2009) using the same material composition found that OMC was 5.82% and MDD 2.161 Mg/m^3. In comparison, there was a slight difference in these values, which could be due to material factors. The origin and the properties of the material used in the mix might contribute to the difference. In addition, the gradation (particle size distribution) and shape of aggregate particles could also influence the compaction behavior.

Table 6 presents the bulk density of the design mix with the addition of stabilizing agents and accelerator. It can be observed that the highest value recorded was on the mix with 4% of OPC and 1% of $CaCl_2$ at 2.05 Mg/m^3, while the lowest recorded on the mix with 3% GCBA and 1% OPC at 1.93 Mg/m^3. Although the increased percentage of GCBA in the mix did not increase the bulk density, the addition of $CaCl_2$ in the stabilizing agent, however, was able to slightly enhance the bulk density compared to the mix without it. This has proven that GCBA needs to be combined with other chemicals apart from OPC in aiding the mix for a better compaction.

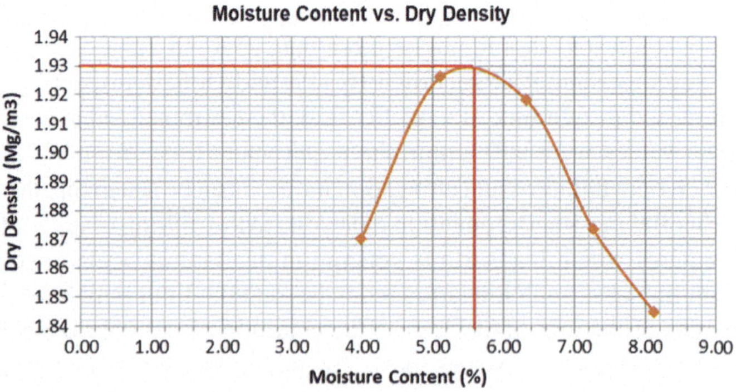

Fig. 4 Moisture content versus dry density

Table 6 Effect of stabilizing agent on bulk density

CA:RAP ratio	Stabilizing agent composition		Accelerator		
	GCBA (%)	OPC (%)	$CaCl_2$(%)	OMC (%)	Bulk density (Mg/m^3)
50:50	3	1	0	5.60	1.93
	2	2	0		1.95
	1	3	0		2.01
	0	4	0		2.02
	3	1	1		1.95
	2	2	1		1.98
	1	3	1		2.02
	0	4	1		2.05

$CaCl_2$ in general can slightly aid in binding the particles together, allowing for a better compaction and higher bulk density.

6 Conclusion

The aim of this study was to investigate the characteristics and compaction properties of the CIPR base course when stabilized with OPC, GCBA, and $CaCl_2$. The following conclusions can be drawn from this study;

- The particle size distribution of the laboratory specimen for the proposed design mix's composition ratio and the site sample were quite comparable. This indicated that the composition material used from the laboratory can be used for further study.
- From pH value analysis, the pH values for all the proposed design mix showed a value of higher than 12.0. The high pH environment (alkaline conditions) is conducive to the formation and stability of the hydrated products, which contribute to the strength and durability of the stabilized road base.
- The specific gravity of typical road base materials can vary widely depending on the type of material used. The highest specific gravity recorded was OPC and the lowest was RAP.
- The flakiness and elongation index for both CA and RAP as well as the mix of CA and RAP showed a value within the limits of JKR specification requirements.
- The OMC and MDD for the design mix at 50:50 ratio of CA and RAP recorded 5.60% and 1.93 Mg/m^3, respectively.
- The bulk density of the mix decreased as a result of the increased in GCBA percentage. However, the addition of $CaCl_2$ affected the mix by slightly increasing the bulk density. Although the GCBA itself when combined with OPC did not enhance the bulk density, further study needs to be conducted on the strength

performance to verify its suitability of the material in partial replacement of the OPC.

Based on the conclusions, it can be stated that the use of OPC, GCBA, and $CaCl_2$ has significantly affected the characteristics properties of the proposed design mix which eventually affects the strength performance of the mix. It is important to understand that using GCBA has many benefits, not only from a technical perspective but also from an environmental viewpoint.

References

Arun NR, Singh P, Gupta S (2020) Utilisation of ground bottom ash in concrete. Mater Today Proc 32:663–669

Baghini MS, Ismail A, Naseralavi SS, Firoozi AA (2016) Performance evaluation of road base stabilized with styrene–butadiene copolymer latex and Portland cement Performance evaluation of road base. Int J Pavement Res Technol 9(4):321–336. https://doi.org/10.1016/j.ijprt.2016.08.006

Dungca JR, Jao JAL (2017) Strength and permeability characteristics of road base materials blended with fly ash and bottom ash. GEOMATE J 12(31):9–15

Ghanizadeh AR, Rahrovan M, Bafghi KB (2018) The effect of cement and reclaimed asphalt pavement on the mechanical properties of stabilized base via full-depth reclamation. Constr Build Mater 161:165–174. https://doi.org/10.1016/j.conbuildmat.2017.11.124

Gonzalo-Orden H, Linares-Unamunzaga A, Pérez-Acebo H, Díaz-Minguela J (2019) Advances in the study of the behavior of full-depth reclamation (FDR) with cement. Appl Sci 9(15). https://doi.org/10.3390/app9153055

JKR Malaysia (2008) Standard specification for roadworks section 4: flexible pavement (JKR/SPJ/2008-S4). JKR Malaysia, Kuala Lumpur

Jun L, Zhenlin L, Weizhuo Z, Hesong J, Feng X, Chaoyun C, Luping T, Yanshuai W (2022) Valorization of municipal solid waste incineration bottom ash (MSWIBA) into cold-bonded aggregates (CBAs): feasibility and influence of curing methods. Sci Total Environ 843:157004. ISSN 0048-9697. https://doi.org/10.1016/j.scitotenv.2022.157004

Li X, Yin X, Ma B et al (2013) Cement-fly ash stabilization of cold in-place recycled (CIR) asphalt pavement mixtures for road bases or subbases. J Wuhan Univ Technol-Mat Sci Edit 28:298–302. https://doi.org/10.1007/s11595-013-0682-9

Marto A, Tan CS (2016) Properties of coal bottom ash from power plants in Malaysia and its suitability as geotechnical engineering material. Jurnal Teknologi (Sci Eng) 78:8–5, 1–10

Mohamad N, Muthusamy K, Embong R, Kusbiantoro A, Hashim MH (2021) Environmental impact of cement production and solutions: a review. Mater Today Proc 48:741–746. https://doi.org/10.1016/j.matpr.2021.02.212

Nguyen Thi N, Phi Hong T, Bui Truong S (2019) Utilizing coal bottom ash from thermal power plants in Vietnam as partial replacement of aggregates in concrete pavement. J Eng 2019:1–11

Suárez-Macías J, Terrones-Saeta, JM, Iglesias-Godino FJ, Corpas-Iglesias FA (2021) Development of cold in-place recycling with bitumen emulsion and biomass bottom ash. Crystals 11:384. https://doi.org/10.3390/cryst11040384

Sufian Z, Aziz N, Matori MY, Hussain MZ (2008) Mix design for cold-in-place pavement recycling; does it guarantee performance? Mixing view project selected physical properties of binary mixtures of crude glycerol and methanol at various temperatures view project. https://www.researchgate.net/publication/229041319

Sufian Z, Aziz NA, Yazip M, Mat M, Hussain Z, Azmi A (2009) Full depth reclaimed pavements using cement and hydrated lime as stabilizing agents

Sutanto M, Bala N, Al Chodri F, Sunarjono S (2018) Effect of hydrated lime and Portland cement on asphalt binders using DSR. MATEC Web Conf 203:05006. https://doi.org/10.1051/matecconf/201820305006

Tan P-C, Chan C-M (2021) Unconfined compressive strength of rehabilitated road base via polymer-modified cold in-place recycling (CIPR). IOP Conf Ser Mater Sci Eng 1150:012014. https://doi.org/10.1088/1757-899X/1150/1/012014

Wirtgen GmbH (2004) In: Group W (ed) Wirtgen cold recycling manual, 2nd edn. Wirtgen GmbH, Windhagen, Germany, p 251

Management of Metal Scrap from Construction Site for a Sustainable Construction Practice in Malaysia: A Pilot Case Study

Umol Syamsyul Rakiman, Raja Zuraidah Raja Mohd Rasi, and Ainol Mardhiyah Rahmat

Abstract This paper is a pilot study on the value creation from the construction site waste management. The pilot study is to explore any sustainable construction site waste management practice such as reuse, reduce or/and recycle in construction site in Malaysia. Proper management of construction site metal scrap is one of the steps to ensure sustainable environment as some of it can be dangerous and hazardous. Furthermore, a proper onsite waste management is one of the requirements to acquire a green building status. Despite this, not many constructions site adopted this approach. Therefore, the main objective of this research is to explore the flow of metal scrap disposal on the construction site. By exploring this flow, one can check whether a proper and sustainable way of managing the metal scrap on the construction site is being practised. This research adopted a semi-structured in-depth interview method to explore the flow of metal scrap disposal on a construction site. This is done to trace the material flow whether it is being reduced, reused or recycled onsite. The result of the research revealed that some of the metal scrap is actually did not end up on the recycling centre but rather being disposed on random land. This indicated an existence of poor metal scrap management on the construction site which signals crucial involvement from the government and law enforcer. Thus, the economy generation from being sustainable is an angle of no stress as sustainability is regarded as welfare practice but not profit making on the construction site which collapses the sustainability pillars of society, environment and economy.

Keywords Construction site management · Sustainability · Metal scrap · Recycling · Green building

U. S. Rakiman (✉) · R. Z. R. M. Rasi · A. M. Rahmat
Department of Business Management, Universiti Teknologi MARA, Perak Branch, Tapah Campus, Tapah, Perak, Malaysia
e-mail: umolsyamsyul@uitm.edu.my

R. Z. R. M. Rasi
e-mail: rzuraida@uthm.edu.my

A. M. Rahmat
e-mail: ainolmardhiyarahmat@uitm.edu.my

E. Shaffie et al. (eds.), *Sustainable Green Infrastructure*,
https://doi.org/10.1007/978-981-96-1486-8_18

1 Introduction

Construction is an important economic generator in Malaysia, contributed 8.1% of the country's Gross Domestic Products (GDP) and provided jobs to 10% (1.4 million) of people in Malaysia (Department of Statistics 2016). In the 11th Malaysia Plan (11MP) 2016–2020, as much as USD 62 billion had been allocated, with the main focus to improve public transportation and rural infrastructure in Malaysia (Economic Planning Unit 2015a). From the 11MP, the construction industry is had a boost of 8.7% in budget allocation compared from the 10th Malaysia Plan (10MP) 2011–2015 which signals the growth and importance of this industry.

Among the main focuses of the 11MP is sustainability, with the intend, action plan and budget to protect the environment. In addition, Construction Industry Development Board (CIDB), the regulating body for the Malaysian construction industry, launched the *Construction Industry Transformation Plan* (CITP) 2016–2020, with the objective to empower and strengthen the construction industry in Malaysia (Esa 2018). There are four strategic thrusts in CITP, which are *quality, safety and professionalism, environmental sustainability, productivity* and *internationalization.* A reduction of irresponsible waste during the construction cycle is one of the focuses under environmental sustainability (CIDB 2015).

Construction wastes will create adverse impacts on environment, economy and society; these impacts will diminish the productivity level and thus the overall performance of the construction projects as well (Poon et al. 2004). In the recent years, the construction industry had increased the focus on environmental, social and economic challenges within this industry (Moshood et al. 2024) which is the pillar of sustainability. Jabbour et al. (2020) stated that to focus on the environment, society and economy (sustainability) will need to new strategic changes.

The main drive within this move is to encourage businesses including the construction industry to reduce the environmental impact (Wang et al. 2019). Construction wastes are produced either directly or unintentionally by the construction activities (Nagapan et al. 2012). Construction wastes can be hazardous and could pollute the areas close to the construction site (Esa 2018). Among the waste being produced in a construction is metal scrap such as steel and aluminium (Alsamarraie 2021).

Considering the increasing budget for the construction industry in Malaysia, the possibility of increasing number of wastes is unavoidable issue to be discussed within Malaysia, being added with rapid economic development and growing populations (Rajendran 2020). As much as 26,000 tonnes of construction wastes were generated daily in Malaysia (Zulzaha 2014) and this will continue to increase as the financial support for this industry increases. One of the highest numbers of construction waste occurred in developed countries like the United States, with staggering number of 534 million tonnes of construction wastes generated in 2014 (EPA 2014) and increasing.

Furthermore, 40% of industrial wastes in the world come from the construction industry (Kulatunga et al. 2006). Thus, the rapid growth of construction activities has resulted in overcrowded landfills, especially in urban areas (Bakshan et al. 2015). Many authors signal that the uncontrolled construction waste generation could bring bad effect to the environment such as flash floods, soil erosion and sedimentation (Udawatta et al. 2015). Moreover, construction industry is among the largest consumer of raw materials whereby the depletion of natural resources is a major concern where the need to maximise the resources has become integral as well as minimise the waste generation (Crawford 2011).

Construction wastes are constantly being generated throughout the construction cycle. For instance, Begum et al. (2007) indicated that construction wastes are generated even at the planning and design stages, due to lack of consideration in minimising wastes. Meanwhile, Del et al. (2010) added that construction waste is not only generated at the planning and design stage but also throughout the construction cycle, whereby the construction stage contributes to the largest portion of wastes when the construction activities are not efficiently managed. The inefficient waste management process has led to too much waste being directed to landfills, including the waste that could be recycled (European Commission 2014). The increasing amount of waste has led to a significant change of perception towards effective waste management, whereby a sustainable waste minimisation strategy needs to be pursued and adopted in order to control and reduce the number of wastes dumped into landfills.

The construction industry is crucially important in supporting the initiatives of the government to promote sustainable development. Therefore, in order to achieve sustainable development while reducing wastes, the adoption of a circular approach to waste management is paramount, thus, the existing practices need to be changed for the better (Crawford 2011). This can be done by focusing more towards the circular approach in managing construction wastes throughout the construction cycle while taking the perspective of turning waste into resources, as the main priority.

The issue where too much waste being generated on a construction site is a concerning issue in developing countries like Malaysia, where waste management is not a priority, which could influence the effort of the government to promote sustainable cities (Esa 2018). The amount of waste will keep increasing in line with the growing population around the world. Currently, 40% of industrial waste in the world comes from the construction industry (Kulatunga et al. 2006). Hence, a proper construction waste management system is now no longer an option but a necessity. Incorrect handling of construction waste will not only affect the environment; it will also become expensive to rectify without proper procedures.

In Malaysia, there is a huge potential to develop a proper construction waste management system, which if successfully developed, will bring a lot of goods to the construction industry such as cleaner and neater construction site, enhance the organisations' image and preserve the environment. Additionally, the government of Malaysia would like to foster sustainable practices by ensuring all types of wastes, including construction wastes, should be managed holistically. A life-cycle approach of waste minimisation should be given a priority, and in fact, by January 2018, all contractors in Malaysia should comply with the waste management programme listed under the "*Environmental Management Systems ISO 14001*" (Economic Planning Unit 2015b).

Given the various issues mentioned above, this research aims to conduct a pilot case study to investigate the flow of metal scrap management on a construction site to check the implementation of proper waste management as part of a sustainable construction practice in Malaysia construction industry.

2 Methodology

This research employs a semi-structured in-depth interview method to identify metal scrap management on the construction site in Malaysia. A semi-structured in-depth interview session was conducted with responsible respondent in charge on site. The respondent (site) was selected based on the use of metal scrap and proper construction waste management activity. This method is chosen due to the explorative nature of this research since the waste management activity might differ depending on site and material heavily involved in the construction process.

This pilot case study consisted of five steps (see Flowchart 1). Firstly, the author conducted a meeting with a representative from Green Building Index Malaysia (GBI) to explore the sustainable construction site waste management practice in Malaysia. Then a semi-structured in-depth interview question was constructed based on the info to check whether the construction waste being recycled or not. In the interview question, respondents were asked questions relating to these constructs of reuse, reduce or/and recycle. After the interview session, the author was assisted to observe the construction site to see how they managed the metal scrap. The interview is manually transcribed; then Nvivo 13 was used to code and build the thematic analysis.

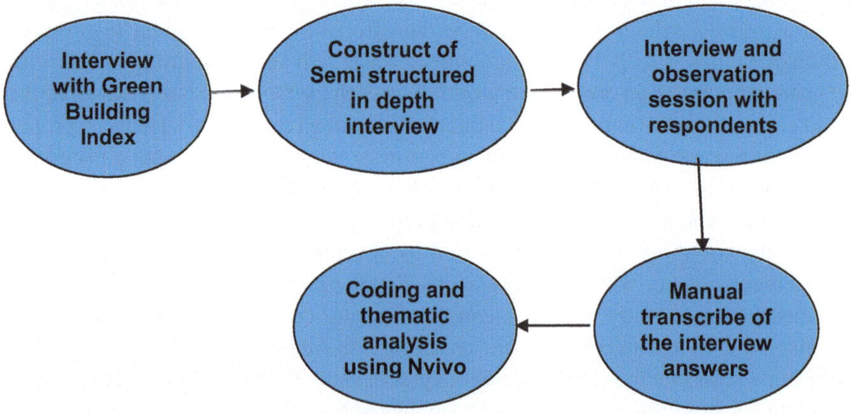

Flowchart 1 Research method for the pilot case study

3 Finding

Case Study Description

Case A was a construction firm. For this particular construction, it was for a mosque for one public university in Malaysia. Case A used mainly aluminium and steel in most part of the construction process. Along the construction process, aluminium and steel scrap was among the by-product. For Case A, the interviewed respondent was the Site Supervisor. The site supervisor had 10 years of experience in construction industry until the time of interview. Case A was a construction project that took 12 months to complete. Roles and responsibilities of the respondent were to monitor the flow of the whole construction site. The construction site was divided into three areas which were Sects. 1, 2 and 3. The distinction between these three areas were location and shape. The respondent has to monitor all three areas. Since different areas of construction started on different date, the respondent had to monitor the due date and its progress and safety of the site. Other than that, the respondent had to report and make necessary claims for the project.

For Case A, the respondent had confirmed that this project used aluminium and steel in the construction process. Based on the documents of the materials being used for construction and researcher observation, aluminium and steel were among the main component in the construction process. Along the interview session, by researcher observation, some of the construction site workers were collecting the metal scrap from the construction activity. This is an excerpt of the interview when the respondent showed that recycling activity was practised on this construction site as part of the practice of the construction waste management:

> Aluminium is the main material for the ceiling. Since aluminium could be sold and we get money. I decided to instruct the worker to collect the aluminium leftover and sell it to metal scrap collector. We also can eliminate it from hurting anyone since this is a mosque and many people come here to pray and some of them is children. (Respondent)

Case A sold the metal scrap from the construction activity to metal scrap collector. The scrap collector was a recycling centre in nearby town. According to the respondent, there was a lot of excess metal in the construction process. The amount of metal scrap being produced varies according to the construction activity being performed. This happened even when workers on the site had processed or cut the metal based on the given measurement being determined. The reason for this, was due to the design complexity and cutting difficulty which inevitably produces the metal scrap.

The operation that produced a lot of scrap was the cutting process. The other process that produced metal scrap was drilling. But this process did not produce scrap as much as the cutting process. The metal was cut based on the required shape. After cutting, metal scrap is produced. Then came the drilling process to make hole. The hole was prepared for the screwing and attachment of the metal as part of the building structures.

The construction activities such as cutting and drilling produced the metal scrap. This metal scrap then will be collected by the same worker who cut or drill it along the construction process. There was no fixed time to do the collection. All workers involved in the construction project have to ensure that the metal scrap being produced is managed properly. Managed properly means they collect the metal scrap being produced then they put it at one spot. After the metal scrap has been collected and gathered at one spot, the scrap collector will be called to take it away from the construction site. The respondent was responsible to contact the scrap collector.

As in the day of the interview, it was estimated as much as 80 kg of metal scrap had been produced. When there were too much metal scrap accumulated on site, then the respondent will contact the scrap collector to collect it. Too much metal scrap means the provided area is fully occupied until it affects the construction works. Since the spot to gather the metal scrap was not specified in terms of width and place as the construction progresses, researcher could not measure the width of the specified area. During the interview, collection had just been done. Making the construction area clear of the metal scrap. The collection of the metal scrap was done by the same worker who cut and drill the metal.

Based on the respondent, there was no specific amount of metal scrap being produced in any specific week along the construction process as it was purely depended on the design, worker's expertise and cutting error of the metal. There was no specific amount of the metal scrap to be collected before it is being sold. There was also no specific operation needed to prepare the metal scrap before they sell it other than collecting it and put it at one spot.

The collection phase was regarded as quite dangerous due to the sharp nature of the metal scrap. It was also the most difficult one. The respondent also did not record the recycling activity, they just simply collect and sell the metal scrap to the scrap collector. One important statement from the respondent was, when one of these two situations happens, most of the time the metal scrap will be dumped on random land to clear the construction site. The first situation is when the scrap collector is willing to buy the metal scrap at a very low price yet they have to send the metal scrap to the scrap collector. Then the respondent would rather dump it on random land. The second situation is when the construction exceeds the construction period and they need to clear the site. Then the metal scrap also will be dumped on random land for a quick clearance of the site. Figure 1 is the simplified form of the construction waste management activities being performed on the construction site (Case A).

4 Discussion and Conclusion

From the finding, the construction site was practising a proper construction waste management in the scope of metal scrap. The steps are metal scrap collection, gathering and disposal on a proper site such as send it to scrap collector for a recycling process. There was no reduce or reuse practise in terms of managing the metal scrap onsite. The next important point from the case study is, metal scrap can be dangerous if not properly managed due to the mechanical or chemical structure of the material. In this specific case, both metal scraps which were aluminium and steel did

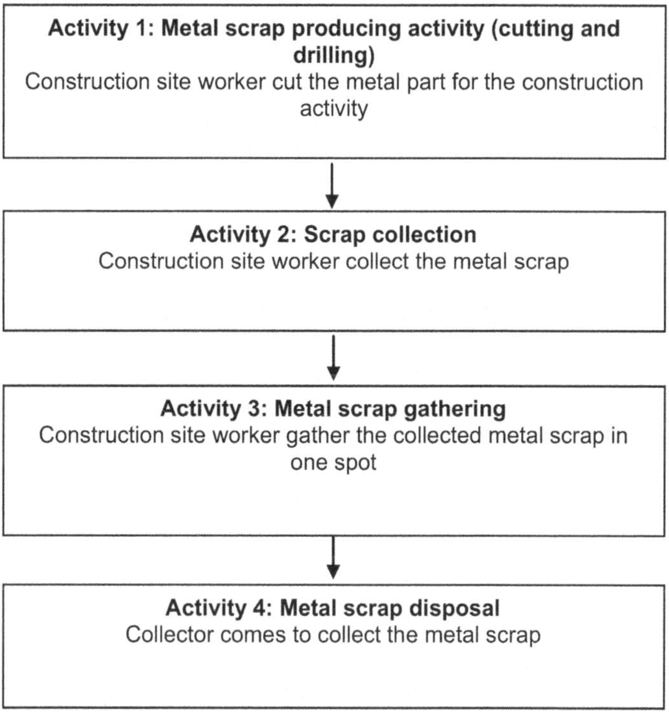

Fig. 1 Metal scrap waste management activity in Case A

not produce dangerous chemicals along its usage. But the structure of both metal scrap can be sharp and cut through most living creature such as human and animal, most of the part unintentionally especially when left unattended. The third important finding is the disposal of the metal scrap from a construction site to random land is a real practice. This is supported by Chen et al. (2015) that reported that the construction waste picked up by contracted waste management companies was only 15%. In contrast, the remaining 85% was left uncollected. This 85% uncollected waste is the candidate to be disposed on random land to clear the construction site before project handling to the client.

Practising proper construction waste management already in practice such as collecting, gathering and disposal. But this practice still can be improved. This is because all of this step is not properly documented such as the date, amount and people responsible involved in the practice. This is important to track how much actual waste is being produced from the construction site. This can tremendously help to keep the correct amount of material usage for future construction project. Hence being closer to sustainable construction waste management practice.

Next, due to the danger that metal scrap as a by-product of construction can impose to the workers and living creature surrounding the area, elimination of waste is difficult to achieve. But a proper management of the waste is key. Illegal dumping of waste materials from a construction site signal for a better construction site waste management as it is unlawful, irresponsible and most importantly dangerous. From the interview excerpt, illegal dumping was partially driven by very low price of metal scrap during the construction period. As fresh metal for construction is of high value yet metal scrap is of a very low value even if it can recycled. If this matter is not being addressed then sustainability can never be achieved as sustainability consists of society, environment and economy. Societal and environmental part is quite obvious, but if the economic part (generate profit) is not being addressed then sustainability will remain a utopia. This is further supported by Chi et al. (2020) that stated that the business to recycle construction waste was said to be a form of public welfare rather than a money-making trade.

One of the ways to mitigate this problem is through proper documentation and reporting of the waste being produced. But this step will only be effective if it is being monitored by a third party. This calls for the authority intervention and effective mobilisation such as CIDB, Department of Environment, local authorities and police. To remain the integrity of the data and information of the waste being produced, the use of digitalised documentation is necessary such as the Blockchain Technology can be helpful as the data cannot be tampered easily and can be available to related party only or also to the public.

As a conclusion, proper management of construction waste is already taking place but only in recycling part. Onsite reduce and reuse practice is yet to be visible. The second constraint for sustainable construction waste management is due to low economic benefit to the practitioner. An in depth future research in this matter is crucial if sustainability is a target in the construction industry. Final piece from this research is the improper documentation and reporting of construction site waste management makes monitoring of sustainability practice difficult.

Acknowledgements The authors would like to thank Ministry of Higher Education (MOHE), Research Management Center (RMC) and research team from UiTM and UTHM, for supporting this research.

References

Alsamarraie M (2021) Evaluation of practical steel waste disposal techniques in Malaysian construction sites. https://doi.org/10.13140/RG.2.2.22695.88484

Bakshan A, Srour I, Chehab G, El-Fadel M (2015) A field based methodology for estimating waste generation rates at various stages of construction projects. Resour Conserv Recycl 100:70–80. https://doi.org/10.1016/j.resconrec.2015.04.002

Begum RA, Siwar C, Pereira JJ, Jaafar AH (2007) Implementation of waste management and minimisation in the construction industry of Malaysia. Resour Conserv Recycl 51(1):190–202. https://doi.org/10.1016/j.resconrec.2006.09.004

Chen Z, Li H, Wong CTC (2015) An application of bar-code system for reducing construction wastes. Autom Constr 11:521–533

Chi B, Lu W, Ye M, Bao Z, Zhang X (2020) Construction waste minimization in green building: a comparative analysis of LEED-NC 2009 certified projects in the US and China. J Clean Product [Online] 256:120749

CIDB (2015) Construction industry transformation plan (CITP) 2016–2020

Crawford R (2011) Life cycle assessment in the built environment. Taylor & Francis

Del M, Izquierdo P, Weis S (2010) Sustainable construction: construction and demolition waste reconsidered. Waste Manage Res 28(2):118–129. https://doi.org/10.1177/0734242X09103841

Department of Statistics (2016) Quarterly construction statistics, second quarter 2016. Retrieved from https://www.statistics.gov.my/index.php?r=column/cthemeByCat&cat=77&bul_id=TEYvVFNZYUhtdElqUXgwZnlqbUdYQT09&menu_id=OEY5SWtFSVVFVUpmUXEyaHppMVhEdz09

Economic Planning Unit (2015a) Eleventh Malaysia plan. Malaysia. Retrieved from http://rmk11.epu.gov.my/index.php/en/

Economic Planning Unit (2015b) Strategy paper 18—transforming services sector. Malaysia. Retrieved from http://rmk11.epu.gov.my/pdf/strategy-paper/Strategy%20Paper%2018.pdf

EPA (2014) Advancing sustainable materials management: facts and figures report from https://www.epa.gov/smm/advancing-sustainable-materialsmanagement-facts-and-figures-rep

Esa MR (2018) Moving towards sustainable construction in Malaysia: a holistic model for construction and demolition waste management. Ph.D. thesis. School of Earth and Environmental Sciences, The University of Queensland. https://doi.org/10.14264/uql.2018.212

European Commission (2014) Waste as a resource. Publications Office of the European Union, Luxembourg, pp 1–8. https://doi.org/10.2777/61395

Jabbour CJC, Seuring S, de Sousa Jabbour ABL, Jugend D, Fiorini PDC, Latan H, Izeppi WC (2020) Stakeholders, innovative business models for the circular economy and sustainable performance of firms in an emerging economy facing institutional voids. J Environ Manage 264:110416

Kulatunga U, Amaratunga D, Haigh R, Rameezdeen R (2006) Attitudes and perceptions of construction workforce on construction waste in Sri Lanka. Manag Environ Qual Int J 17(1):57–72. https://doi.org/10.1108/14777830610639440

Moshood TD, Rotimi JO, Shahzad W (2024) Enhancing sustainability considerations in construction industry projects. Environ Dev Sustain

Nagapan S, Rahman IA, Asmi A (2012) Construction waste management Malaysian perspective. In: The international conference on civil and environmental engineering sustainability (IConCEES 2011). Johor Bahru, Malaysia, 3–5 Apr 2012

Poon CS, Yu T, Wong S, Cheung E (2004) Management of construction waste in public housing projects in Hong Kong. Constr Manag Econ 22(7):675–689. https://doi.org/10.1080/0144619042000213292

Rajendran N (2020) Research thesis—construction waste management-concrete—construction waste management in. Studocu. https://www.studocu.com/my/document/universiti-kuala-lumpur/final-year-project-1/research-thesis-construction-waste-management-concrete/20769926

Udawatta N, Zuo J, Chiveralls K, Zillante G (2015) Improving waste management in construction projects: an Australian study. Resour Conserv Recycl 101:73–83. https://doi.org/10.1016/j.resconrec.2015.05.003

Wang C, Ghadimi P, Lim MK, Tseng ML (2019) A literature review of sustainable consumption and production: a comparative analysis in developed and developing economies. J Clean Prod 206:741–754

Zulzaha FF (2014) New plan to manage solid waste systematically. The Star

Strategic Entry Location Decisions by Using of Entry Location Decisions Modeling (ELDeMo)

Norizzati Binti Ibrahim and Che Maznah Mat Isa

Abstract Firms often ask several questions before entering the international market, ranging from "Where are the markets with potential?" to "What are the factors that influence the firm's decision in choosing the location?". It is important for researchers to continue investigating and developing strategies to address this issue. In this regard, models are less developed for helping construction firms choose strategic locations by continent. Hence, the scant knowledge and access model related to the global market make it difficult for Malaysian companies to win projects abroad. Additionally, local construction firms lack an in-depth understanding of the overseas market. Moreover, firms are unable to identify effective entry location strategies when selecting appropriate entry location decisions in international markets. In this regard, this study fills this knowledge gap by developing Entry Location Decisions Modeling (ELDeMo), which is expected to help Malaysian construction firms (Mcf) reach the international level. Subsequently, ELDeMo is a combination of key factors that influence Mcf to enter international markets and also six main continents based on hierarchy scales from poor to excellent decision-making that suit their respective companies' capabilities to succeed in the host country. ELDeMo incorporates the six main factors, including firm-specific, firm-resource commitment, project-specific, target country, home country, and market-specific, as depicted through an integrative literature review and synthesized, and six continents comprised of Europe, Australia, the United States, Asia, Africa, and ASEAN. This study points the way forward for future work on developing strategic entry location models.

Keywords Assessment · Entry location · International · Construction · Modeling · Factors

N. B. Ibrahim
School of Civil Engineering, College of Engineering, Universiti Teknologi MARA Pasir Gudang Campus, Masai, Johor, Malaysia

C. M. Mat Isa (✉)
Civil Engineering Centre of Studies, College of Engineering, Universiti Teknologi MARA Pulau Pinang Branch, Permatang Pauh Campus, Permatang Pauh, Pulau Pinang, Malaysia
e-mail: chema982@uitm.edu.my

E. Shaffie et al. (eds.), *Sustainable Green Infrastructure*,
https://doi.org/10.1007/978-981-96-1486-8_19

1 Introduction

The global growth was expected to fall from an estimated 3.4% in 2022 to 2.9% in 2023, then rose to 3.1% in 2024 (IMF Data 2023). It is a known fact that the construction industry has a big effect on economic growth around the world (Alaloul et al. 2021). In this way, developing countries like Malaysia depend on the construction industry to carry out their plans for long-term growth. Evidently, the construction industry is one of Malaysia's most important industries. In 2023, the construction industry is expected to make up 4.7% of Malaysia's gross domestic product (Ministry of Finance Malaysia 2022). In this way, highways, power plants, and local building projects that are part of global construction projects also help the national economy (Construction Industry Development Board 2022). This highlights the significance of the construction industry in sustaining the country's economy. Furthermore, the Strategic Trust 4 objectives, as highlighted in the Construction 4.0 Strategic Plan (2021–2025), emphasize the role of foreign direct investment (FDI) or collaboration for local firms to enter foreign markets.

Competition and a high number of projects at home have led many Malaysian construction companies to take on projects abroad (Sarpin et al. 2019). International construction projects that involve a lot of different international companies are called "intercultural" projects (Kerur and Marshall 2012). In other words, the employer, contractor, and consultant are not of the same nationality, and at least one of the construction teams is working outside of the home country. It is a trend: international opportunities for Mcf to expand their business outside of Malaysia to ensure long-term survival at the international level. Since the early 1980s, Malaysian construction firms have been venturing into the global construction market. However, their ventures were not always successful. Mcf's participation in a foreign country is a challenging task that requires strategies that yield success abroad. Statistical data released by CONVINCE CIDB (2023) shows the lack of involvement of local players in acquiring construction projects at a global level. This situation is considered critical when the statistics of project decline for four international projects managed by Mcf are recorded in 2022, compared to 26 projects and 17 projects in 2020 and 2021, respectively. The financial statements also showed a decrease in the value of international projects, from 23.01 billion in 2021 to 8.94 billion in 2022.

Obviously, there are some constraints in the Malaysian construction sector (MCS), such as a lack of framework for guiding the MCF to win construction projects internationally (Alaloul et al. 2021). This weakness was identified due to poor educational readiness and insufficient data from local researchers. The high level of uncertainty about political risks and the instability of financial resources (Sarpin et al. 2019; Preece et al. 2015) is another reason why MCF doesn't take part in international projects. Other challenges faced by construction companies for international construction projects include competitiveness in the host country, a shortage of resources, an unstable global economy, and uncertainty in the host country (Sarpin et al. 2019). Accordingly, this problem has resulted in a diminished number of successful local construction firms involved in the global market. There are local

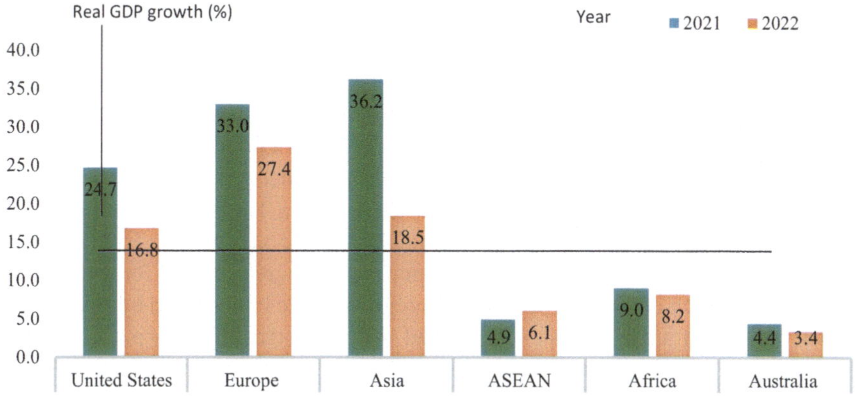

Fig. 1 World economic outlook growth projections by continents. *Source* International Monetary Fund (2023)

players who have gained years of experience and expertise in foreign countries, boosting their confidence to offer and win construction projects internationally (Ibrahim et al. 2021b). Malaysian construction firms have indeed ventured into ASEAN countries such as Singapore, Indonesia, and Cambodia, as well as non-ASEAN countries such as Saudi Arabia, Dubai, the United States, and Australia (Sarpin et al. 2019). The involvement of such companies can drive the construction sector, which is the backbone of economic growth. The involvement of construction companies internationally is of paramount importance. It affects the role of each sector at all levels of the economy (Alaloul et al. 2021; Ibrahim et al. 2021b). World Economic Outlook growth projections by continents through the International Monetary Fund (IMF) are depicted in Fig. 1.

Based on the IMF (2023), the United States and Europe are classified as advanced economies. Asia, ASEAN, Africa, and Australia, on the other hand, are considered emerging markets and developing economies. In this regard, the development of MEDIF is focused on six continents. Before conducting the analysis in this current research, Fig. 1 predicts that Asia will have the most strategic ELs, followed by Europe, the United States, Africa, ASEAN, and Australia. However, as the world prepares for the COVID-19 pandemic, construction firms must consider the market potential for 2023. According to Han et al. (2007), top contractors use a variety of methods to maintain their international performance. However, there is still a lack of a scale-based strategic framework in the matrix to decide whether a firm needs to enter the international market or not based on its performance. For instance, inadequate measurements associated with complex entry results have resulted in a negative perception for most construction companies conducting operations in other countries (Low et al. 2015). In addition, there is a lack of exposure to effective market entry strategies that can be considered by Mcf, while other different factors cannot be measured (Papadopoulos and Martín 2011).

Therefore, the most strategically estimated location in ELDeMo has been developed based on Mcf's experience running construction projects internationally. Later, ELDeMo was formulated as a result of the combination of continents and significant factors that influenced the company to enter the international market. The choice of the most strategic location can be measured by the scale of poor to brilliant decision-making based on ELDeMo measurements to ensure the success of the company when entering the international market at the selected location. This study also shows the gaps identified in the literature for developing ELDeMo.

Being a young player in international competition, there is a need for a deep and core understanding of unique strategies to integrate all the issues faced by Malaysian industry players. Later, the models developed are expected to help new players determine strategic locations based on continents so that they can win project offers in foreign countries. This research is also expected to be useful as a guideline and reference source for researchers to assist in increasing the number of local industry players in the global market. Malaysian financial planners and policymakers can use the findings of this research to improve the construction sector through the assessment guidelines, allowing Malaysia to become a progressive and prosperous country by 2030.

From this angle, this paper conducts an integrative literature review to expose the basic elements involved in developing the framework named "ELDeMo". Three research objectives have been planned to help in ELDeMo's formation process: (1) identify the key factors that attract Malaysian construction firms to carry out construction projects abroad; (2) study the relationships between the main factors and the location selected by the firm to conduct construction operations; and (3) develop a conceptual framework for the most strategic entry location aligned with the company's capabilities and subsequently suggest direction for its future research. The findings from this study offer a new way to apply based analysis methodology using Smart PLS analysis to identify relationships in strategic location selection. The originality of this study is in the development of ELDeMo) as a novel contribution. This model integrates multiple factors influencing entry decisions and offers a structured approach to help Malaysian construction firms (Mcf) make informed choices about international market entry.

1.1 Key Terms and Definitions Related to Internationalization

Internationalization is the process by which organizations gradually increase their international involvement (Ibrahim et al. 2021b; Durmaz and Taşdemir 2014). Improvement of skills at the domestic level, whether services, technology, or products, is used in the internationalization process to gain opportunities in the international market (Che Ibrahim et al. 2009; Ramayah et al. 2010; Asgari and Ahmad 2010). The relationship between entry location and internationalization is important where internationalization is seen as a contracting mechanism for companies leaving the local market to seek opportunities abroad (Ibrahim et al. 2021b). In addition, internationalization aims to increase a company's ideas, products, and funds, as well as investment opportunities.

According to Ibrahim et al. (2021b), countries with developing economies focus on international development. Construction firms operating overseas look for opportunities in the host country to earn high returns. In addition, the competitiveness of a firm is also important in determining the entry location (EL) into the host country. Many firms start as indirect or direct exporters and then move on to licensing, joint ventures, and finally direct investment. This evolution of the business is called the "process of internationalization" (Durmaz and Taşdemir 2014). This view is very similar to the study by Dawei (2008), where firms move in the process, taking into account time and knowledge. However, multinational companies in emerging markets internationally are at high risk in terms of market volatility due to national political and economic instability (Dawei 2008; Han et al. 2007). In addition, there are several external factors such as culture, politics, economic distance, and legal distance that can be barriers to firms' involvement in the international market (Dawei 2008; Swoboda et al. 2015; Tan et al. 2007; Dacko 2002).

Construction companies face major challenges in selecting strategic business entry locations for the international market. The international market strategy related to the entry location decisions for construction companies has been under research, more so in respect of developing countries. According to Kaur and Sandhu (2014), company internationalization has generally been conceived as a gradual and sequential process that takes place in stages. The company is assumed to build a secure domestic position before increasing its commitment to international operations and foreign markets.

Several questions of interest to previous researchers and practitioners will be increasingly asked, including: "What are the factors that influence the firm's decision in choosing the location?" and "Where are the markets with potential?". However, the answers to these and other related questions in international construction are still not very clear. Furthermore, previous studies on entry location decisions are more focused on the contribution factors that prevent firms from entering international markets, with a lack of solutions in terms of developing models or frameworks that serve as guidelines to help firms enter the international market. Discussions about comparisons between previous studies will be discussed further in the next subtopic in order to provide novelty for current studies. Finally, it is important to help construction companies choose a strategic entry location in the field of construction engineering based on empirical data and models, or frameworks.

1.2 Knowledge Gap of Entry Location Decisions in the Construction Literature

Entry location decisions (ELDs) leading to construction projects have gained growing interest among scholars in both developed and developing countries. Previous research, however, has revealed a lack of modeling of the combination of entry location decisions with the factors that assist construction firms in gaining opportunities for construction projects abroad. This issue is also highlighted in the Construction 4.0 Strategic Plan (2021–2025), which demonstrates the scarcity of a model or strategic framework to assist Malaysian construction firms in expanding their business abroad.

For example, Mat Isa (2016) has published a model named "An Integrated Model for International Market Entry Location, Entry Timing, and Entry Mode (ELETEM)". ELETEM consists of 27 mutually inclusive and significant factors (MISFs) that attract the attention of firms to the international market. Further, the 27 MISFs are divided into four main factors: country, market, firm, and project factors, to guide construction firms in avoiding poor decision-making, which can lead to failure in their new international ventures. The model is published based on logit values resulting from analysis using the Rasch model.

While this paper was designed specifically for entry location decisions, it examined the relationship between 41 significant factors influencing entry decisions (SIFIEDs) under the seven main factors that influence firms to enter the international market: firm-resource commitment, project-specific, target country, home country, firm-resource commitment, firm-specific, and market-specific factors. These factors were derived primarily from the previous studies (Ibrahim et al. 2021b). Next, in order to produce an ELDeMo model, these factors have been analyzed using Smart PLS to find the relationship between these factors and the entry locations consisting of ASEAN, Europe, Australia, the United States, Asia, and Africa.

In addition, limitations of the study by Alashwal et al. (2017) focus on success factors based on empirical data based on respondent profiles (such as profession and background), project location (e.g., ASEAN, Middle East, etc.), and type of project (e.g., infrastructure projects, residential projects, mixed development, etc.). For instance, a study by Ibrahim et al. (2021a) evaluated the performance of Malaysian construction firms using weighted scores characterized by three (3) dimensions of the country with taxonomic criteria under the ELAC model. The first dimension includes entry location results to ASEAN, non-ASEAN, and both regions, while the second dimension includes entry location results to ASEAN, Asia, Africa, the United States, and Europe, and the third dimension includes entry location results to countries classified as Developing Countries (DLC), Less Developed Countries (LDCs), New Industrial States (NIC), Developed Countries (DC), and Highly Developed Countries (HDC). However, the model does not analyze the factors that help Malaysian construction firms be sustainable in the international market.

Although the studies on ELDs were conducted based on previous research (see Table 1), the limitations of the studies only involved key factors that help Malaysian construction firms in the international market as well as score-based measurements to identify sustainable Malaysian construction firms in the international market. As a result, this research has been renewed based on the study by introducing ELDeMo. This model is a measurement tool that can predict the likelihood of Malaysian construction firms being sustainable in the international market based on a combination of factors influencing Malaysian construction firms in the international market

Table 1 Research gaps related to entry location decisions for Malaysian construction firms (from 2017 onwards)

Examples of authors	Focus of entry location decisions	Focus of data analysis method	Findings
Mat Isa et al. (2017)	ASEAN, non-ASEAN, both region	Rasch model analysis	The ELETEM decision model consists of 27 mutually inclusive and significant factors (MISFs)
Alashwal et al. (2017)	ASEAN, the Middle East, etc.	Regression analysis	The success factors of international construction projects include three main components, namely team power and skills, availability of resources, and external environment. Other components that influence international project success but to a limited extent include organization capability, project support, and project organization
Ibrahim et al. (2021a)	First dimension: ASEAN, non-ASEAN, both region Second dimension: ASEAN, Asia, Africa, the United States and Europe Third dimension: Developing Country. (DLC); Less Developed Countries (LDCs); New Industrial State (NIC); Developed Countries (DC); and Highly Developed Countries (HDC)	Weighted scores analysis	The ELAC score shows that three (3) Malaysian construction firms have sustained international operations in various locations by obtaining an ELAC score of more than 50% by venturing to ASEAN countries, developing countries and developed countries

(also known as independent determinants) and six entry locations spread across six continents (Europe, Asia, the United States, ASEAN, Australia, and Africa).

1.3 Entry Location Decisions (ELDs)

The researchers define internationalization as expanding across country borders into geographic locations that are new to the firm (Sousa 2017). It also creates new opportunities by providing access to new foreign stakeholders, resources, and institutions. The entry location decisions (ELDs) of firms, which are the organizational structure of how a firm establishes itself in a foreign market, are one of the important factors of internalization. However, internalization firms must choose an appropriate EL. According to Che Ibrahim et al. (2009), internationalization is a continuous development in terms of involvement as firms are expected to target gradually more distant markets.

It is important to consider the strategic EL of local construction firms entering a newly opened country market. The decision process on where to enter is a critical strategic decision that requires extensive determination of weaknesses and strengths and environmental scanning (Che Ibrahim et al. 2009). Hence, it is essential for Malaysian construction firms to identify the entry locations involved in the internationalization process and obtain appropriate strategies in order to ensure the success of constructing projects in foreign markets. Next, researchers (see Hutzschenreuter et al. 2016) studied the location of entry in six dimensions: cultural, institutional, economic, physical, geographical, and others. The majority of previous studies indicate that as cultural distance increases, firms are more likely to choose a joint venture than wholly owned subsidiaries (Hutzschenreuter et al. 2016; Driscoll and Paliwoda 1997; Magnusson et al. 2009). If a company has a lot of resources and a lot of experience, it is willing to take massive risks in the international market (Johanson and Vahlne 1990).

1.4 The Determinants Influencing Firms' Entry Decisions to Enter International Markets

Assessing the ELD is a continuous affair and has significant implications for the firms' long-term and short-term financial positions. In general, once a construction firm chooses a particular market to enter (EL decisions), it must consider other suitable entry decisions. A review of the available literature was conducted in order to identify the significant factors (independent determinants) that influence Malaysian construction firms' international market entry decisions. Firm-specific advantages toward internationalization refer to a firm's specific advantages toward internationalization that aim to reveal something new, such as a firm establishing a design center in a new location known as a new resource or introducing technologies, capabilities, and products that offer different and better growth (Zander 2015). Several independent determinants related to firm-specific characteristics obtained from previous literature reviews are firm ability to assess market signals and opportunities (Preece et al. 2016), firm experience related to overseas projects (Majocchi, Bacchiocchi, and Mayrhofer 2005), and firm product differentiation such as a strong brand name (Fahy 2002). In other words, firms must have competent leaders or employees to lead the firm. Leaders or employees who are always vigilant see opportunities for profit for the company based on innovation (Preece et al. 2016). Next, Ozorhon (2013) emphasizes the long-term strategic orientation or objectives of construction firms through innovation. Ozorhon (2013) highlighted that project-based businesses must manage innovation across organizational boundaries and within a network of interdependent suppliers, customers, and regulatory bodies. According to Chiao et al. (2008), large-sized firms have an advantage when it comes to being able to engage in overseas purchases, such as evaluating potential resources, negotiating with suppliers, checking material quality, and managing contract enforcement, in order to be competitive in the host country. Firm-resource commitment refers to the advancement of a firm's commitment to internationalization in order to ensure effective decisions and profitable actions in the international market (see Bianchi et al. 2018) or to the extent to which organizations and management resources are allocated to internationalization (Lages et al. 2008). Internationalization is dependent on having strong resources (employees) in terms of research and development knowledge (Daszkiewicz 2014). Next, firms with stronger financial and productive infrastructure have a greater opportunity to enter the international market (Arndt et al. 2009). The project also encourages firms to enter the international market. However, the technical or physical complexity of a project that requires the use of advanced technology and related issues in the environment in which the project is carried out is referred to as project complexity (Gidado 1996). As a result, companies with expertise in these technologies will seize the opportunity to enter the international market. Furthermore, the target country influences the decision of construction firms to enter the international market. According to Zeqiri and Angelova (2011), the target country's legal environment (including legal issues such as foreign exchange rates, jurisdiction, corruption, the presence of strict time limits, strict quality requirements, etc.) must be considered before entering the country. Next, in the international market, similarity to the host country in terms of social, cultural, and religious factors is an important factor to consider (Ozorhon et al. 2010). Furthermore, environmental factors in the home country are identifiable elements in the economic, demographic, or technological environment that affect an organization's growth, operations, and survival, resulting in domestic competition in terms of market segmentation. As a result, the forty-one (41) identified factors are derived primarily from the previous studies, and the ELDeMo incorporates the six main independent determinants, including firm-specific, firm-resource commitment, project-specific, target country, home country, firm-resource commitment, and market-specific, as depicted by previous research as tabulated in Table 2. Meanwhile, the dependent determinants comprising firm entry locations used in ELDeMo development are comprised of Europe, Australia, the United States, Asia, Africa, and ASEAN.

2 Methodology

This research adopts a quantitative method using a survey questionnaire administered to managers from construction firms to solicit their opinions to determine the significant determinants influencing their internationalization process.

Table 2 Key of independent determinants

The key of independent determinants	Significant independent determinants/authors
Firm-specific	Firm ability is to assess market signals and opportunities (Preece et al. 2016), firm experience related to overseas project (Majocchi et al. 2005), long-term and strong management strategic orientation/ objectives (Ozorhon 2013), large-sized firms have advantages in terms of the ability to engage in overseas purchases (Chiao et al. 2008), experience of firm in similar works in the international market (Souchon and Diamantopoulos 1999), international business network (strong relationship with foreign partners in host countries) Cavusgil 1998), product differentiation (strong brand name) (Fahy 2002), availability of strategic alliance through research and development partnerships and technical exchanges toward innovation (Abdul Ghani et al. 2011), the firm has employees who are skilled at interacting (employees who can speak the language of the host country, such as English, Arabic, Mandarin, Japanese, etc.) to facilitate discussions between different parties in the host country (Adsit et al. 1997)
Firm resource-commitment	Larger firms with greater resources as resources refer to number of employees and total annual sales (Abdul-Talib et al. 2011), superior management and having dynamic organizational capabilities in terms of production skills and refining business processes (Teece 2017), technical competence (Canavan et al. 2012), a firm has strong resources (employees) in terms of the level of knowledge related to research and development (Daszkiewicz 2014), strong financing capacity (as to cover labor costs, materials, machinery, and equipment costs) (Arndt et al. 2009)
Project-specific	The technical or physical complexity of a project that requires the use of highly advanced technology and related issues in the environment in which the project is carried out (Gidado 1996), project size project types such are building, highway, bridge, road etc. (Kostopoulos et al. 2019), type of clients such are public versus private projects (Othman 2011), contract types or procurement methods such as lump sum, cost-plus, design and build projects (Jeyakumar 2016), and availability of fund for projects (Othman 2011)

(continued)

Table 2 (continued)

The key of independent determinants	Significant independent determinants/authors
Target country	The legal environment of the target country (legal law related to such issues as foreign exchange rates, jurisdiction, corruption, existence of strict time limitations, strict quality requirements etc. (see Zeqiri and Angelova 2011), similarity to host country/market in terms of social, cultural, and religious (Ozorhon et al. 2010), attitude and intervention of host governments (Ozorhon et al. 2010), geographical distance to host country (Dachs and Pyka 2010), host government control on licensing, restrictions, and other FDI requirements (Ozorhon et al. 2010), similarity of host country/market (interest rate, per capita income has close relationship with size of market of target country (Braunerhjelm et al. 2005), availability of facility (Zeqiri and Angelova 2011), system similarity used in host country and home country (Zeqiri and Angelova 2011), a firm is contacted by foreign investors and/or the government of the host country to handle the projects (Zeqiri and Angelova 2011)
Home country	Promotion of export efforts of home government (Cuervo-Cazurra 2018), financial support from home country banks (Cardoza and Fornes 2011), domestic competition is due to changes in the environment in terms of technology (Pearce and Robinson 2013)
Market-specific	Market intensity of competition (Greening et al. 1996), market profit potential/attractiveness (Greening et al. 1996), market entry barriers in host country such as the existence of local competitors in the host country (Asgari and Ahmad 2010), product/service market growth (Hollensen 2007), construction market demand such as the host country also offers financial resources, labor, materials, transportation, and other utilities for the purpose of construction projects (Hollensen 2007), availability of innovative and entrepreneurial opportunities in market (Gunhan and Arditi 2005), current construction market size (Ibrahim et al. 2021b)

2.1 Respondent Selection Method and Target Respondents

The methodology for selecting respondents in this study involved a structured and systematic approach. The target population consisted of managers from Malaysian construction firms with experience in international projects. This specific focus ensured that the data collected was relevant to the study's objectives. The sampling frame was based on the list of global players provided by the Construction Industry Development Board (CIDB), Malaysia for the year 2020. This provided a reliable and comprehensive source of potential respondents who were actively engaged in international construction projects.

The data was obtained from a survey study of 74 Malaysian construction firms with experience of operating internationally. The selection of seventy-four respondents for this study can be considered sufficient and representative for several reasons. First is in terms of sample size adequacy. Generally, a sample size of more than

30 is acceptable for studies involving structural equation modeling (SEM). This study's sample size of seventy-four respondents exceeds this threshold, providing a robust basis for the analysis. Next is based on the diversity of respondents which represent a broad spectrum of roles within their respective companies, including general managers, senior project managers, project managers, senior engineers, engineers, project executives, and QA/QC engineers. This diversity ensures that the data collected encompasses a wide range of perspectives and experiences relevant to international construction projects. Third is based on their international experience in international operations, which is critical for the study's focus on entry location decisions in the international market. Their firsthand experience ensures that the data reflects practical insights and real-world challenges faced by Malaysian construction firms in global markets.

The determinants for the international markets selected by Malaysian construction firms were analyzed using Smart PLS software to obtain the relationship between the determinants and the entry location decisions into the international market as used in the development of ELDeMo.

Next, structural equation modeling (SEM) through Smart PLS software is used to obtain a significant relationship between each determinant for the purpose of constructing ELDeMo.

2.2 *Design of Questionnaire Survey*

In this study, the questionnaires were refined and reviewed by designing the initial instrument through piloting and enhancing the tool by carrying out the final check. The questionnaires based on the framework for the research area were used to answer the research questions regarding international market entry decisions made by construction firms. The questionnaire is divided into several sections, including Section A, which inquires about the respondents' background, designation, and international experience, and Section B, which is related to the global business market ELD of the firms. Next, Part C deals with market entry strategies, whereas value was measured using a five-point Likert scale (1: not important; 2: a little important; 3: important; 4: very important; 5: extremely important). Entry location decisions are divided into continents like ASEAN, Asia, Africa, the United States, Europe, and Australia.

Section D solicits the respondents' opinions to evaluate the determinants of the ELD of their firms based on the level of agreement. Each response was measured using a five-point Likert scale (1: not critical; 2: a little critical; 3: critical; 4: very critical; and 5: extremely critical). Section D divides into firm-specific (FSP), firm-resource commitments (FRC), project-specific (PS), home country (HC), target country (TC), and market-specific (MS). The following section explains the analysis and discussion based on the descriptive analysis and using statistical analysis tools (SPSS 20) and Smart PLS analysis. The purpose of each method and its results are explained in the following section.

2.3 Data Collection for Final Data

The primary or main data were collected through the pilot study, the questionnaire survey, and the interview sessions. Questionnaires were sent through postal mail and e-mail to 132 Malaysian construction firms as global players based on the CIDB Malaysia list in the year 2020. The total final response rate was approximately 62% at the end of this study. The response rate of the current study via online is considered to acceptable, i.e., an average of 30%, as stated by Wu et al. (2022). Questionnaires were sent via postal mail and e-mail to the identified firms. This dual approach maximized the response rate by accommodating the preferences and capabilities of different firms in terms of communication. The high response rate enhances the reliability and generalizability of the findings.

2.3.1 Mounting the Measuring and Structural Equation Models on the Smart PLS

The dependent determinants refer to entry location (i.e., ASEAN, Asia, Africa, Europe, the United States, and Australia), while the independent variables were the determinants to enter the international market that new entrants needed to consider (i.e., target country (TC), home country (H), market-specific (Ms), firm-resources (FRC), projects (PSF), and firm-specific (FSP). Various aspects need to be considered to produce quality and valid structural equation modeling (SEM), and to ensure that the equation obtained is valid. The first thing to look at in model measurement is the convergent validation obtained by observational average variance extracted (AVE). Furthermore, after all variables are analyzed, the criteria average value (AVE) must be greater than 0.50. If there is a variable that has an AVE of less than 0.50, then the variable must be deleted from the equation (Fornell and Larcker 1981). If the value AVE > 0.5, the equation is valid (Fornell and Larcker 1981).

The internal consistency values (Cronbach's Alpha) and the composite reliability (CR) were then investigated. The CA value must be above 0.60 or 0.70, and the values of CR of 0.70 and 0.90 are considered reliable (Fornell and Larcker 1981) to ensure that the samples are free of bias. Next, observe the value of the standard linear regression (Rsquare) for linear correlation as ($p < 0.05$), and the value of R^2 should be above 50%. The third step is that the value of the t-test for each correlation relation between variables should be calculated (between − 1.96 and + 1.96, corresponding to the probability of 95% and 5% outside of this interval, in a normal distribution) and correspond to p-values of 0.05. An assessment of the model adjustment indicator by the mean weighted AVE (goodness of fit for the measuring model) shows that all that is constructive is reflexive (Ramayah et al. 2010). A value of AVE of 0.36 is adequate for the areas of the social and behavioral sciences.

Table 3 shows six (6) main analyses (convergent validity, discriminant validity, model reliability, *T*-test, Pearson closure coefficient (R^2) evaluation, and path coefficient) that were emphasized during data analysis using Smart PLS, referring to statements or studies that have been done based on previous researchers.

The *T*-ratio tests for evaluation of the significance of the correlations and regressions are important for this study as they make sure the threshold values are + 1.96.

Table 3 Summary of the steps of analysis by using Smart PLS

Indicator/analysis	Purpose	Referential values/criteria	References
AVE measurement	Convergent validity	AVE > 0.50	Henseler et al. (2009)
Criteria of Fornell and Larcker	Discriminant validity	Compare the R^2 of the AVE values of each item construct with the correlation between the construct (latent variables). Next, the R^2 of AVE values should be > correlation of the construct	Fornell and Larcker (1981)
Cronbach alpha and composite reliability	Model reliability	CA > 0.70, CR > 0.70	Hair et al. (2014)
T-test or *T*-analysis	Evaluation of significance of the correlations and regressions	Beta coefficient, $\beta > \pm 1.96$	Hair et al. (2014)
Evaluation of the coefficient of the Pearson's determination (R^2)	Evaluate the range of R^2	$R^2 = 2\%$(small effect); $R^2 = 13\%$ (median effect); $R^2 = 26\%$ (large effect)	Cohen (1988)
Path coefficient/ *P*-value analysis	Evaluation of relation	Interpretation of values. The *P*-value test is used to conduct a test of the hypothesis that is 0. At the 0.05 significance level (i.e.,1–95%), we calculate the one-tailed *P*-value associated with the path coefficient. Generally, this quantity could be interpreted as the probability that belongs to a distribution with a mean of zero. The hypothesis is accepted if $P \leq 0.05$; otherwise, it is rejected	Cohen (1988)

3 Data Analysis

The data for the main study were analyzed using two approaches, including SPSS analysis (SPSS Software, version 20) and SEM analysis (Smart PLS software). Initially, demographic data were analyzed using SPSS 20 software to describe the basic features of the data. This analysis is very useful in analyzing the background of the respondents and firms, the percentage frequency of the respondents' designations, international experience, and the ELD of firms in the international market. The following subsections will explain through descriptive analysis using percentages in the form of bar or pie charts, histograms, and tables.

Meanwhile, Smart PLS analysis was used to obtain the relationship between independent and dependent determinants using Smart PLS 2.0 M3 software. This analysis is very useful for this study because it can solve complex data (many variables were observed) and models with less consecrated theoretical support (Ringle et al. 2014). Next, several linear equations show the connection between dependent and independent determinants developed through this structural model. The following subsections describe in more detail the analysis of the data using SPSS and Smart PLS software.

3.1 Descriptive Analysis

A total of seventy-four respondents participated in the survey study, each representing 74 construction companies with experience in international projects. The respondents consisted of a general manager ($n = 6$), a senior project manager ($n = 8$), a project manager ($n = 15$), a senior engineer ($n = 19$), an engineer ($n = 20$), a project executive ($n = 3$), a QA/QC engineer ($n = 2$), and other managerial posts ($n = 1$). One respondent for the "other managerial post" is a senior manager. In relation, Fig. 2 shows the percentage of respondents by designation.

The results, as shown in Fig. 2, show the highest percentage of respondents consisting of engineers (27%), followed by senior engineers (25.7%), project managers (20.3%), senior project managers (10.8%), general managers (8.1%), QA/QC engineers (2.7%), and lastly other managerial posts (1.4%). All respondents have international experience and are also directly involved in the operations of their

Fig. 2 Percentage of the respondent's designation

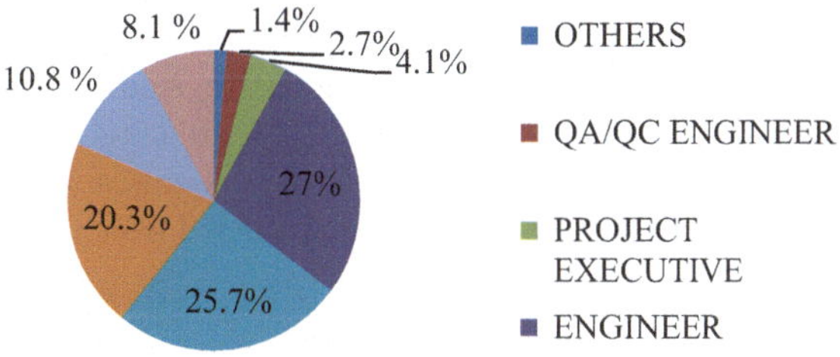

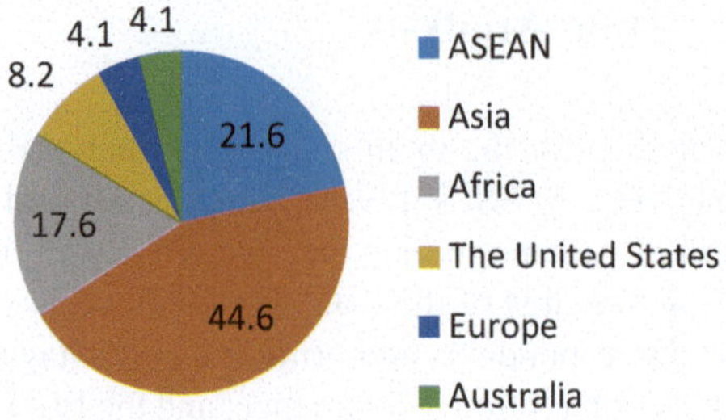

Fig. 3 Company's entry location decisions

firm, whether in the management, planning, or control of construction projects in the international market.

Thus, this profile indicates a diverse background of top managers and professionals involved in international operations in this study. Thus, the findings of this study indicate the diversity of backgrounds among top managers and professionals involved in international operations.

Next, the following Fig. 3 shows the location of the company's operations by continent, covering ASEAN, Asia, Africa, the United States, Europe, and Australia.

Based on the findings in Fig. 3, the majority of firms choose to conduct international operations in Asian countries (44.6%). The second continent chosen by the firm to conduct operations internationally is ASEAN (21.6%), followed by Africa (17.6%), the United States (8.1%), Europe, and Australia (4.1%), respectively.

3.2 The Relationship Between EL Decisions and Sustainable Performance in the International Market

CR values are used in SEM analysis applications to identify the relationship between the levels of sustainability of firms with operations at six locations. The path (arrow) and its coefficient are present in Fig. 4, which measure the correlation significance of every item construct. For example, the relationship between Asia and sustainability indicates that at every 1.495, the firms' survival increases in the international market.

Notably, all items are valid (> 0.50) based on the algorithm analysis shown in Fig. 4 previously. SEM shows all items under (all continents) are significant, influencing firms to sustain in international markets. Moreover, the equations are valid as long as the average value (AVE) is greater than 0.5 for all items (Fornel and Larcker 1981). Next, all the CR values are valid, exceeding 0.70, as similarly exhibited by the CA value at more than 0.70 (Hair et al. 2014). Additionally, the results indicated that all the *P*-values are significant (0.000) for all items. All items were found to influence firms' survival in international markets significantly.

Next, the regression Eq. (1) constructed from the above findings is as follows:

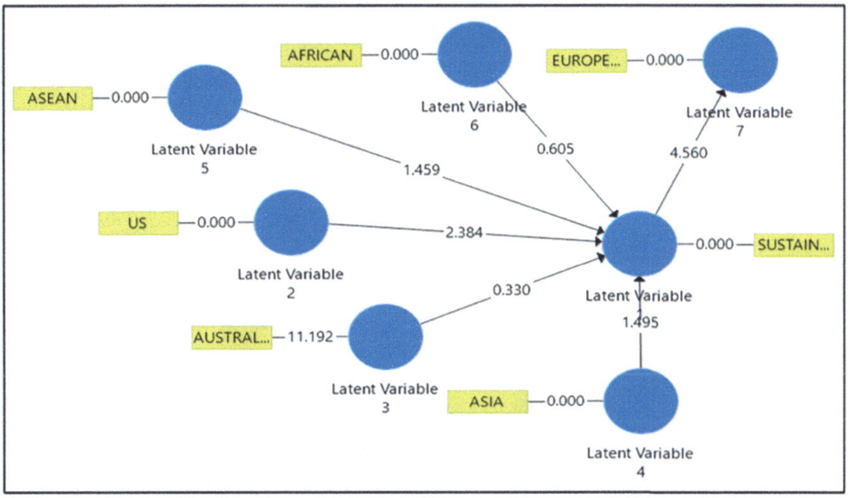

Fig. 4 Standardized regression weights of the relationship between EL decisions and sustainable performance in the international market

$$\begin{aligned}\text{ELDeMo} &= 4.560\,\text{Europe} + 2.384\,\text{US} + 1.495\,\text{Asia} \\ &\quad + 1.459\,\text{ASEAN} + 0.605\,\text{Africa} + 0.330\,\text{Australia}. \qquad (1)\end{aligned}$$

Overall, Europe exhibited the most significant beta coefficient, 4.560, influencing firms' sustainability in international markets. Accordingly, the values indicate that the European countries are a suitable EL in influencing the firm's success or sustain internationally. This result is followed by the United States, Asia, ASEAN, Africa, followed by Australia and Oceania.

3.3 *Measurement of ELDeMo*

ELDeMo models are built to rank the most significant determinants that lead to success in the international market, increasing companies' potential to enter the international market. Firstly, the EL used by Malaysian construction firms to enter the international market is tested through a questionnaire survey. On a scale of one to six, the order is labeled for each determinant to represent from poor to excellent entry decisions made into the international market, ensuring the firms' long-term sustainability in the international market. The higher the ß value, the more significant the EL decision that enables firms to sustain for an extended period in the international market. Equation (2) presents the AoELD calculation, where the X label represents each ET decision (Awang et al. 2018).

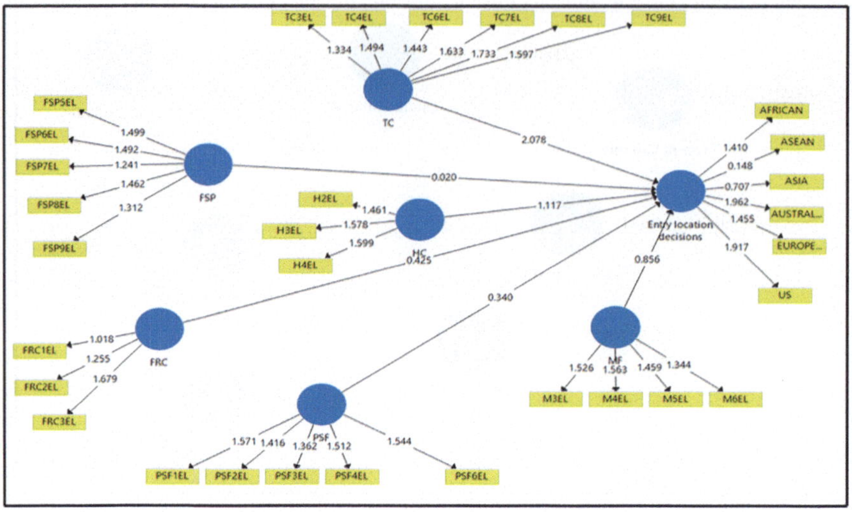

Fig. 5 Relationship between Malaysian construction firms' determinants and EL decisions in international markets using bootstrapping modules

$$\text{ELDeMo} = \beta_1 X_1 + \beta_2 X_2 + \beta_3 X_3 + \cdots \beta_4 X_n. \tag{2}$$

Meanwhile, Fig. 5 presents the path (arrow) and its coefficient that measures the correlation significance of every item constructed after bootstrapping.

The findings from Smart PLS in Fig. 5 show that all items (TC, FSP, PS, HF, FRC, and MF) significantly influence the EL decisions (> 0.50).

Finally, the regression Eq. (3) derived from the above findings for independent determinant and firm sustain in international markets is as follows:

$$\begin{aligned}\text{ELDeMo (independent determinants)} &= 2.078\,\text{TC} + 1.117\,\text{HC} + 0.856\,\text{MF} \\ &\quad + 0.425\,\text{FRC} + 0.340\,\text{PSF} + 0.002\,\text{FSP}.\end{aligned} \tag{3}$$

The TC significantly presented the most significant beta coefficient, 2.078, influencing EL decisions. The value also means that TC is a critical determinant influencing the firm's EL decision to enter the international market. The second determinant influencing the EL decision was the HC (1.117), followed by MS (0.856), FRC (0.425), PS (0.340), and FSP (0.002). The next section discussed the measurement of ELDeMo.

3.4 *Measurement of EL Decisions*

The measurement of ELDeMo is based on the relationship of critical independent determinants (MS, HC, TC, FSP, FRC, and PS). Meanwhile, dependent determinants include the EL decisions of six continents (Europe, ASEAN, Asia, Africa, the United States, and Australia).

3.4.1 Assessment of the Key Determinants of EL Decisions as to Analyze the Firms' Ability to Enter the International Market

Firstly, the firm can test whether it has the potential or ability to sustain itself in the international market. Firm capabilities can be tested based on the following Eq. (3) resulting from Smart PLS analysis.

3.5 *Assessment of ELDeMo Based on the Relationship Between the Dependent Determinants and EL Decisions*

Firstly, measurements for all EL are made based on equation below (1).

$$\begin{aligned}&\text{ELDeMo (Dependent determinants)}\\&\quad = 4.560\,\text{Europe} + 2.384\,\text{the United States} + 1.495\,\text{Asia}\\&\qquad + 1.459\,\text{ASEAN} + 0.605\,\text{Africa} + 0.330\,\text{Australia}.\end{aligned}$$

Measurements for ELDeMo are as follows:

$$\begin{aligned}&\text{ELDeMo (Dependent determinants)}\\&\quad = 4.560\,\text{Europe} + 2.384\,\text{the United States} + 1.495\,\text{Asia}\\&\qquad + 1.459\,\text{ASEAN} + 0.605\,\text{Africa} + 0.330\,\text{Australia}.\end{aligned}$$

Hence, for Option 1:

$$\begin{aligned}\text{Final score, } X &= 4.560(6) + 2.384(5) + 1.495(4) + 1.459(3)\\&\quad + 0.605(2) + 0.330(1) = 51.56,\end{aligned}$$

$$\begin{aligned}X \text{ in percentage} &= (51.56/51.56) * 100\\&= 100.\end{aligned}$$

Y refers to the score for the independent determinant. For example, if score Y is 100%.

Hence,

$$\begin{aligned}\text{Total Score } Z &= ((Y + X)/200) * 100 \\ &= ((100 + 100)/200) * 100 \\ &= 100.\end{aligned}$$

Based on the assessment (ELDeMo), Europe is the best EL to enter the international market with an EL decision score of 76.7, or a *B* score, which ensures the firm's sustainability in the global market. It is followed by the United States, Asia, ASEAN, Africa, and Australia. Europe also showed that the firm's highly sustainable in foreign markets was evident.

The ELDeMo model incorporates several key and significant independent determinants derived from previous research to predict the likelihood of Malaysian construction firms' success in the international market. For instance, the model's application can be seen in the successful international expansion of Gamuda Berhad into the Australian market. Gamuda's strategic entry was influenced by firm-specific advantages, resource commitment, and market-specific factors, as detailed in the ELDeMo framework as described below.

3.6 A Case Study: Gamuda Berhad's Successful Expansion into the Australian Market

Gamuda Berhad, a leading Malaysian engineering and construction firm, ventured into the Australian market as part of its international expansion strategy. The company's entry into Australia is a noteworthy case of successful market penetration using strategic entry location decisions.

Strategic Entry Factors: Gamuda Berhad's expansion was driven by several key factors identified in the ELDeMo model:

1. **Firm-Specific Factors**: Gamuda's extensive experience and strong reputation in large-scale infrastructure projects provided a solid foundation for its international ventures. The firm's established track record in Malaysia and other ASEAN countries showcased its capabilities to potential Australian partners and clients.
2. **Firm-Resource Commitment**: Gamuda committed significant resources to its Australian operations, including financial investments and the deployment of experienced personnel. The firm's willingness to allocate resources to research and development, as well as innovation in construction techniques, positioned it favorably in the competitive Australian market.
3. **Project-Specific Factors**: The technical complexity and scale of Australian infrastructure projects matched Gamuda's expertise. Projects such as the Sydney Metro West and the Melbourne Metro Tunnel required advanced engineering solutions, which Gamuda was well-equipped to provide.

4. **Target Country Factors**: Australia's stable legal and regulatory environment, coupled with a transparent procurement process, made it an attractive destination for Gamuda. The country's robust economic growth and commitment to infrastructure development presented numerous opportunities for international construction firms.
5. **Home Country Factors**: Support from the Malaysian government and financial institutions facilitated Gamuda's overseas expansion. The government's promotion of export efforts and the availability of financing options from Malaysian banks played a crucial role in enabling Gamuda's entry into the Australian market.
6. **Market-Specific Factors**: The high demand for infrastructure development in Australia, driven by urbanization and population growth, created a favorable market environment. Gamuda identified and capitalized on the market's potential, leveraging its innovative and entrepreneurial capabilities to secure key projects.

Gamuda Berhad successfully secured contracts for major infrastructure projects in Australia. The firm's strategic approach included forming joint ventures with local partners, which helped navigate the cultural and regulatory nuances of the Australian market. For instance, Gamuda partnered with Laing O'Rourke, a prominent UK-based construction company, for the Sydney Metro West project. This collaboration combined Gamuda's technical expertise with Laing O'Rourke's local market knowledge, ensuring project success. Berhad's entry into the Australian market exemplifies the application of the ELDeMo model's strategic entry location decisions. By aligning its capabilities with market opportunities and leveraging firm-specific, resource-commitment, project-specific, target country, home country, and market-specific factors, Gamuda achieved significant success in the international construction arena. This case study highlights the importance of a well-structured approach to market entry, ensuring sustainable performance and growth in foreign markets.

4 Conclusion

Assessment of ELDeMo is developed to provide guidelines to local construction firms in making strategic decisions in making the right ELDs (Asia, ASEAN, Europe, the United States, Africa, and Australia). Thus, ELDeMo highlighted six key determinants for entering the international market, arranged by scale; the most important determinants chosen by firms to enter the global market were the TC, followed by HC, MS, FRC, PS, and FSP. Hence, ELDeMo can facilitate firms' decisions on entry according to the firm's capabilities through the scale. The AoELD was made based on the scale with coefficient values from Smart PLS analysis. Based on the overall assessment of ELDeMo, firms can sustain in the international market and it could provide indicators or guidelines for local firms to operate globally, from poor to excellent decision-making. ELDeMo also assists firms in strategic location selection by considering six significant independent determinants influencing entry decisions to ensure firms can sustain long-term strategic goals in the international market. ELDeMo has identified the most strategic continent to ensure the firm's sustainability in the international market, which is Europe, followed by the United States, Asia, ASEAN, Africa, and Australia.

Limitation of the Study

This study covered Malaysian construction firms' ELDs to the global market. However, the ELDs in AoELD were limited to continents such as Asia, ASEAN, Europe, the United States, Africa, and Australia, which will be chosen by Malaysian construction firms to enter the international market for this research. Hence, recommendations for future studies should divide the ELDs into developed, developing, and highly developed countries.

Acknowledgements The authors gratefully acknowledge the Ministry of Higher Education Malaysia for providing the Fundamental Research Grant Scheme (600-IRMI/FRGS 5/3 (025/2019) and also to the School of Civil Engineering, College of Engineering, UiTM Shah Alam for their support in carrying out this research. We would like to thank all the managers who have participated in this study.

References

Abdul Ghani AB, Ishak MM, Tull M (2011) An empirical case study of international strategic alliances in Malaysia. Int J Bus Soc Sci 2(6):194–209

Abdul-Talib A-N, Salleh MFM, Shamsuddin FM, Ashari H (2011) The effects of firm size and international business experience on export attitudes. Am Coll Radiol 19(1&2):312–347

Adsit JD, London M, Crom S, Jones D (1997) Cross-cultural differences in upward ratings in a multinational company. Int J Hum Resour Manag 8(4):385–401

Alaloul WS, Musarat MA, Rabbani MBA, Iqbal Q, Maqsoom A, Farooq W (2021) Construction sector contribution to economic stability: Malaysian GDP distribution. Journal of Sustainability 13:5012. https://doi.org/10.3390/su13095012

Alashwal AM, Fareed NF, Al-obaidi KM (2017) Determining success criteria and success factors for international construction projects for Malaysian contractors. Constr Econ Build 17(2):62–80. https://doi.org/10.5130/AJCEB.v17i2.5319

Arndt C, Claudia MB, Mattes A (2009) Barriers to internationalisation: firm-level evidence from Germany. In: EFIGE working paper

Asgari M, Ahmad SZ (2010) Internationalization of Malaysian service firms: business strategy and choice of foreign market entry mode. In: International conference on business, economics and tourism management (CBETM 2010), vol 1, pp 1–6

Awang Z, Hui LS, Zainudin NFS (2018) Pendekatan Mudah structural equation modelling. MPWS Rich Publication

Bianchi C, Carneiro J, Wickramasekera R (2018) Internationalisation commitment of emerging market firms. J Small Bus Enterpr Dev 25(2):201–221. Retrieved from https://doi.org/10.1108/jsbed-07-2017-0221

Braunerhjelm P, Oxelheim L, Thulin P (2005) The relationship between domestic and outward foreign direct investment: the role of industry-specific effect. Int Bus Rev 14(6):677–694. https://doi.org/10.1016/j.ibusrev.2005.09.004

Canavan D, Sharkey-Scott P, Mangematin V (2012) Aligning strategy and talent in creative professional service firms. J Bus Strateg. https://doi.org/10.21427/D7S03G

Cardoza G, Fornes G (2011) The Internationalisation of SMEs from China: the case of Ningxia Hui autonomous region. Asia Pac J Manag 28(4):737–759
Cavusgil ST (1998) International partnering: a systematic framework for collaborating with foreign business partners. J Int Mark 6(1):91–107. Retrieved from http://www.jstor.org/stable/25048719
Che Ibrahim KI, Osman MH, Baki AM, Ghani MK (2009) Location consideration in international construction markets by Malaysian firms: specific local perspective. In: Conference on scientific and social research
Chiao Y, Yu JC, Li P, Chen Y (2008) Subsidiary size, internationalization, product diversification, and performance in an emerging market. Int Mark Rev 25(6):612–633
Cohen J (1988) Statistical power analysis for the behavioral sciences, 2nd edn. Lawrence Erlbaum Associates, Hillsdale
Construction 4.0 Strategic Plan (2021–2025) Next revolution of the Malaysian construction industry. Published by Construction Industry Development Board Malaysia in year 2020
Construction Industry Development Board (2022) International construction project. Retrieved from https://convince.cidb.gov.my
CONVINCE CIDB (2023) International construction project. Retrieved on 3 Mar 2023 from https://convince.cidb.gov.my/(S(ulpzk5t4eqteqeti0cqhao3g))/CMS/Page/2
Cuervo-Cazurra A (2018) The evolution of business groups' corporate social responsibility. J Bus Ethics 153:997–1016. https://doi.org/10.1007/s10551-018-3912-4
Dachs B, Pyka A (2010) What drives the internationalisation of innovation? Evidence from European patent data. Econ Innov New Technol 19(1):71–86. Retrieved from https://doi.org/10.1080/10438590903016476
Dacko SG (2002) Understanding market entry timing decisions: the practitioner-academic gap. Market Intell Plann 20(2):70–81. https://doi.org/10.1108/02634500210418482
Daszkiewicz N (2014) Small and medium-sized enterprises in Visegrad Countries towards internationalisation challenges in the European Union. In: Patterns of business internationalisation in Visegrad Countries—in search for regional specifics, pp 179–192
Dawei G (2008) Entry strategy of enterprises: a case study of Chinese firm. Perspective 53:1–53
Driscoll AM, Paliwoda SJ (1997) Dimensionalizing international market entry mode choice. J Mark Manag. https://doi.org/10.1080/0267257X.1997.9964459
Durmaz Y, Taşdemir A (2014) A theoretical approach to the methods introduction to international markets. Int J Bus Soc Sci 5(6(1)):47–54
Fahy J (2002) A resource-based analysis of sustainable competitive advantage in a global environment. Int Bus Rev 11(1):57–78. Retrieved from https://doi.org/10.1016/S0969-5931(01)00047-6
Fornell C, Larcker DF (1981) Evaluating SEM models with unobservable variables and measurement error. J Mark Respect 18(1):39–50
Gidado K (1996) Project complexity: the focal point of construction production planning. Constr Manag Econ 14:213–225
Greening DW, Barringer BR, Macy G (1996) A qualitative study of managerial challenges facing small business geographic expansion. J Bus Ventur II 11(4):233–256
Gunhan S, Arditi D (2005) Factors affecting international construction. J Constr Eng Manag 131(3):273–282. Retrieved from https://doi.org/10.1061/(ASCE)0733-9364(2005)131:3(273)
Hair JF, Hult TM, Ringle CM, Sarstedt MA (2014) Primer on partial least squares structural equation modeling (PLS-SEM). SAGE, Los Angeles
Han SH, Park SH, Kim DY, Kim H, Kang YW (2007) Causes of bad profit in overseas construction projects. J Constr Eng Manag 133(12):932–943. https://doi.org/10.1061/(ASCE)0733-9364(2007)133:12(932)
Henseler J, Ringle CM, Sinkovics RR (2009) The use of partial least squares path modeling in international marketing. Adv Int Mark 20:277–319
Hollensen S (2007) Global marketing: a decision-oriented approach, 4th edn. Pearson Education, Harlow

Hutzschenreuter T, Kleindienst I, Lange S (2016) The concept of distance in international business research: a review and research Agenda. Int J Manag Rev 18(2):160–179. Retrieved from https://doi.org/10.1111/ijmr.12065

Ibrahim N, Mat Isa CM, Preece CN, Mustaffa NK, Najwa A (2021a) Development of entry location assessment criteria (ELAC) model for Malaysian construction firms in international markets. J Acad 9(1):21–38

Ibrahim N, Mat Isa CM, Mohd Nusa FN, Preece CN, Ghani MK (2021b) Significant factors influencing entry decisions (SEFIED) into international market. In: 2021 IEEE 12th control and system graduate research colloquium (ICSGRC 2021), pp 138–143. Retrieved from https://doi.org/10.1109/ICSGRC53186.2021.9515240

International Monetary Fund (2023) Global inflation will fall in 2023 and 2024 amid subpar economic growth. Retrieved from https://www.imf.org/en/Publications/WEO/Issues/2023/01/31/world-economic-outlook-update-january-2023

Jeyakumar J (2016) Basic civil engineering: what are the type of contracts?. Retrieved from Basic Civil Engineering. Retrieved from https://basiccivilengineering.com/2016/05/type-of-contracts.html

Johanson J, Vahlne JE (1990) The mechanism of internationalisation. Int Mark Rev 7. https://doi.org/10.1108/02651339010137414

Kaur S, Sandhu MS (2014) Internationalisation of born global firms: evidence from Malaysia. J Asia Pac Econ 19(1):101–136. https://doi.org/10.1080/13547860.2013.818426

Kerur S, Marshall W (2012) Identifying and managing risk in international construction projects. Int Rev Law

Kostopoulos KC, Spanos Y, Soderquist KE (2019) Market- firm-, and project-level effects on the innovation impact of collaborative R&D projects. J Knowl Econ 10:1384–1403. Retrieved from https://doi.org/10.1007/s13132-015-0342-8

Lages LF, Jap SD, Griffith DA (2008) The role of past performance in export ventures: a short-term reactive approach. J Int Bus Stud 39(2):304–325

Low WW, Abdul-Rahman H, Zakaria N (2015) The impact of organizational culture on international bidding decisions: Malaysia context. Int J Project Manage 33(4):917–931. https://doi.org/10.1016/j.ijproman.2014.10.010

Magnusson P, Westjohn SA, Boggs DJ (2009) Order-of-entry effects for service firms in developing markets: an examination of multinational advertising agencies. J Int Mark 17(2):23–41. https://doi.org/10.1509/jimk.17.2.23

Majocchi A, Bacchiocchi E, Mayrhofer U (2005) Firm size, business experience and export intensity in SMEs: a longitudinal approach to complex relationships. Int Bus Rev 14(6):719–738. https://doi.org/10.1016/j.ibusrev.2005.07.004

Mat Isa CM (2016) An integrated model for international market entry location, entry timing and entry mode (ELETEM) decisions of Malaysian construction firms. Ph.D. thesis. Universiti Teknologi Mara Malaysia. Retrieved from https://ir.uitm.edu.my/id/eprint/18563/1/18563.pdf

Mat Isa CMB, Saman HM, Preece CN (2017) Development of OLI+S entry decision model for construction firms in international markets. Constr Econ Build 17(4):66–91. https://doi.org/10.5130/AJCEB.v17i4.5573

Ministry of Finance Malaysia (2022) Malaysia's GDP expected to grow 4–5 percent in 2023, 2022's revised upwards. Published on 7 Oct 2022. Retrieved from https://www.mof.gov.my/portal/en/news/press-citations/malaysia-s-gdp-expected-to-grow-4-5-pct-in-2023-2022-s-revised-upwards-mof

Othman AA (2011) Modelling marketing resources, procurement process coordination and firm performance in the building construction industry: the integration of resource based view and coordination theories. Ph.D. thesis. Retrieved from https://doi.org/10.13140/RG.2.1.3529.7362

Ozorhon B (2013) Analysis of construction innovation process at project level. J Manag Eng 29(4):455–463. Retrieved from https://doi.org/10.1061/(ASCE)ME.1943-5479.0000157

Ozorhon B, Arditi D, Dikmen I, Birgonul M, Talat (2010) Performance of international joint ventures in construction. J Manag Eng 26(4):209–222. Retrieved from https://doi.org/10.1061/(ASCE)ME.1943-5479.0000022

Papadopoulos N, Martín OM (2011) International market selection and segmentation: perspectives and challenges. Int Mark Rev 28(2):132–149. https://doi.org/10.1108/02651331111122632

Pearce J, Robinson R (2013) Strategic management: strategy formulation implementation and control, 13th edn. McGraw-Hill/Irwin, New York

Preece CN, Ibrahim CK, Mat Isa CM, Saman HM (2015) Development of entry location, entry timing and entry mode decision model for construction firms in international markets. Constr Manag Econ. https://doi.org/10.1080/01446193.2015.1084429

Preece CN, Mat Isa CM, Saman HM, Che Ibrahim CK (2016) Development of entry location, entry timing and entry mode decision model for construction firms in international markets. Constr Manag Econ 34(4–5). Retrieved from https://doi.org/10.1080/01446193.2015.1084429

Ramayah T, Mohamad O, Jaafar M, Aziz ARA, Wong SS (2010) Internationalisation of Malaysian contractors. J Int Bus Entrepr Dev 5(1):18–27

Sarpin N, Yi Ling C, Kasim N, Mohd Noh H (2019) Key strategies in undertaking international construction project: contractors' perspective. MATEC Web Conf 266:03019. Retrieved on https://doi.org/10.1051/matecconf/201926603019

Souchon AL, Diamantopoulos A (1999) Export information acquisition modes: measure development and validation. Int Mark Rev 16(2–3):143–168

Sousa ML (2017) Internationalization of construction professional service firms-overcoming challenges in mature market. Master thesis. University Catalica Portuesa

Swoboda B, Elsner S, Olejnik E (2015) How do past mode choices influence subsequent entry? A study on the boundary conditions of preferred entry modes of retail firms. Int Bus Rev. https://doi.org/10.1016/j.ibusrev.2014.10.008

Tan A, Brewer P, Liesch PW (2007) Before the first export decision: internationalisation readiness in the pre-export phase. Int Bus Rev 16(3):294–309. https://doi.org/10.1016/j.ibusrev.2007.01.001

Teece DJ (2017) Business models and dynamic capabilities, long range planning. Retrieved from https://doi.org/10.1016/j.lrp.2017.06.007

Wu Meng-Jia, Kelly Zhao and Francisca Fils-Aime (2022). Response rates of online surveys in published research: A meta-analysis. Comput Human Behav 7 (2022) 100206

Zander I (2015) Internationalization: patterns in business firms. In: International encyclopedia of the social & behavioral sciences, pp 580–586. Retrieved from https://doi.org/10.1016/b978-0-08-097086-8.73008-x

Zeqiri J, Angelova B (2011) Factors that influence entry mode choice in foreign markets. Eur J Soc Sci 22(4):572–584

Experimental and Numerical Investigation of Concrete-Bamboo Sandwich Panel (BSP) as Load-Bearing Wall Under Compression Loading

Mohammad Afif Hasnizam, Rohana Hassan, Mohd Hanizan Bahari, Ezahtul Shahreen Ab Wahab, Ali Awaludin, and Ahmad Mazlan Othman

Abstract Studies on recent tropical bamboo development as the structural component are very limited. Therefore, an experimental programme was conducted to investigate the possible application of the infilled concrete with bamboo as a load-bearing wall. This study observed effective and practical techniques of concrete-bamboo sandwich panels (BSP) constructed as load-bearing walls with and without openings to withstand compression loads. A nonlinear 3D finite element model (FEM) of the BSP wall was developed using the ABAQUS 2022 to simulate the entire BSP component. As a result, the BSP without openings exhibits higher compressive strength than the BSP with openings. The stress predictions from the finite element model align closely with experimental data, with error rates of 16.98% for walls with openings and 1.38% for walls without openings. The failure mechanism of load-bearing walls is effectively modelled, and the predicted failure loads are in close agreement with the experimental results. Furthermore, an increase in the percentage of longitudinal bamboo reinforcement significantly enhances the compression strength and load-carrying capacity of the BSP walls.

M. A. Hasnizam · R. Hassan (✉)
School of Civil Engineering, College of Engineering, Universiti Teknologi Mara (UiTM), 40450 Shah Alam, Selangor, Malaysia
e-mail: rohan742@uitm.edu.my

R. Hassan
Institute for Infrastructure Engineering and Sustainable Management (IIESM), Universiti Teknologi MARA, 40450 Shah Alam, Selangor, Malaysia

M. H. Bahari
Kolej Kemahiran Tinggi MARA (KKTM), 17000, Pasir Mas, Kelantan, Malaysia

E. S. A. Wahab
Faculty of Civil Engineering Technology, Universiti Malaysia Pahang Sultan Abdullah, Lebuh Persiaran Tun Khalil Yaakob, 26300 Gambang, Pahang, Malaysia

A. Awaludin
Faculty of Engineering, Universitas Gadjah Mada, Bulaksumur, 555281 Yogyakarta, Indonesia

A. M. Othman
Iramo Bamboo Resources, 31120, Sungai Siput Utara, Perak, Malaysia

E. Shaffie et al. (eds.), *Sustainable Green Infrastructure*,
https://doi.org/10.1007/978-981-96-1486-8_20

Keywords Bamboo strand panel (BSP) · Compressive strength · Finite element model · Load-bearing walls · Longitudinal reinforcement

1 Introduction

The usage of tropical bamboo as a material in the construction field has yet to be appropriately introduced. The lack of available information and lack of proper management of bamboo throughout the countries has contributed to the lower percentage of bamboo applications, especially in the construction industry. The advantage of bamboo being matured in shorter life growth compared to timber makes bamboo one of the potential materials as a structural component. The installation of bamboo is fairly simple. The bamboo is easy to cut, with no bark to peel during processing, and its lightness makes it easy to handle, transport, and store. Due to its lightness, the bamboo wall system does not require sophisticated technical skills; only jointing at the site is needed (Vishal et al., 2017).

As reported by Anshari et al. 2010, 3D finite element modelling (FEM) has been created to improve the size, quantity, and position of the elements. The advantage of FEM is that it allows for a comprehensive analysis of stress distribution in bamboo. The FEM's reliability is crucial in establishing the correctness of the behaviour of the concrete with bamboo under a range of loading scenarios. FEM is an effective technique for use in building projects because of its versatility, which allows designers to spend hours using software to generate new designs of models and materials continuously. FEM may also be adjusted to satisfy specific accuracy criteria, avoiding the need for complicated geometries throughout the design phase. Among other modelling techniques, FEM is widely used for its dependability in simulating and verifying the behaviour of structural elements precisely as they perform in experimental tests because it is a cost-effective and efficient tool for achieving the desired result with acceptable accuracy (Emmanuel et al., 2018; Kumar and Ashish, 2018; Xuesong et al., 1999).

Few researchers reported on bamboo strength behaviour, such as the test results of compressive strength for dry and treated Dendrocalamus asper and Bambusa vulgaris (Mansor et al., 2019) and behaviour and strength of bearing wall strengthening by CFRP (Ibrahim et al., 2015). Ma et al. (2014) studied an investigation of industrialized building system performance compared to conventional construction methods. Esti, 2016 demonstrated that bamboo has the potential to be used as a construction material for organic structures. Based on the limitation of the available reports in this field, this study investigates the compression strength of concrete-bamboo sandwich panel (BSP) load-bearing walls. The primary goal is to experimentally test the BSP and simulate the BSP in a 3D model using ABAQUS.

2 Literature Review

Bamboo contains a lot of carbohydrates and sugar. This nutrient in bamboo will attract insects to eat the bamboo. With treatment and protection, the durability of bamboo will be at least five years. Bamboo degradation will affect bamboo products' strength, usage, and value, leading to decay and disintegration, splits and cracks, and unsightly stains and blotches. The bamboo needs to be treated with safe preservatives so that it can be preserved and used for a long time.

According to Ghavami (2005), the relationship between concrete and bamboo is primarily due to the water absorption behaviour. The bamboo's water absorption might produce fissures in hardened concrete. The bamboo will expand and push away from the concrete. The bamboo then loses moisture and shrinks to its former proportions after curing, creating spaces around itself. In order to address this issue, bamboo must be treated before it can be utilized in buildings.

Concrete and treated bamboo have a stronger connection than concrete, steel, or untreated bamboo. The link between untreated bamboo and cement mortar frequently weakens due to bamboo's expanding behaviour. The water in the new cement mortar may be absorbed by the bamboo during the casting and curing of the wall panel, changing the bamboo's physical characteristics (Ghavami, 2005).

Bamboo walls are extensively employed as building separations due to their low maintenance costs, financial efficiency, and architectural attractiveness. These walls have much better axial strength than brick infill walls. Because bamboo is light, it favours earthquake-prone places because the chance of falling is much reduced owing to flexibility, and even if it does fall, it can be readily re-established with low work and money (Raj and Agarwal, 2014). Bamboo walls can be built using entire bamboo culms, vertical halved culms, or split bamboo woven mats, and mortar can be placed on one or both sides of the bamboo frame. Bamboo grown vertically was proven more durable than bamboo grown horizontally (Raj and Agarwal, 2014).

In a review of experimental observation on the structural performance of bamboo-reinforced concrete beams reported by Azuwa 2024, the author clarifies that bamboo has diverse applications, including enhancing the performance, properties, and behaviour of reinforced concrete beams as internal reinforcement. It summarizes the impact of bamboo on the concrete beam properties and discusses its use as reinforcement. Azuwa, 2024 concluded that the previous experimental results indicate that incorporating bamboo into reinforced concrete beams significantly improves flexural strength, stress–strain behaviour, load–deflection response, failure modes, crack patterns, tensile strength, compression, and shear modulus. Due to its superior strength, force resistance, and anchoring properties, bamboo can serve as an alternative to traditional steel reinforcing bars in reinforced concrete beams. Most of the relations between concrete and bamboo were tested on concrete beams studies such as the flexural performance of bamboo-reinforced foamed aerated concrete beams (Ikponmwosa et al. 2014), bond-slip improvement of bamboo reinforcement in concrete beams using hose clamps (Muhtar Dewi et al., 2016), and a comparative study of bamboo-reinforced concrete beams using different stirrup materials for rural construction, (Mark and Russell, 2011).

3 Methodology

3.1 Experimental Set-Up

For the material preparation, Bambusa vulgaris bamboo was used. All bamboo specimens were cut into 1.2 m each. The bamboo specimens were treated in a chemical solution and placed in a specific box layered with canvas. The chemical solution was diluted using boric acid, borax, and water in a ratio of 1:1:50. The specimens were soaked in the chemical solution for two weeks. Then, the specimens were dried at ambient temperature for one week.

Two (2) types of bamboo sandwich panel (BSP) were prepared, BSP with and without opening. In total, 6 panels were prepared, 3 each with and without an opening panel. Both BSPs are cast in a size of 1.2 m × 0.6 m, and the panel thickness is 0.15 m. For the panel with an opening, the opening size is 0.15 m × 0.15 m in a square shape at the centre of the panel (Fig. 1).

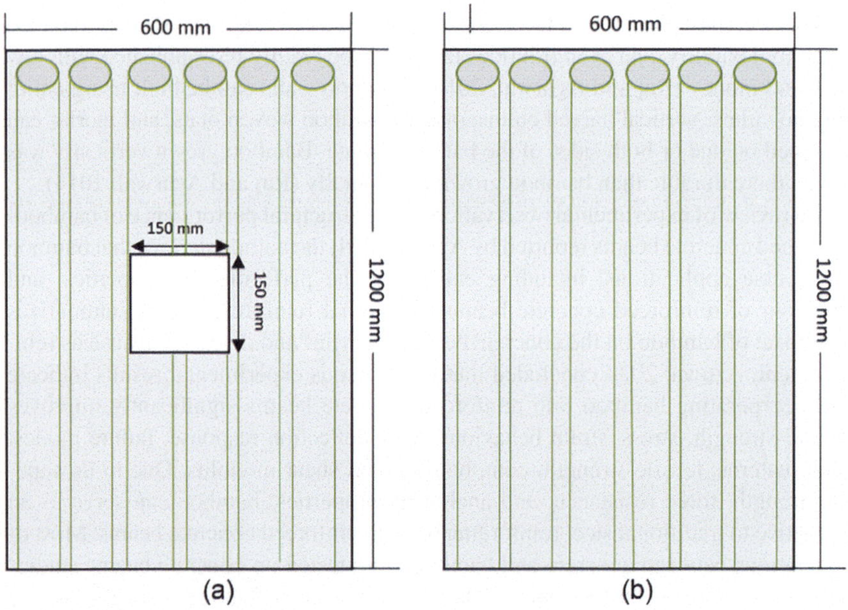

Fig. 1 Schematic diagram of bamboo sandwich panel (BSP); **a** BSP with opening and **b** BSP without opening

The preparation of BSP involved arranging 1.2 m-long specimens together in vertical alignment (Fig. 2). The formwork was prepared according to the planned size of the wall panel, which is 1.2 m × 0.6 m and 0.15 m thick. The gap between each specimen is ± 75 mm, and the top and bottom of the specimens were covered with C-channels.

Concrete is mixed with fine and coarse aggregate, a binder, and water. Portland cement is used as the binder. The mixture proportions of cement, fine aggregate, and coarse aggregate are in the ratio of 1:2:4. Concrete that is too dry will not spread appropriately while the concrete is too wet; it will settle quickly and will not be easy to trowel. A good concrete mix has features such as possessing good workability, spreading quickly, and sticking to vertical surfaces easily. The concrete was applied on the bamboo panel within a duration of 90 min. Figure 3 shows the casting of the BSP.

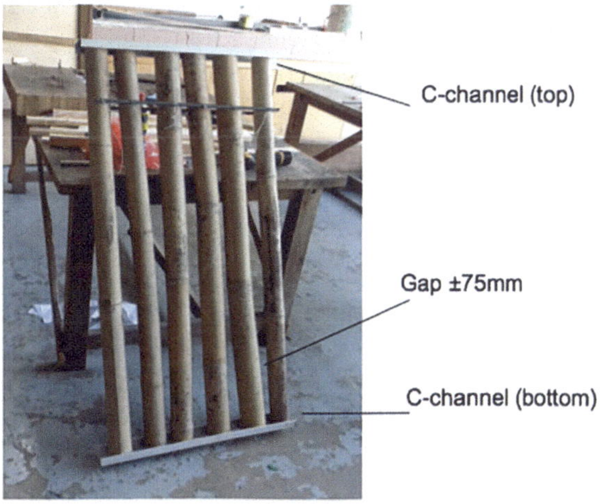

Fig. 2 Round bamboo arrangement for the BSP

Fig. 3 Casting of wall panel

The bamboo panel is sandwiched by concrete on both sides of the BSP. Then, the BSP is cured by wet fabrics and placed in an open space. After 28 days, the BSP is compressed using a Universal Testing Machine (UTM). The experimental test set-up is prepared according to the BS EN 1990 standard.

3.2 *Finite Element Modelling*

The elastic behaviour of the concrete is modelled in ABAQUS 2022 using basic linear elasticity. The Young's modulus and Poisson's ratio were adjusted to 3.0e6 N/cm^3 and 0.2, respectively. The concrete damage-plasticity model was utilized to define the concrete's multilinear isotropic elastic behaviour. It captures plasticity by utilizing compressive and tensile strength. The parameters used to model and depict the behaviour of the bamboo Bambusa vulgaris are the density, Young's modulus, and Poisson's ratio as 0.6 g/cm^3, 2.0 e7 N/cm^2, and 0.15, respectively. The mild steel c-channel is an elastic material with density, Young's modulus, and Poisson's ratio set to 7.85 g/cm^3, 2.1e7 N/cm^2, and 0.3, respectively.

The assembly of the different pieces (concrete, bamboo, and c-channel) developed remains independent in the model; they are joined during experimental testing by replicating instances in the model and reorienting the parts to portray the beam. Meshing is one of the most crucial procedures in any finite element analysis technique; it essentially divides the entire volume being modelled into more minor elements so that all applied loads on the structure are distributed equally throughout. The loading was supplied to the steel loading plate nodes, and the failure load was distributed throughout the steel and put on the nodes after meshing (Fig. 4).

The analysis used influences both the calculation time and the degree of difference between the laboratory experiment and the ABAQUS software. The static method

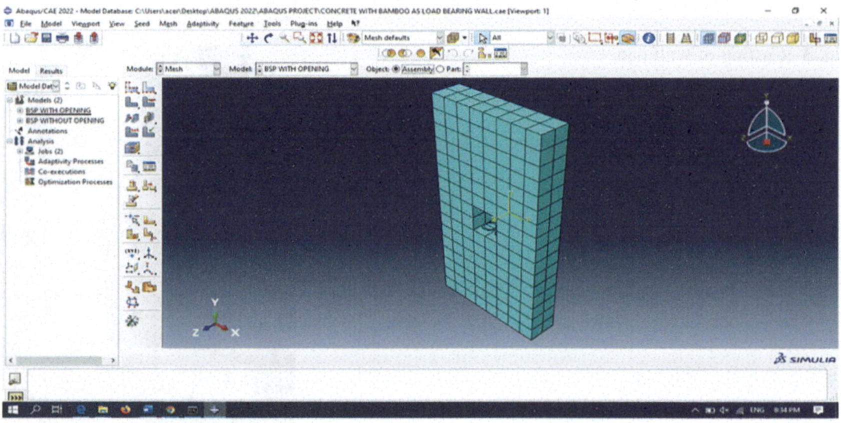

Fig. 4 BSP meshing model

and incremental loading were used for analysis. The experimental concrete with bamboo as load-bearing walls was simulated with load stages and time periods.

4 Results and Discussion

4.1 Experimental Work

The result of the experimental work is presented in Fig. 5.

The provided graph illustrates the relationship between load (kN) and displacement (mm) for concrete BSP walls with and without openings. The graph includes multiple series, each representing different conditions: The W01, W02, and W03 kN series represent walls without openings, while the With1, With2, and With3 series represent walls with openings.

The observations of load–displacement behaviour show that the walls without openings (W01, W02, and W03) show a generally smooth and continuous increase in load with displacement. The peak loads for these walls are higher, with W01 reaching around 800 kN, W02 around 700 kN, and W03 slightly lower but still substantial. The displacement at peak load is also higher for these walls, indicating better performance and ductility.

The walls with openings (With 1, With 2, and With 3) exhibit a more irregular load–displacement behaviour. The peak loads for walls with openings are significantly lower than those without openings. BSP With1 peaks around 350 kN, With2 around 450 kN, and With3 around 300 kN. The displacement at peak load is lower for walls with openings, suggesting reduced ductility and structural performance.

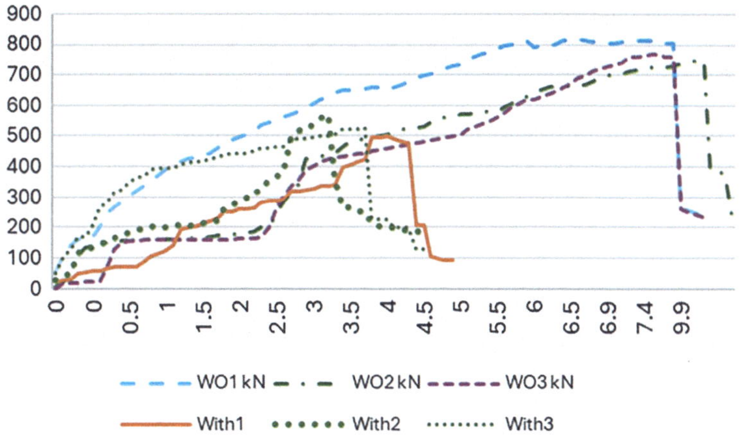

Fig. 5 Relationship between load (kN) and displacement (mm) for BSP walls with and without openings

The peak load comparison between the BSP walls without openings (W01, W02, and W03) consistently shows higher peak loads than those with openings (With1, With2, and With3). This behaviour indicates that the presence of openings reduces the overall load-bearing capacity of the walls. For the displacement at failure, walls without openings reach higher displacement values before failure, indicating they can undergo more deformation and absorb more energy before collapsing. At the same time, the BSP walls with openings show sudden drops in load after reaching peak values, which may indicate brittle failure modes due to stress concentrations around the openings.

The ductility performances show the smoother curves of the walls without openings, suggesting more ductile behaviour and allowing them to withstand more significant deformations. The walls with openings show more abrupt changes in the load–displacement relationship, indicating less ductility and a higher likelihood of sudden failure. The presence of openings significantly compromises the structural integrity of the walls, as evidenced by the reduced peak loads and lower displacement at failure. The walls with openings must be more capable of carrying loads, especially under large deformations. Openings create stress concentrations that lead to premature failure. The abrupt drops in the load for walls with openings suggest that once the stress exceeds a specific limit, the walls fail suddenly. In practical applications, designing walls with openings requires careful consideration of the reduced load-bearing capacity and ductility. Additional reinforcement around openings might be necessary to mitigate the adverse effects observed. Walls without openings can absorb more energy due to their higher ductility, making them more suitable for dynamic or impact loads. In contrast, walls with openings might be more vulnerable in such scenarios due to their reduced energy absorption capacity.

In summary, the provided graph demonstrates that openings in walls significantly reduce their load-bearing capacity, ductility, and overall structural performance. The design and reinforcement of walls with openings need to address these issues to ensure safety and functionality.

4.2 Comparison of Numerical and Experimental

The comparison of numerical and experiment results is tabulated in Table 1. It shows a percentage difference of 16.98% for the maximum load from BSP with the opening and 1.38% for BSP without the opening. In this study, the simulation of the with and without opening could be affected due to the FEM modelling process. The percentage difference calculation is ((Experimental − FEM) / Experimental) × 100%. Figure 6 shows the stress model distribution for BSP without and with opening.

Figure 6 shows the maximum stress model for Fig. 6a, BSP without opening, and Fig. 6b; BSP with the opening is 505.3 kN and 957.2 kN, respectively. The load chosen is 498.41 kN for BSP without opening, while 818.28 kN for BSP with opening taken from the maximum load applied to the experimental data. In comparing

Table 1 Comparison between experimental and FEM

Specimen	Average load (KN)		Percentage difference (%)
	Experimental data (Max load)	FEM data (Max load)	
BSP with opening	818.28	957.2	16.98
BSP without opening	498.41	505.3	1.38

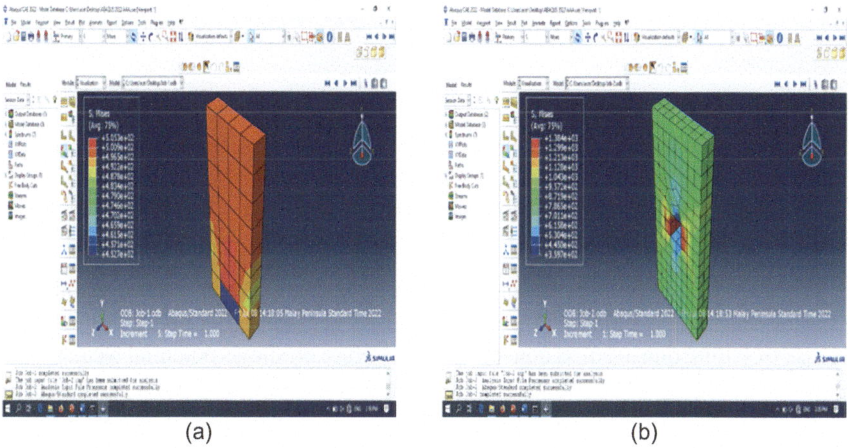

Fig. 6 Stress model distribution, **a** BSP without opening, **b** with opening

Fig. 6a (BSP without opening) and Fig. 6b (BSP with opening), the observations are as follows:-

Figure 6a shows the BSP without an opening for the stress distribution. It shows that the stress distribution is more uniform throughout the wall. The highest stress regions are concentrated at the bottom and decrease upwards. No significant stress concentrations indicate a continuous load-bearing capacity without interruptions.

For the BSP with an opening, the stress distribution shows a significant concentration around the opening. The stress is more localized around the opening area, especially at the edges of the opening, indicating a potential weak spot. There are visible high-stress regions around the opening, suggesting that the wall with the opening might experience higher localized stresses, which could affect its structural integrity.

For the stress magnitude, in Fig. 6a BSP without opening, the stress values are relatively lower and more evenly spread, which is typical for a solid structure without openings. The maximum stress is observed at the bottom of the wall, which is expected due to the higher load at the base. Figure 6b (BSP with opening) shows the maximum stress values around the opening are higher than the rest of the wall. The presence of the opening introduces stress concentrations that result in higher localized stresses.

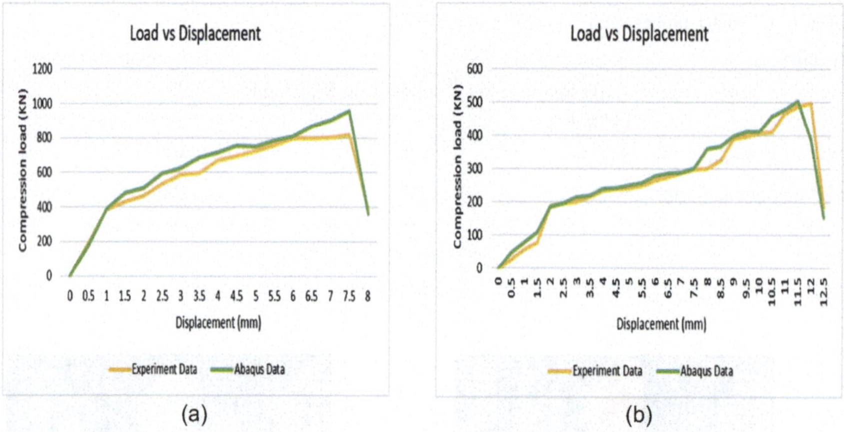

Fig. 7 Compression load vs displacement graph, **a** BSP with opening, **b** BSP without opening

The overall structural behaviour shown in Fig. 6a exhibits a more stable and predictable stress distribution. The wall will likely perform better under load without the risk of failure due to stress concentrations. Figure 6b shows potential weak points around the opening where failure could initiate. The wall's performance under load is compromised by the presence of the opening, which introduces significant stress concentrations.

The comparison data and performance between the numerical and the experimental are explained in Fig. 7.

According to Fig. 7, the greatest compression load specimen with opening using ABAQUS data is 957.2 kN, whereas the experiment data is 818.28 kN, while the highest compression load specimen without opening using ABAQUS data compared to experiment data are 505.3 kN and 498.41 kN, respectively. Compression strength is described as a material's resistance to breaking under compression or a structure's ability to sustain a load that tends to shrink in size. The results clearly illustrate that compression strength from modelling is consistent and error-free when the proper data is used instead of experimental data. As seen in Table 2, FEM results varied greatly from experimental testing due to various parameters in the ABAQUS software that impact simulation modelling. From Fig. 7, the BSP without opening data has higher compression compressive strength compared to the BSP with opening. The experimental testing did not offer several parameters, like stress and strain rate, which might influence the FEM results. The result of stress distribution from the FEM is shown in Fig. 8.

Figure 8 illustrates a graph of stress versus strain for BSP with opening and BSP without opening. The ABAQUS graph shows linear graphs compared to the theoretical graph due to input limitations in generating the parts during the meshing process. The lack of input data provides the flaws. The improper boundary conditions, kNown as discretization errors, are more visible in 2D and 3D settings; however, they may be reduced by appropriately increasing the number of components (SimScale, 2021).

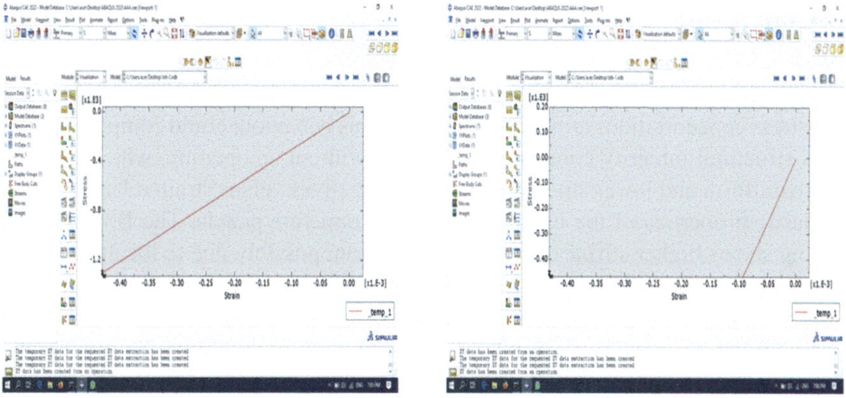

(a) (b)

Fig. 8 Stress vs strain graph, **a** BSP with opening, **b** BSP without opening

The provided figures show the stress versus strain graphs for bamboo-reinforced concrete panels (BSP), with Fig. 8a representing the BSP with an opening and Fig. 8b representing the BSP without an opening. From the observations of stress–strain behaviour, Fig. 8a (BSP with opening) shows a linear relationship between stress and strain, indicating elastic behaviour up to a point shown. The slope of the line is steeper, suggesting a higher stiffness in the initial elastic region. In Fig. 8b (BSP without opening), the graph also shows a linear relationship between stress and strain, but the slope is less steep than that of Fig. 8a. This behaviour indicates a lower stiffness in the initial elastic region compared to the panel with an opening. Both graphs show strain values ranging from − 0.45 to 0, indicating compressive strain. The negative stress values are consistent with compressive stress, ranging from − 1 to 0.

The slope of the stress–strain curve in the linear region represents the material's elastic modulus. For the BSP with opening, it shows a steeper slope, indicating a higher elastic modulus. This suggests that the material in this region is stiffer compared to the BSP without an opening. The BSP without opening shows a less steep slope, indicating a lower elastic modulus, suggesting more flexibility or less stiffness in the initial elastic region.

The higher stiffness in the BSP with an opening could be due to localized stress concentrations around the opening, causing a higher apparent stiffness in the initial elastic region. The lower stiffness in the BSP without an opening indicates a more uniform stress distribution, allowing for more flexibility. The stress–strain graphs should be compared with the load–displacement behaviour analyzed previously. The higher stiffness observed in the BSP with an opening might explain the sudden load drops in the load–displacement graph, indicating brittle behaviour due to stress concentrations. The lower stiffness in the BSP without an opening correlates with the smoother load–displacement curves, indicating more ductile behaviour.

5 Conclusion

An opening in the BSP wall significantly affects the stress distribution, creating high-stress concentrations around the opening. This behaviour could compromise the wall's structural integrity compared to the wall without an opening, which exhibits a more uniform and lower stress distribution. The stress versus strain relations show the elastic properties of the bamboo-reinforced concrete panels. The BSP with an opening shows higher stiffness in the elastic region, possibly due to localized stress concentrations. The BSP without an opening exhibits a more uniform stress distribution, resulting in lower stiffness but potentially higher ductility. These observations align with the analyzed load–displacement behaviour, highlighting the importance of considering openings' impact on structural performance.

Acknowledgements The authors gratefully acknowledge the help of Universiti Teknologi MARA, Shah Alam and SEGi University, Kota Damansara, which contributed to this research.

References

Anshari B, Guan Z, Komatsu K (2010) Finite element modeling of the pre-camber of glulam beams reinforced by compressed wood. In: The 11th world conference on timber engineering, Italy, pp1654–1659

Bai X, Lee AWC, Thompson LL, Rosowsky DV (1999) Finite element analysis of moso bamboo-reinforced southern pine OSB composite beams. Wood Fiber Sci 31:403–415

Boakye EO, Osei JB, Adom-Asamoah M (2018) Finite element modelling of bamboo reinforced concrete beams. J Constr Build Mater Eng 4(2)

EN 1990 (2002) Eurocode—basis of structural design

Errors in FEA and understanding singularities. SimScale. https://www.simscale.com/blog/2016/06/errors-in-fea-and-singularities/

Nurdiah EA (2016) The potential of bamboo as building material in organic shaped buildings. Procedia—Social Behav Sci 216:30–38

Ghavami K (2005) Bamboo as reinforcement in structural concrete elements. Cem Concr Compos 27(6):637–649

Ibrahim AM, Salman WD, Ahmed QW (2015) Behavior and strength of bearing wall strengthening by CFRP. Int J Eng Res Technol (IJERT) 4(06), ISSN: 2278–0181

Ikponmwosa E, Falade F, Fapohunda C, Aransiola J (2014) Flexural performance of bamboo-reinforced foamed aerated concrete beams. Am J Mater Sci 4:56–63

Kumar G, Ashish DK (2018) Analysis of stress dispersion in bamboo reinforced wall panels under earthquake loading using finite element analysis. Comput Concr 21(4):451–461

Mansor H, Wahab NMAA, Hamid YS, Kamarudin MK (2019) A mockup unit of the an-eco budget bamboo chalet: design and cost estimation analysis. MATEC Web Conf 258:01010

Mark AA, Russel AO (2011) A comparative study of bamboo reinforced concrete beams using different stirrup materials for rural construction. Int J Civ Struct Eng 2(1):407–423

Muhtar M, Dewi SM, Wisnumurti, As'ad Munawir (2016) Bond-slip improvement of bamboo reinforcement in concrete beam using hose clamps. In: Proceedings of the 2th international multidisciplinary conference 2016, vol 1, no 1

Othuman Mydin MA, Sani NM, Phius AF (2014) Investigation of industrialised building system performance in comparison to conventional construction method. MATEC Web Conf 10:04001

Puri V, Chakrabortty P, Anand S, Majumdar S (2017) Bamboo reinforced prefabricated wall panels for low cost housing. J Build Eng 9:52–59.

Raj AD, Agarwal AB (2014) Bamboo as a building material. J Civ Eng Environ Technol 1(3):56–61, Print ISSN: 2349–8404; Online ISSN: 2349–879X

Azuwa SB (2024) A review on experimental observation on structural performance of bamboo reinforced concrete beam. Heliyon, 10(2):e24628

Cold In-place Recycling in Malaysia: Implementation Problems and Solutions

Ahmad Kamil Arshad, Ekarizan Shaffie, Mohd Izzat Asyraf Mohamad Kamal, Noor Azreena Kamaluddin, Wardati Hashim, and Mohd Yazip Hj Matori

Abstract Pavement recycling is an environmentally sustainable pavement rehabilitation technique which reuses the existing materials in the pavement. Various pavement recycling techniques have been carried out as a method of pavement rehabilitation in Malaysia for more than 20 years. These techniques include cold in-place and plant recycling and hot in-place recycling. Early recycling methodologies and construction techniques were adopted from other countries such as South Africa and Germany. However, there are challenges and shortcomings in carrying out these recycling techniques due to the lack of experience of the local contractors. This paper details the problems of implementing the cold in-place recycling (CIPR) technique in Malaysia. Among these problems include unavailability of specific local recycling structural and mix design guidelines, inconsistency of the in situ pavement materials, limited availability of machinery to carry out the recycling works and lack of experience workers and finally, the difficulty in the construction and implementation of quality assurance/control procedures. Suggestions on overcoming the problems and improving the implementation of this rehabilitation technique are then discussed. It is hoped that the suggestions and solutions presented in this paper will encourage and further increase the use of CIPR as a sustainable pavement rehabilitation technique.

Keywords Cold in-place recycling · In-situ pavement recycling · Pavement recycling · Pavement rehabilitation

A. K. Arshad (✉) · E. Shaffie
Institute for Infrastructure Engineering and Sustainable Management, Universiti Teknologi MARA, Shah Alam, Selangor, Malaysia
e-mail: drahmadkamil@uitm.edu.my

M. I. A. Mohamad Kamal · N. A. Kamaluddin · W. Hashim
School of Civil Engineering, College of Engineering, Universiti Teknologi MARA, Shah Alam, Selangor, Malaysia

M. Y. Hj Matori
Kumpulan IKRAM Sdn Bhd, Kajang, Selangor, Malaysia

E. Shaffie et al. (eds.), *Sustainable Green Infrastructure*,
https://doi.org/10.1007/978-981-96-1486-8_21

1 Introduction

One of the recommended practices in the pavement sustainability is to incorporate pavement recycling in the pavement life cycle (FHWA 2015). Pavement recycling is an environmentally sustainable pavement rehabilitation technique which reuses the existing materials in the pavement, thus minimizing the use of virgin aggregate materials and avoids the landfilling of old pavement materials. Pavement recycling can be categorized into four types, namely, hot plant recycling, hot in-place recycling, cold in-place recycling and full depth reclamation (ARRA 2015; Xiao et al. 2018). Pavement recycling has been practiced in Malaysia particularly, since the early 2000s, where several pavement recycling have been carried out throughout the central region states in Peninsular Malaysia (Arshad 2004). Two of the most used recycling methods in Malaysia are hot in-place recycling and cold in-place recycling (Arshad 2005). In Malaysia, the term cold in-place recycling method refers to the full depth reclamation method used elsewhere (Arshad 2006). The full depth reclamation method involves the reclamation of all the asphalt pavement layer and a predetermined amount of the underlying base material treated with either cement, bitumen emulsion or foamed bitumen to produce a stabilized base layer, whereas for cold in-place recycling, only the top asphalt layer is recycled. Figure 1 shows the typical set of equipment (CIPR Equipment Train) used in CIPR construction of roads and highways in Malaysia.

Fig. 1 CIPR machineries train using foamed bitumen used in Malaysia

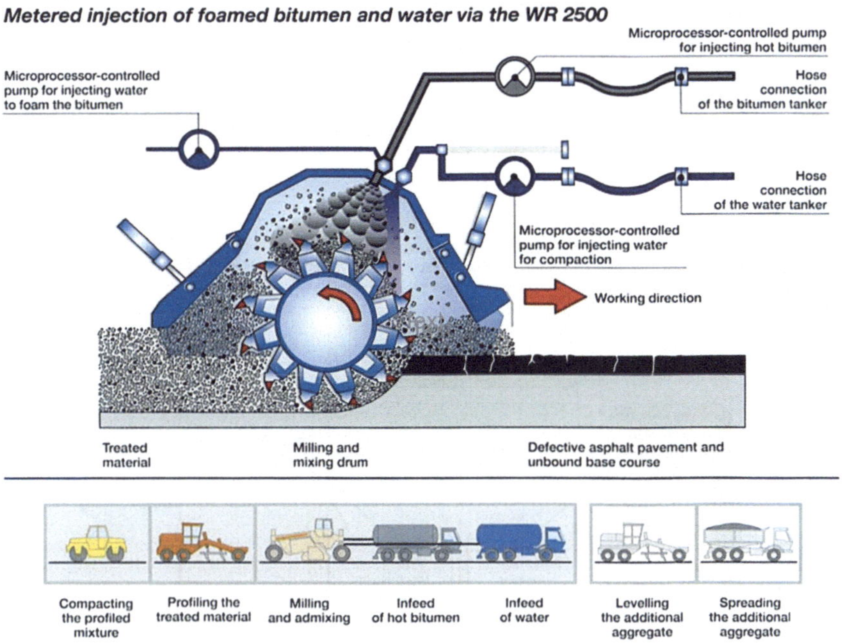

Fig. 2 CIPR process using foamed bitumen and cement [9]

The cold in-place recycling (CIPR) method in Malaysia involves the pulverization of the asphalt layer and part of the existing road-base layer and at the same time mixes with the stabilizing agent to create a new pavement base material consisting of reclaimed asphalt pavement (RAP) and aggregate mixture. This recycled pavement material is then compacted to form a new high strength pavement base layer followed by the application of a new asphalt surfacing (Lu and Sun 2023). Figure 2 shows the CIPR process using foamed bitumen and cement.

The use of foamed bitumen and cement enhanced the properties of the pavement, in particular improved flexibility (foamed bitumen) and enhanced or improved resistance to pavement deformation due to the cement producing a much stiffer material (Aker and Ozer 2023). Figure 3 shows the benefits of using foamed bitumen and cement to improve the properties of the recycled material in CIPR.

The use of CIPR therefore provides an approach that offers significant benefits for pavement rehabilitation compared with other techniques, where applicable, among which are:

- conservation of high-quality natural aggregates
- reduction of traffic disruption and road user time delays
- significant construction time savings
- reduction in heavy construction traffic.

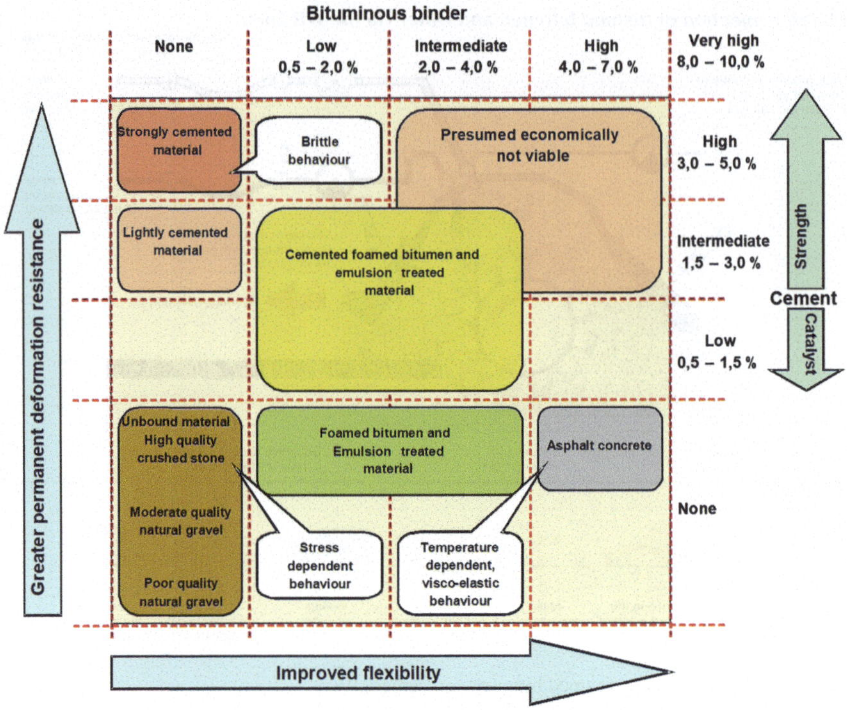

Fig. 3 Improvements in pavement properties using foamed bitumen and cement

These benefits all contribute to minimizing environmental impact, and clearly address the problems identified as being of major concern to the road owners and operators.

There are numerous guidelines and specifications available for pavement recycling and for CIPR. Early recycling methodologies and construction techniques were adopted in Malaysia from other countries such as Germany and South Africa (Wirtgen 2012; SABA 2020; Tabakovic et al. 2016). However, there are many initial problems in applying the CIPR rehabilitation technique to Malaysia based on the available guidelines and specifications as these were developed elsewhere and may have to be modified to local conditions (JKR 2008). Most of the challenges in CIPR relate to in-situ condition of existing pavements and construction problems such as curing and compaction (Redelius et al. 2015; Ogbo et al. 2022). Among the problems include the unavailability of specific local CIPR design guidelines, inconsistency of the in situ pavement materials, limited availability of machinery and experienced workers, the difficulty in spreading of cement during the construction of CIPR and difficulty in the implementation of quality assurance/control procedures for CIPR. These initial challenges and problems in the implementation of the CIPR technique in Malaysia and the methods to overcome the challenges are discussed in this paper.

Table 1 Layer stiffness values for mechanistic design

Material	Layer stiffness (MPa)	Poisson's ratio
Bitumen-treated base	1500–2000	0.40
Cement/lime-treated base	1200–2000	0.30

2 Unavailability of Specific Local Recycling Design Guidelines

Pavements in Malaysia has been traditionally designed using component analysis method based on the Public Work's Department of Malaysia (PWD Malaysia) Guideline for Pavement Design 5/85, developed in 1985, which was based on the AASHTO Interim Guide for the Design of Rigid and Flexible Pavements, 1972. There is a need to change the way the structure of the pavements is designed for CIPR pavements from the present empirical method toward the more rational mechanistic design method. To apply the mechanistic design method, key design parameters like layer stiffness or resilient modulus must be determined for local materials and conditions. The Public Works Department of Malaysia built test sites using CIPR and measured the layer stiffness with a falling weight deflectometer. Table 1 provides the recommended values for layer stiffness and Poisson's ratio for foamed bitumen and cement-treated base.

In terms of mix design, initially there were no established procedures to determine the appropriate type and content of stabilizing agent, strength, RAP/aggregate proportions, optimum moisture content and maximum density. Guidelines and specifications on recycling and construction techniques were adopted and modified to suit local conditions from other countries such as South Africa and Germany. Since the establishment of the JKR Specifications for cold in-place recycling, mix design report has been made mandatory for all CIPR projects under PWD Malaysia.

Another problem is in terms of mix design which specifically relates to how the material samples were obtained from the site. The recommended practice elsewhere was to use pulverized material from the site. In Malaysia, trial pits (Fig. 4) were used to obtain the material samples. However, the research has shown that material samples obtained from trial pits gave no significant difference in the mix design results.

3 Inconsistency of the In Situ Pavement Materials

In terms of construction materials, a problem faced by the local road industry was the inappropriate existing material encountered on-site to be used for CIPR. Roads that were constructed in the 1970s in Malaysia used large stones measuring more than 75 mm for the road-base (Fig. 5). The use of large stones is not suitable for the CIPR process to be carried out. Since the recycling machines available in Malaysia

Fig. 4 Sampling of CIPR materials from the test pit

are only capable of recycling aggregates of 50 mm size or less, these roads could not be properly recycled as the stones had the tendency to get stuck in between the cutting teeth of the recycler. For these types of roads, it is recommended that only the RAP portion of the existing pavement layer be utilized with addition of imported aggregates prior to recycling. In addition, there were occasions where materials from sites were found to be of marginal quality (Fig. 6).

In the existing pavement, inconsistency can occur with regard to the proportion of RAP to crushed aggregate, layer thickness and material properties. RAP content not only varies from site to site, but it may also vary within the same road stretch. The material design depends on the proportion of RAP to the existing crushed aggregate from the site. The main reason for the inconsistencies was that the road had been overlaid several times throughout its service life, thus making the asphalt layer very thick (Fig. 7). To ensure that these inconsistencies are considered, more trial pits, coring and dynamic cone penetrometer (DCP) tests should be carried out. From these test results, uniform discrete sections are identified, and different treatment are

Fig. 5 Large stones measuring more than 75 mm for the road-base

Fig. 6 Marginal quality materials for CIPR from a site

applied for each section. However, for each type of treatment, a minimum length of 100 m should be maintained for practical and economic reasons.

Fig. 7 Very thick asphalt layer due to numerous overlaying of asphalt layers

Fig. 8 Typical CIPR recycling machine used in Malaysia

4 Availability of Machinery and Experienced Workers

Currently, CIPR projects could only be undertaken by few specialist contractors as it requires high capital to purchase purpose-built equipment needed for recycling. Figure 8 shows the typical CIPR machine used in Malaysia, which is very costly and must be imported from overseas such as Germany. To encounter this situation, there must be changes in the Government policies and establish incentives to the contractors. This is because CIPR as rehabilitation method is proven to be more cost-effective, sustainable and environmentally friendly. Depending on the severity of the existing pavement, CIPR method provides a cost saving of up to 40%, in addition to other benefits such as conservation of natural resources and reduced material transportation, thus minimizing environmental impact.

5 Spreading of Cement During Construction

For CIPR using cement as the stabilizing agent, the cement is first spread on the existing roadway manually. Greater accuracy can be achieved using a mechanical spreader or cement slurry machine. Instead of using these costly machines, manual spreading of powdered cement carried out on selected sites (Fig. 9) was found with

Fig. 9 Manual spreading of cement prior to recycling

no major problems on the completed CIPR layer, other than the tendency of the cement to be blown away by the passing traffic in the adjacent lane. This problem can be minimized by controlling the length of the section to be spread ahead of the recycler. The length of section to be spread by cement depends on the geometry of the road.

6 Implementation of Quality Assurance/Control Procedures

Curing time is defined as the duration after the recycling process and before laying of asphalt layer on top of the recycled layer. One of the problems when using CIPR is the duration after curing time after recycling before the laying of asphalt layer on top of the recycled layer. Current specifications require at least 72 h for cement stabilizing agents and 48 h for bituminous stabilizing agents. Previous research carried out in the laboratory showed that with RAP contents less than 25%, the required strength was achieved as early as one day.

The current specification also requires testing of unconfined compressive strength (UCS) after seven days. It was common that within this seven-day period, the recycled layer was already overlaid with the asphalt surfacing (Fig. 10). Thus, if the CIPR

Fig. 10 Overlaying the recycled layer with asphalt surfacing

layer does not achieve the specified strength at seven days, then the asphalt layer would have to be removed. This may induce contractors to impose higher unit rates and in turn dissuade industry and road authorities from using the technology.

To overcome this problem and to suit the 72-h requirement, the UCS strength test was carried out at three days. From research and the actual CIPR works with cement, it was found that the CIPR layer achieved the minimum specified strength of 2 MPa.

7 Conclusion

CIPR is a sustainable solution to pavement rehabilitation works as it reused the existing materials on-site and therefore reduces the material requirements and transportation to site. However, there are problems of implementing the cold in-place recycling (CIPR) technique in Malaysia. Among these problems include unavailability of specific local recycling structural and mix design guidelines, inconsistency of the in situ pavement materials, limited availability of machinery to carry out the recycling works and lack of experienced workers, and finally, the difficulty in the construction and implementation of quality assurance/control procedures. Suggestions on overcoming the problems and improving the implementation of this rehabilitation technique have been discussed. It is hoped that the suggestions and solutions presented in this paper will encourage and further increase the use of CIPR as a sustainable pavement rehabilitation technique.

References

Aker SNA, Ozer H (2023) Cold recycling mix design approach targeting permanent deformation resistance. Constr Build Mater 400:1–12

Arshad AK (2004) Pavement recycling: an overview. Jurutera, 14–17

Arshad AK (2005) Hot In-Place Recycling: An Introduction, Jurutera, 10–14

Arshad AK (2006) Cold in-place recycling: an introduction. Jurutera, 30–35

Asphalt Recycling and Reclaiming Association (2015) Basic asphalt recycling manual, 2nd edn. Asphalt Recycling and Reclaiming Association, Glen Ellyn, IL

Federal Highway Administration (2015) Towards sustainable pavement systems. FHWA, Washington, DC

JKR Malaysia (2008) Standard Specification for Road Works: Section 4: Flexible Pavement (JKR/SPJ/2008-S4), JKR Malaysia

Liu Z, Sun L (2023) A review of compaction methods on cold recycling asphalt mixtures. Constr Build Mater 401:1–20

Ogbo C, Dave EV, Sias JE, Zegeye E (2022) Correlating field and laboratory evolution of curing in cold in-place recycled (CIR) materials. Constr Build Mater 345:1–12

Redelius P, Ostlund J, Soenen H (2015) Field experience of cold mix asphalt during 15 years, road materials and pavement design. Taylor & Francis

South African Bitumen Association (2020) A guideline for the design and construction of bitumen emulsion and foamed bitumen stabilised materials. TG2, 3rd edn. Howard Place, South Africa

Tabakovic A, McNallY C, Fallon E (2016) Specification development for cold in-situ recycling of asphalt. Constr Build Mater 102:318–328

Wirtgen GmbH (2012) Wirtgen cold recycling technology, 1st edn. Wirtgen GmbH, Windhagen, Germany

Xiao F, Yao S, Wang J, Li X, Amirkhanian S (2018) A literature review on cold recycling technology of asphalt pavement. Constr Build Mater 180:579–604

The manufacturer's authorised representative in the EU is Springer Nature Customer Service Centre GmbH, Europaplatz 3, 69115 Heidelberg, Germany. If you have any concerns regarding our products, please contact ProductSafety@springernature.com

Printed and bound by CPI Group (UK) Ltd, Croydon, CR0 4YY

01/10/2026

02225941-0009